T0310387

**Spline Collocation Methods
for Partial Differential Equations**

Spline Collocation Methods for Partial Differential Equations

With Applications in R

William E. Schiesser

Lehigh University
Bethlehem, PA, USA

Registered Office
John Wiley & Sons, Inc., 111 River Street, Hoboken, NJ 07030, USA

Editorial Office
111 River Street, Hoboken, NJ 07030, USA

For details of our global editorial offices, customer services, and more information about Wiley products visit us at www.wiley.com.

Wiley also publishes its books in a variety of electronic formats and by print-on-demand. Some content that appears in standard print versions of this book may not be available in other formats.

Library of Congress Cataloging-in-Publication Data

Names: Schiesser, W. E., author.
Title: Spline collocation methods for partial differential equations : with
    applications in R / William E. Schiesser.
Description: Hoboken, NJ : John Wiley & Sons, 2017. | Includes
    bibliographical references and index.
Identifiers: LCCN 2017000481 (print) | LCCN 2017007223 (ebook) | ISBN
    9781119301035 (cloth) | ISBN 9781119301059 (pdf) | ISBN 9781119301042
    (epub)
Subjects: LCSH: Differential equations, Partial–Mathematical models. |
    Spline theory. Classification: LCC QA377 .S355 2017 (print) | LCC QA377 (ebook) | DDC
    515/.353–dc23 LC record available at https://lccn.loc.gov/2017000481

Cover image: Wiley
Cover design by (Background) © hakkiarslan/iStockphoto;
(Equations) courtesy of the author

Set in 10/12pt WarnockPro by SPi Global, Chennai, India

Printed in the United States of America

10 9 8 7 6 5 4 3 2 1

To
Alan Foust, Gary Kohler, Douglas Leith, Everett Pitcher, Gilbert Stengle,
Richard Wilhelm

Contents

Preface

This book is an introduction to the use of spline collocation (SC) for the numerical analysis of partial differential equations (PDEs). SC is an alternative to finite differences (FDs), finite Volumes (FVs), finite elements (FEs), spectral methods, weighted residuals, and least squares.

The features and advantages of SC are demonstrated through a series of PDE examples, including the use of SC library routines that are part of the R programming system [1, 2]. Applications of the PDEs are mentioned only briefly. Rather, the emphasis is on the SC methodology, particularly in terms of PDE examples developed and discussed in detail.

The papers cited as a source of the PDE models generally consist of (i) a statement of the equations followed by (ii) a reported numerical solution. Generally, little or no information is given about how the solutions were computed (the algorithms), and in all cases, the computer code that was used to calculate the solutions is not provided.

In other words, what is missing is (i) a detailed discussion of the numerical methods used to produce the reported solutions and (ii) the computer routines used to calculate the reported solutions. For the reader to complete these two steps to verify the reported solutions with reasonable effort is essentially impossible.

A principal objective of this book is therefore to provide the reader with a set of documented R routines that are discussed in detail, and they can be downloaded and executed without having to first master the details of the relevant numerical analysis and then code a set of routines.

The example applications are intended as introductory and open ended. They are based mainly on legacy PDEs. Since each of the legacy equations is well known, a search on the equation name will give an extensive set of references, generally including many application studies. Rather than focus on the applications, which would require extended discussion, the emphasis in each chapter is on the following:

1. A statement of the PDE system, including initial Conditions (ICs), boundary conditions (BCs), and parameters

2. The algorithms for the calculation of numerical solutions, with particular emphasis on SC
3. A set of R routines for the calculation of numerical solutions, including a detailed explanation of each section of the code
4. Discussion of the numerical solution
5. Summary and conclusions about extensions of the computer-based analysis

In summary, the presentation is not as formal mathematics, for example, theorems and proofs. Rather, the presentation is by examples of SC analysis of legacy PDEs, including the details for computing numerical solutions, particularly with documented source code. The author would welcome comments, especially pertaining to this format and experiences with the use of the R routines. Comments and questions can be directed to (wes1@lehigh.edu).

References

1 Hiebeler, D.E. (2015), *R and Matlab*, CRC/Taylor and Francis, Boca Raton, FL.
2 Soetaert, K., J. Cash, and F. Mazzia (2012), *Solving Differential Equations in R*, Springer-Verlag, Heidelberg, Germany.

William E. Schiesser
Bethlehem, PA, USA
March 1, 2017

About the Companion Website

This book is accompanied by a companion website:

www.wiley.com/go/Spline_Collocation

The website includes:
R Routines

1

Introduction

The principal objective of this book is to present the method of lines (MOL) numerical integration of partial differential equations (PDEs), with spline collocation (SC) approximation of the PDE boundary-value derivatives. This approach is therefore termed *spline collocation method of lines* (SCMOL).

The details of SCMOL computer implementation are presented in terms of a series of applications, first for one-dimensional (1D) PDEs, then for two-dimensional (2D) PDEs, and finally for a series of legacy PDEs to illustrate the broad applicability of SCMOL. The approach is not with formal mathematics, for example, theorems and proofs, but rather, by examples of SCMOL discussed in detail, including routines in R.[1]

In this introduction, some basic properties of splines are reviewed, including example applications of the utilities (functions) for splines in R. The R spline utilities are then applied to PDEs in the following chapters within the SCMOL setting.

Splines are polynomials that can be used for the functional approximation of a set of numerical pairs

Table 1.1 Data pairs for spline interpolation.

x_1	y_1
x_2	y_2
$\vdots$	$\vdots$
x_{n-1}	y_{n-1}
x_n	y_n

---

1 R is a quality open-source scientific programming system that can be easily downloaded from the Internet (http://www.R-project.org/). In particular, R has (i) vector–matrix operations that facilitate the programming of linear algebra, (ii) a library of quality ordinary differential equation (ODE) integrators, and (iii) graphical utilities for the presentation numerical ODE/PDE solutions. All of these features and utilities are demonstrated through the applications in this book.

Spline Collocation Methods for Partial Differential Equations: With Applications in R, First Edition.
William E. Schiesser.
© 2017 John Wiley & Sons, Inc. Published 2017 by John Wiley & Sons, Inc.
Companion website: www.wiley.com/go/Spline_Collocation

A cubic spline is of the form

$$p_3(x) = a_0 + a_1 x + a_2 x^2 + a_3 x^3 \tag{1.1a}$$

The coefficients a_0, a_1, a_2, a_3 are evaluated to (i) return the original data of Table 1.1 and (ii) provide continuity in the computed first and second derivatives of $p_3(x)$ at $x_1, x_2, \ldots, x_{n-1}, x_n$. The sequence in x does not have to be uniformly spaced, which is a particularly useful feature in the SCMOL solution of PDEs, that is, the points can be placed as required to achieve good resolution in x.

$p_3(x)$ can be differentiated to give the first and second derivatives

$$\frac{dp_3(x)}{dx} = a_1 + 2a_2 x + 3a_3 x^2 \tag{1.1b}$$

$$\frac{d^2 p_3(x)}{dx^2} = 2a_2 + 6a_3 x \tag{1.1c}$$

$$\frac{d^3 p_3(x)}{dx^3} = 6a_3 \tag{1.1d}$$

The derivatives of Eqs. (1.1a)–(1.1d) can then be used to approximate first, second, and third derivatives in PDEs.

All of the operations reflected in Eqs. (1.1) are implemented in the R function `splinefun`,[2] which returns the coefficients a_0, a_1, a_2, a_3 for a set of n data as listed in Table 1.1. a_0, a_1, a_2, a_3 are computed by the solution of a tridiagonal algebraic system of $n - 1$ equations. This set of $n + 1$ coefficients therefore requires the specification of two additional conditions.[3]

A variety of additional conditions can be specified. For example, if the two second derivatives at the end points of the data set are set to zero, so-called *natural cubic splines* result. The details of the splines resulting from various sets of two additional conditions as implemented in `splinefun` are given in Appendix A1 and in [1].

The use of `splinefun` is illustrated with a series of examples that follows.

1.1 Uniform Grids

Application of `splinefun` to the function $y = \sin(\pi x)$ is illustrated with the following code:

```
#
# Previous workspaces are cleared
```

---

2 Details of `splinefun`, including the programming options and example applications, are available from the online documentation accessed by `help(splinefun)` entered at the R prompt. Excerpts from this documentation are given in Appendix A1.
3 This gives a set of $n + 1$ equations in the $n + 1$ unknown coefficients.

```
  rm(list=ls(all=TRUE))
#
# Define uniform grid
  xl=0;xu=1;n=11;
  x=seq(from=xl,to=xu,by=(xu-xl)/(n-1));
#
# Define function to be approximated
  u=sin(pi*x);
#
# Set up spline table
  utable=splinefun(x,u);
#
# Compute spline approximation
  us=utable(x);
#
# Display comparison of function and its spline
# approximation
  cat(sprintf("\n      x              u            us
              diff"));
  for(i in 1:n){
    cat(sprintf("\n%5.2f%10.5f%10.5f%12.7f",
      x[i],u[i],us[i],us[i]-u[i]));
  }
```

Listing 1.1 Spline approximation of $y = \sin(\pi x)$.

We can note the following details about this listing:

- Previous workspaces are cleared to avoid the unintended use of out-of-date files.

```
#
# Previous workspaces are cleared
  rm(list=ls(all=TRUE))
```

- A uniform grid is defined with 11 points, $0 \le x \le 1$ so that $x = 0, 0.1, \ldots, 1$.

```
#
# Define uniform grid
  xl=0;xu=1;n=11;
  x=seq(from=xl,to=xu,by=(xu-xl)/(n-1));
```

The seq utility is used for this purpose.
- The function $y = \sin(\pi x)$ is defined on the grid in x.

```
#
```

```
# Define function to be approximated
  u=sin(pi*x);
```

This definition of u[4] illustrates the vectorization available in R, that is, u is an *n*-vector since *x* is an *n*-vector. Also, the function sin can operate on a vector to produce a vector.

- splinefun is used to define a table of spline coefficients, for example, a_0, a_1, a_2, a_3 in Eq. (1.1a).

```
#
# Set up spline table
  utable=splinefun(x,u);
```

splinefun is part of the basic R and does not have to be accessed from an external library. Here the default spline mmf is used in which an exact cubic polynomial is fitted to the four points at each end of the data of Table 1.1 [1], p. 73.

- The table utable is used to compute spline approximations to *u* at the values of *x* in x.

```
#
# Compute spline approximation
  us=utable(x);
```

Other values of *x* within the interval x1 to xu could be used for interpolation between the grid points.

- The values of u, their spline approximations us, and the difference between the two are displayed.

```
#
# Display comparison of function and its spline
# approximation
  cat(sprintf("\n    x           u          us
                diff"));
  for(i in 1:n){
    cat(sprintf("\n%5.2f%10.5f%10.5f%12.7f",
      x[i],u[i],us[i],us[i]-u[i]));
  }
```

The use of the combination cat(sprintf()) provides detailed formatting of the output.

Execution of the code in Listing 1.1 gives the following output:

---

4 The function $y = \sin(\pi x)$ is named u in this and subsequent code since it will ultimately be the dependent variable of a PDE. That is, in accordance with the usual convention in the literature, the dependent variables of PDEs are designated with *u*.

Table 1.2 Output from Listing 1.1.

x	u	us	diff
0.00	0.00000	0.00000	0.0000000
0.10	0.30902	0.30902	0.0000000
0.20	0.58779	0.58779	0.0000000
0.30	0.80902	0.80902	0.0000000
0.40	0.95106	0.95106	0.0000000
0.50	1.00000	1.00000	0.0000000
0.60	0.95106	0.95106	0.0000000
0.70	0.80902	0.80902	0.0000000
0.80	0.58779	0.58779	0.0000000
0.90	0.30902	0.30902	0.0000000
1.00	0.00000	0.00000	0.0000000

This output demonstrates that the spline in `splinefun` returns the original data (an important feature of splines in general).

Equations (1.1a)–(1.1d) can also give numerical approximations to the first to third derivatives, as demonstrated by the following routine:

```
#
# Previous workspaces are cleared
  rm(list=ls(all=TRUE))
#
# Define uniform grid
  xl=0;xu=1;n=11;
  x=seq(from=xl,to=xu,by=(xu-xl)/(n-1));
#
# Define function to be approximated, and its
# derivatives
     u=sin(pi*x);
   ux=pi*cos(pi*x);
  uxx=-pi^2*sin(pi*x);
 uxxx=-pi^3*cos(pi*x);
#
# Set up spline table for function, derivatives
  utable=splinefun(x,u);
#
# Compute spline approximation of function
# derivatives
     us=utable(x);
   usx=utable(x,deriv=1);
```

```
    usxx=utable(x,deriv=2);
    usxxx=utable(x,deriv=3);
#
# Display comparison of function and its spline
# approximation, and its derivatives
#
# u
  cat(sprintf("\n      x           u          us
               diff"));
  for(i in 1:n){
    cat(sprintf("\n%5.2f%10.5f%10.5f%12.7f",
      x[i],u[i],us[i],us[i]-u[i]));
  }
#
# ux
  cat(sprintf("\n      x           ux         usx
               diff"));
  for(i in 1:n){
    cat(sprintf("\n%5.2f%10.5f%10.5f%12.7f",
      x[i],ux[i],usx[i],usx[i]-ux[i]));
  }
#
# uxx
  cat(sprintf("\n      x           uxx        usxx
               diff"));
  for(i in 1:n){
    cat(sprintf("\n%5.2f%10.5f%10.5f%12.7f",
      x[i],uxx[i],usxx[i],usxx[i]-uxx[i]));
  }
#
# uxxx
  cat(sprintf("\n      x           uxxx       usxxx
               diff"));
  for(i in 1:n){
    cat(sprintf("\n%5.2f%10.5f%10.5f%12.7f",
      x[i],uxxx[i],usxxx[i],usxxx[i]-uxxx[i]));
  }
```

Listing 1.2 Spline approximation of $y = \sin(\pi x)$ and the first to third derivatives.

We can note the following details about Listing 1.2:

- Previous workspaces are cleared and the same uniform grid in x as in Listing 1.1 is defined.

- The function $y = \sin(\pi x)$ and its first three derivatives are computed. Again, u, ux, uxx, uxxx are n-vectors.

```
#
# Define function to be approximated, and its
# derivatives
    u=sin(pi*x);
   ux=pi*cos(pi*x);
  uxx=-pi^2*sin(pi*x);
 uxxx=-pi^3*cos(pi*x);
```

The derivatives are used to evaluate the numerical approximations from splinefun.

- The table of spline coefficients is defined as in Listing 1.1.

```
#
# Set up spline table for function, derivatives
   utable=splinefun(x,u);
```

- The function us and its first three derivatives are computed from utable.

```
#
# Compute spline approximation of function
# derivatives
      us=utable(x);
     usx=utable(x,deriv=1);
    usxx=utable(x,deriv=2);
   usxxx=utable(x,deriv=3);
```

us, usx, usxx, usxxx are n-vectors. The argument deriv specifies the order of the derivative to be computed.

- The exact values of $y = \sin(\pi x)$ and the numerical approximations are compared and displayed.

```
#
# Display comparison of function and its spline
# approximation, and its derivatives
#
# u
   cat(sprintf("\n     x          u          us
              diff"));
   for(i in 1:n){
     cat(sprintf("\n%5.2f%10.5f%10.5f%12.7f",
       x[i],u[i],us[i],us[i]-u[i]));
   }
#
```

```
# ux
  cat(sprintf("\n    x           ux          usx
            diff"));
  for(i in 1:n){
    cat(sprintf("\n%5.2f%10.5f%10.5f%12.7f",
      x[i],ux[i],usx[i],usx[i]-ux[i]));
  }
#
# uxx
  cat(sprintf("\n    x           uxx         usxx
            diff"));
  for(i in 1:n){
    cat(sprintf("\n%5.2f%10.5f%10.5f%12.7f",
      x[i],uxx[i],usxx[i],usxx[i]-uxx[i]));
  }
#
# uxxx
  cat(sprintf("\n    x           uxxx        usxxx
            diff"));
  for(i in 1:n){
    cat(sprintf("\n%5.2f%10.5f%10.5f%12.7f",
      x[i],uxxx[i],usxxx[i],usxxx[i]-uxxx[i]));
  }
```

The output from Listing 1.2 is in Table 1.3.

Table 1.3 Output from Listing 1.2.

x	u	us	diff
0.00	0.00000	0.00000	0.0000000
0.10	0.30902	0.30902	0.0000000
0.20	0.58779	0.58779	0.0000000
0.30	0.80902	0.80902	0.0000000
0.40	0.95106	0.95106	0.0000000
0.50	1.00000	1.00000	0.0000000
0.60	0.95106	0.95106	0.0000000
0.70	0.80902	0.80902	0.0000000
0.80	0.58779	0.58779	0.0000000
0.90	0.30902	0.30902	0.0000000
1.00	0.00000	0.00000	0.0000000

Table 1.3 (Continued)

x	ux	usx	diff
0.00	3.14159	3.14930	0.0077113
0.10	2.98783	2.98556	-0.0022759
0.20	2.54160	2.54203	0.0004266
0.30	1.84658	1.84633	-0.0002520
0.40	0.97081	0.97079	-0.0000154
0.50	0.00000	0.00000	0.0000000
0.60	-0.97081	-0.97079	0.0000154
0.70	-1.84658	-1.84633	0.0002520
0.80	-2.54160	-2.54203	-0.0004266
0.90	-2.98783	-2.98556	0.0022759
1.00	-3.14159	-3.14930	-0.0077113

x	uxx	usxx	diff
0.00	-0.00000	-0.27309	-0.2730880
0.10	-3.04988	-3.00187	0.0480096
0.20	-5.80121	-5.86869	-0.0674824
0.30	-7.98468	-8.04528	-0.0606045
0.40	-9.38655	-9.46551	-0.0789615
0.50	-9.86960	-9.95029	-0.0806842
0.60	-9.38655	-9.46551	-0.0789615
0.70	-7.98468	-8.04528	-0.0606045
0.80	-5.80121	-5.86869	-0.0674824
0.90	-3.04988	-3.00187	0.0480096
1.00	-0.00000	-0.27309	-0.2730880

x	uxxx	usxxx	diff
0.00	-31.00628	-27.28778	3.7184973
0.10	-29.48872	-27.28778	2.2009421
0.20	-25.08460	-21.76592	3.3186866
0.30	-18.22503	-21.76592	-3.5408860
0.40	-9.58147	-4.84776	4.7337109
0.50	-0.00000	-4.84776	-4.8477555
0.60	9.58147	14.20231	4.6208422
0.70	18.22503	14.20231	-4.0227235
0.80	25.08460	28.66824	3.5836399
0.90	29.48872	28.66824	-0.8204768
1.00	31.00628	27.28778	-3.7184973

We can note the following details about this output:

- The output for the function is repeated as in Table 1.2.
- As expected, the computed derivatives are not exact, and the errors increase with the order of the derivative. The increasing errors are expected when considering Eqs. (1.1a)–(1.1d) (the approximations vary from a second-order quadratic of Eq. (1.1a) to a constant of Eq. (1.1d)).
  The errors tend to be largest at the end points in x, for example, ±0.0077113 for ux, -0.2730880 for uxx. The exception is for the errors in uxxx, but since the approximation is the constant $6a_3$ of Eq. (1.1d), the error is large, even within the interval. In other words, the use of the third derivative in the PDE applications to follow should be avoided.

One method to reduce the errors in Table 1.3 would be to use a smaller grid spacing (more points in the grid). This is accomplished by changing n=11 in Listing 1.2 to, for example, n=21 (everything else in the code remains unchanged). The numerical output follows.

Table 1.4 Output from Listing 1.2, n=21.

x	u	us	diff
0.00	0.00000	0.00000	0.0000000
0.05	0.15643	0.15643	0.0000000
0.10	0.30902	0.30902	0.0000000
0.15	0.45399	0.45399	0.0000000
0.20	0.58779	0.58779	0.0000000
0.25	0.70711	0.70711	0.0000000
0.30	0.80902	0.80902	0.0000000
0.35	0.89101	0.89101	0.0000000
0.40	0.95106	0.95106	0.0000000
0.45	0.98769	0.98769	0.0000000
0.50	1.00000	1.00000	0.0000000
0.55	0.98769	0.98769	0.0000000
0.60	0.95106	0.95106	0.0000000
0.65	0.89101	0.89101	0.0000000
0.70	0.80902	0.80902	0.0000000
0.75	0.70711	0.70711	0.0000000
0.80	0.58779	0.58779	0.0000000
0.85	0.45399	0.45399	0.0000000
0.90	0.30902	0.30902	0.0000000

Table 1.4 (Continued)

x	ux	usx	diff
0.95	0.15643	0.15643	0.0000000
1.00	0.00000	0.00000	0.0000000

x	ux	usx	diff
0.00	3.14159	3.14209	0.0004935
0.05	3.10291	3.10277	-0.0001456
0.10	2.98783	2.98786	0.0000261
0.15	2.79918	2.79916	-0.0000192
0.20	2.54160	2.54160	-0.0000060
0.25	2.22144	2.22143	-0.0000082
0.30	1.84658	1.84658	-0.0000061
0.35	1.42625	1.42625	-0.0000049
0.40	0.97081	0.97080	-0.0000033
0.45	0.49145	0.49145	-0.0000017
0.50	0.00000	-0.00000	-0.0000000
0.55	-0.49145	-0.49145	0.0000017
0.60	-0.97081	-0.97080	0.0000033
0.65	-1.42625	-1.42625	0.0000049
0.70	-1.84658	-1.84658	0.0000061
0.75	-2.22144	-2.22143	0.0000082
0.80	-2.54160	-2.54160	0.0000060
0.85	-2.79918	-2.79916	0.0000192
0.90	-2.98783	-2.98786	-0.0000261
0.95	-3.10291	-3.10277	0.0001456
1.00	-3.14159	-3.14209	-0.0004935

x	uxx	usxx	diff
0.00	-0.00000	-0.03493	-0.0349290
0.05	-1.54395	-1.53776	0.0061820
0.10	-3.04988	-3.05866	-0.0087840
0.15	-4.48071	-4.48926	-0.0085487
0.20	-5.80121	-5.81333	-0.0121181
0.25	-6.97886	-6.99318	-0.0143132
0.30	-7.98468	-8.00112	-0.0164442
0.35	-8.79388	-8.81197	-0.0180931
0.40	-9.38655	-9.40587	-0.0193171
0.45	-9.74809	-9.76815	-0.0200599

(Continued)

Table 1.4 (Continued)

x			
0.50	-9.86960	-9.88991	-0.0203103
0.55	-9.74809	-9.76815	-0.0200599
0.60	-9.38655	-9.40587	-0.0193171
0.65	-8.79388	-8.81197	-0.0180931
0.70	-7.98468	-8.00112	-0.0164442
0.75	-6.97886	-6.99318	-0.0143132
0.80	-5.80121	-5.81333	-0.0121181
0.85	-4.48071	-4.48926	-0.0085487
0.90	-3.04988	-3.05866	-0.0087840
0.95	-1.54395	-1.53776	0.0061820
1.00	-0.00000	-0.03493	-0.0349290

x	xuxx	usxxx	diff
0.00	-31.00628	-30.05671	0.9495706
0.05	-30.62454	-30.05671	0.5678318
0.10	-29.48872	-28.61192	0.8768047
0.15	-27.62679	-28.61192	-0.9851220
0.20	-25.08460	-23.59703	1.4875760
0.25	-21.92475	-23.59703	-1.6722802
0.30	-18.22503	-16.21706	2.0079710
0.35	-14.07656	-16.21706	-2.1405061
0.40	-9.58147	-7.24569	2.3357781
0.45	-4.85045	-7.24569	-2.3952380
0.50	-0.00000	2.43523	2.4352332
0.55	4.85045	2.43523	-2.4152171
0.60	9.58147	11.87787	2.2964074
0.65	14.07656	11.87787	-2.1986812
0.70	18.22503	20.15889	1.9338574
0.75	21.92475	20.15889	-1.7658589
0.80	25.08460	26.48141	1.3968090
0.85	27.62679	26.48141	-1.1453810
0.90	29.48872	30.41790	0.9291820
0.95	30.62454	30.41790	-0.2066345
1.00	31.00628	30.05671	-0.9495706

We can conclude generally that the errors in Table 1.4 have been reduced with the change n=11 to n=21. This suggests the question of how the errors vary with n. This question can be addressed empirically (from the observed numerical output for increasing n).

For n=11, 21, 31, 41, 51, selected errors are summarized in Table 1.5.

Table 1.5 Summary of errors for n=11,21,31,41,51.

derivative	x	n	error
usx	0	11	0.0077113
		21	0.0004935
		31	0.0000979
		41	0.0000310
		51	0.0000127
	0.5	11	-0.0000000
	0.5	21	-0.0000000
	0.5	31	0.0000000
	0.5	41	0.0000000
	0.5	51	0.0000000
usxx	0	11	-0.2730880
		21	-0.0349290
		31	-0.0103934
		41	-0.0043912
		51	-0.0022499
	0.5	11	-0.0806842
	0.5	21	-0.0203103
	0.5	31	-0.0090227
	0.5	41	-0.0050744
	0.5	51	-0.0032474

We can note the following details about Table 1.5:

- Quantitative relationships between the error and the number of points n are not apparent. For example, an order relationship such as error $= O(\Delta x^p)$, where Δx is the grid spacing, does not apply for ux, x=0.5 (the error does not vary with grid spacing).
- Generally, some form of convergence appears to occur in the sense that the errors decrease with increasing n.
- The accuracy decreases with successive differentiation, for example, the errors for usxx are greater than for usx.

- The errors at the boundaries, for example, $x = 0$, are larger than at the interior points, for example, $x = 0.5$.

These results, although specific to the function $y = \sin(\pi x)$, suggest that some experimentation with the grid interval (value of n) can be used to infer accuracy of the spline approximation, even when exact values are not available (the usual case in PDE applications). For example, for x=0.5, n=11, usxx=-9.95029 (Table 1.3) and for x=0.5, n=21, usxx=-9.88991 (Table 1.4) suggests an accuracy of -9.95029 - (-9.88991)= -0.06038, which can be inferred merely by changing *n* and comparing numerical results.[5] Exact values of the function and its derivatives and the associated exact errors are not required for an error analysis (and are generally unavailable anyway). Rather, the errors can be *estimated* from approximate results.

To summarize, the preceding results indicate the calculation of derivatives of tabulated data pairs using splinefun is straightforward. This is further demonstrated next with the test function $y = e^{ax}$.

The routine for this case is the same as in Listing 1.2 except that

```
#
# Define function to be approximated, and its
# derivatives
      a=1;
      u=exp(a*x);
     ux=a*exp(a*x);
    uxx=a^2*exp(a*x);
   uxxx=a^3*exp(a*x);
```

is used in place of

```
#
# Define function to be approximated, and its
# derivatives
      u=sin(pi*x);
     ux=pi*cos(pi*x);
    uxx=-pi^2*sin(pi*x);
   uxxx=-pi^3*cos(pi*x);
```

The rate of change of $y = e^{ax}$ is defined by the parameter a, which can be varied to test the spline approximations as discussed next.

The output is in Table 1.6.

Since the function and its derivatives are the same (for $a = 1$), the reduction in the accuracy of successive derivatives is clear from Table 1.6. Also, the error

---

5 This form of error analysis is termed *h refinement* since the grid spacing in the numerical analysis literature is frequently designated as *h*.

Table 1.6 Output for the function $y = e^{ax}, a = 1$.

x	u	us	diff
0.00	1.00000	1.00000	0.0000000
0.10	1.10517	1.10517	0.0000000
0.20	1.22140	1.22140	0.0000000
0.30	1.34986	1.34986	0.0000000
0.40	1.49182	1.49182	0.0000000
0.50	1.64872	1.64872	0.0000000
0.60	1.82212	1.82212	0.0000000
0.70	2.01375	2.01375	0.0000000
0.80	2.22554	2.22554	0.0000000
0.90	2.45960	2.45960	0.0000000
1.00	2.71828	2.71828	0.0000000

x	ux	usx	diff
0.00	1.00000	1.00026	0.0002555
0.10	1.10517	1.10510	-0.0000692
0.20	1.22140	1.22142	0.0000177
0.30	1.34986	1.34985	-0.0000056
0.40	1.49182	1.49182	0.0000003
0.50	1.64872	1.64872	-0.0000005
0.60	1.82212	1.82212	-0.0000038
0.70	2.01375	2.01376	0.0000095
0.80	2.22554	2.22550	-0.0000408
0.90	2.45960	2.45975	0.0001463
1.00	2.71828	2.71773	-0.0005526

x	uxx	usxx	diff
0.00	1.00000	0.99030	-0.0097019
0.10	1.10517	1.10663	0.0014559
0.20	1.22140	1.21975	-0.0016548
0.30	1.34986	1.34891	-0.0009520
0.40	1.49182	1.49053	-0.0012956
0.50	1.64872	1.64739	-0.0013349
0.60	1.82212	1.82050	-0.0016196
0.70	2.01375	2.01244	-0.0013094
0.80	2.22554	2.22232	-0.0032249
0.90	2.45960	2.46267	0.0030664
1.00	2.71828	2.69693	-0.0213550

(Continued)

Table 1.6 (Continued)

x	uxxx	usxxx	diff
0.00	1.00000	1.16329	0.1632873
0.10	1.10517	1.16329	0.0581164
0.20	1.22140	1.29159	0.0701854
0.30	1.34986	1.29159	-0.0582707
0.40	1.49182	1.56857	0.0767480
0.50	1.64872	1.56857	-0.0801486
0.60	1.82212	1.91944	0.0973222
0.70	2.01375	1.91944	-0.0943117
0.80	2.22554	2.40353	0.1779934
0.90	2.45960	2.40353	-0.0560688
1.00	2.71828	2.34257	-0.3757088

in the computed derivatives is greater at the boundaries $x = 0, 1$ than at the interior points, as before with $\sin(\pi x)$. Since the function changes more rapidly at $x = 1$ than at $x = 0$, the errors are greater at $x = 1$.

The last point of greater variation at $x = 1$ can be demonstrated by using a larger value of a. For $a = 2$, the numerical output is

Table 1.7 Output for the function $y = e^{ax}$, $a = 2$.

x	u	us	diff
0.00	1.00000	1.00000	0.0000000
0.10	1.22140	1.22140	0.0000000
0.20	1.49182	1.49182	0.0000000
0.30	1.82212	1.82212	0.0000000
0.40	2.22554	2.22554	0.0000000
0.50	2.71828	2.71828	0.0000000
0.60	3.32012	3.32012	0.0000000
0.70	4.05520	4.05520	0.0000000
0.80	4.95303	4.95303	0.0000000
0.90	6.04965	6.04965	0.0000000
1.00	7.38906	7.38906	0.0000000

Table 1.7 (Continued)

x	ux	usx	diff
0.00	2.00000	2.00460	0.0045955
0.10	2.44281	2.44155	-0.0012576
0.20	2.98365	2.98395	0.0003043
0.30	3.64424	3.64412	-0.0001189
0.40	4.45108	4.45106	-0.0000235
0.50	5.43656	5.43654	-0.0000249
0.60	6.64023	6.64007	-0.0001672
0.70	8.11040	8.11074	0.0003389
0.80	9.90606	9.90444	-0.0016219
0.90	12.09929	12.10491	0.0056194
1.00	14.77811	14.75661	-0.0215021

x	uxx	usxx	diff
0.00	4.00000	3.82687	-0.1731250
0.10	4.88561	4.91217	0.0265622
0.20	5.96730	5.93594	-0.0313577
0.30	7.28848	7.26736	-0.0211146
0.40	8.90216	8.87143	-0.0307319
0.50	10.87313	10.83818	-0.0349520
0.60	13.28047	13.23238	-0.0480843
0.70	16.22080	16.18106	-0.0397388
0.80	19.81213	19.69302	-0.1191094
0.90	24.19859	24.31641	0.1178172
1.00	29.55622	28.71751	-0.8387162

x	uxxx	usxxx	diff
0.00	8.00000	10.85298	2.8529819
0.10	9.77122	10.85298	1.0817599
0.20	11.93460	13.31420	1.3795979
0.30	14.57695	13.31420	-1.2627549
0.40	17.80433	19.66744	1.8631082
0.50	21.74625	19.66744	-2.0788190
0.60	26.56094	29.48678	2.9258419
0.70	32.44160	29.48678	-2.9548224
0.80	39.62426	46.23387	6.6096075
0.90	48.39718	46.23387	-2.1633128
1.00	59.11245	44.01101	-15.1014368

The larger variation in e^{ax} and the resulting larger errors are clear in Table 1.7 when compared with Table 1.6. Again, these errors could be reduced by using a larger value of n. However, a reduction in the errors could also possibly be accomplished by concentrating the grid points near $x = 1$. That is, we could take advantage of the basic feature of splines that the grid spacing can be variable.

1.2 Variable Grids

To investigate this alternative, we consider first an example of how a variable grid can be produced. This is illustrated by the following code:

```
n=11;x=rep(0,n);
x[1]=0;xin=1/(n-1);
cat(sprintf("\n x[1]  = %6.4f   xin = %6.4f",
            x[1],xin));
for(i in 2:n){
  x[i]=x[i-1]+xin/i;
    cat(sprintf("\n i = %2d   x = %6.4f   dx = %6.4f",
              i,x[i],x[i]-x[i-1]));
}
```

Listing 1.3 Generation of a variable grid.

We can note the following details of Listing 1.3:

- A grid of n=11 points is declared with the rep utility.

```
n=11;x=rep(0,n);
```

- The first (leftmost) point is defined. Then an increment is computed.

```
x[1]=0;xin=1/(n-1);
cat(sprintf("\n x[1]  = %6.4f   xin = %6.4f",
            x[1],xin));
```

- A for is used to step through points i = 2,3,...,n.

```
for(i in 2:n){
  x[i]=x[i-1]+xin/i;
    cat(sprintf("\n i = %2d   x = %6.4f
                dx = %6.4f",i,x[i],x[i]-
                x[i-1]));
}
```

A particular value of the grid point, x[i], is computed from the previous point, u[i-1], plus an increment, xin/i. This increment decreases with increasing i, thereby giving a grid with a decreasing spacing.

The output from this code follows (Table 1.8).

Table 1.8 Output from Listing 1.3.

```
x[1]  =  0.0000   xin  =  0.1000

i  =   2   x  =  0.0500   dx  =  0.0500
i  =   3   x  =  0.0833   dx  =  0.0333
i  =   4   x  =  0.1083   dx  =  0.0250
i  =   5   x  =  0.1283   dx  =  0.0200
i  =   6   x  =  0.1450   dx  =  0.0167
i  =   7   x  =  0.1593   dx  =  0.0143
i  =   8   x  =  0.1718   dx  =  0.0125
i  =   9   x  =  0.1829   dx  =  0.0111
i  =  10   x  =  0.1929   dx  =  0.0100
i  =  11   x  =  0.2020   dx  =  0.0091
```

We can note the following details about this output:

- The grid spacing varies from 0.1000 at $x = 0$ to 0.0091 at $x = 0.2020$.
- The interval in x is not correct since it should be $0 \leq x \leq 1$. This can be corrected by using a normalizing factor as

```
x[1]=0;xin=(1/0.2020)/(n-1);
           .
           .
           .
x[i]=x[i-1]+xin/i;
```

in Listing 1.3. The resulting output is in Table 1.9.
The grid spacing is again variable (by more than a factor of 1/10, 0.4950 to 0.0450) and the interval in x is $0 \leq x \leq 1$.
This discussion is presented in some detail to illustrate how a variable grid might be generated. The approach can be generalized as

```
x[1]=0;xin=c/(n-1);
           .
           .
           .
x[i]=x[i-1]+xin/(i^p);
```

where c is a normalizing constant (e.g., $1/0.2020$) and p is a power of the grid index i. For $p < 1$, the grid spacing varies more slowly with x than for

Table 1.9 Output from Listing 1.3 with $0 \leq x \leq 1$.

```
x[1]  =  0.0000   xin = 0.4950

i =   2   x = 0.2475   dx = 0.2475
i =   3   x = 0.4125   dx = 0.1650
i =   4   x = 0.5363   dx = 0.1238
i =   5   x = 0.6353   dx = 0.0990
i =   6   x = 0.7178   dx = 0.0825
i =   7   x = 0.7885   dx = 0.0707
i =   8   x = 0.8504   dx = 0.0619
i =   9   x = 0.9054   dx = 0.0550
i =  10   x = 0.9549   dx = 0.0495
i =  11   x = 0.9999   dx = 0.0450
```

$p = 1$, and for $p > 1$, it varies more rapidly. An important detail is that the grid spacing varies smoothly with x (as in Table 1.9), since an abrupt change in the spacing can cause problems with the spline approximation.

The variable grid in Table 1.9 is now used as an input to `splinefun`. This requires only a change in the grid definition code in Listing 1.2 (with the use of e^{ax}).

```
#
# Previous workspaces are cleared
  rm(list=ls(all=TRUE))
#
# Define grid
  xl=0;xu=1;n=11;ng=2;
#
# Uniform grid
  if(ng==1){
    x=seq(from=xl,to=xu,by=(xu-xl)/(n-1));
    for(i in 2:n){
      cat(sprintf("\n i = %2d   x = %6.4f   dx = %6.4f",
                   i,x[i],x[i]-x[i-1]));
    }
  }
#
# Variable grid
  if(ng==2){
```

```
    x=rep(0,n);
    x[1]=0;xin=(1/0.2020)/(n-1);
    cat(sprintf("\n x[1] = %6.4f  xin = %6.4f",
               x[1],xin));
    for(i in 2:n){
      x[i]=x[i-1]+xin/i;
      cat(sprintf("\n i = %2d  x = %6.4f  dx = %6.4f",
                 i,x[i],x[i]-x[i-1]));
    }
  }
#
# Define function to be approximated, and its
# derivatives
    a=2;
    u=exp(a*x);
   ux=a*exp(a*x);
  uxx=a^2*exp(a*x);
 uxxx=a^3*exp(a*x);
#
# Set up spline table for function, derivatives
    utable=splinefun(x,u);
#
# Compute spline approximation of function
# derivatives
     us=utable(x);
    usx=utable(x,deriv=1);
   usxx=utable(x,deriv=2);
  usxxx=utable(x,deriv=3);
#
# Display comparison of function and its spline
# approximation, and its derivatives
#
# u
  cat(sprintf("\n     x          u          us
             diff"));
  for(i in 1:n){
    cat(sprintf("\n%7.4f%10.5f%10.5f%12.7f",
      x[i],u[i],us[i],us[i]-u[i]));
  }
#
# ux
  cat(sprintf("\n     x         ux         usx
             diff"));
```

```
  for(i in 1:n){
    cat(sprintf("\n%7.4f%10.5f%10.5f%12.7f",
      x[i],ux[i],usx[i],usx[i]-ux[i]));
  }
#
# uxx
  cat(sprintf("\n     x         uxx        usxx
              diff"));
  for(i in 1:n){
    cat(sprintf("\n%7.4f%10.5f%10.5f%12.7f",
      x[i],uxx[i],usxx[i],usxx[i]-uxx[i]));
  }
#
# uxxx
  cat(sprintf("\n     x         uxxx       usxxx
              diff"));
  for(i in 1:n){
    cat(sprintf("\n%7.4f%10.5f%10.5f%12.7f",
      x[i],uxxx[i],usxxx[i],usxxx[i]-uxxx[i]));
  }
```

Listing 1.4 Spline approximation of $y = e^{ax}$ and the first to third derivatives, uniform and variable grids.

For ng=1 a uniform grid is used (the output is in Table 1.7) and for ng=2 a variable grid is used. The output for ng=2 follows.

Table 1.10 Output for the function $y = e^{ax}$, $a = 2$, $n = 11$, $ng = 2$.

x	u	us	diff
0.0000	1.00000	1.00000	0.0000000
0.2475	1.64058	1.64058	0.0000000
0.4125	2.28207	2.28207	0.0000000
0.5363	2.92299	2.92299	0.0000000
0.6353	3.56309	3.56309	0.0000000
0.7178	4.20235	4.20235	0.0000000
0.7885	4.84083	4.84083	0.0000000
0.8504	5.47859	5.47859	0.0000000
0.9054	6.11570	6.11570	0.0000000
0.9549	6.75221	6.75221	0.0000000
0.9999	7.38816	7.38816	0.0000000

Table 1.10 (Continued)

x	ux	usx	diff
0.0000	2.00000	2.05688	0.0568850
0.2475	3.28116	3.26995	-0.0112076
0.4125	4.56414	4.56707	0.0029304
0.5363	5.84598	5.84545	-0.0005361
0.6353	7.12617	7.12639	0.0002166
0.7178	8.40470	8.40470	0.0000029
0.7885	9.68166	9.68168	0.0000179
0.8504	10.95719	10.95726	0.0000707
0.9054	12.23141	12.23122	-0.0001861
0.9549	13.50442	13.50513	0.0007133
0.9999	14.77632	14.77382	-0.0025024

x	uxx	usxx	diff
0.0000	4.00000	3.07122	-0.9287777
0.2475	6.56232	6.73035	0.1680358
0.4125	9.12828	8.99071	-0.1375689
0.5363	11.69196	11.66786	-0.0241047
0.6353	14.25234	14.20721	-0.0451373
0.7178	16.80939	16.77906	-0.0303371
0.7885	19.36332	19.33390	-0.0294229
0.8504	21.91438	21.89290	-0.0214782
0.9054	24.46281	24.42823	-0.0345759
0.9549	27.00883	27.03774	0.0289033
0.9999	29.55264	29.34267	-0.2099619

x	uxxx	usxxx	diff
0.0000	8.00000	14.78289	6.7828905
0.2475	13.12464	14.78289	1.6582549
0.4125	18.25655	21.63137	3.3748219
0.5363	23.38392	21.63137	-1.7525503
0.6353	28.50469	31.17082	2.6661338
0.7178	33.61879	31.17082	-2.4479661
0.7885	38.72664	41.35348	2.6268371
0.8504	43.82876	41.35348	-2.4752784
0.9054	48.92562	52.71191	3.7862934
0.9549	54.01766	52.71191	-1.3057497
0.9999	59.10527	51.21573	-7.8895430

A comparison of Tables 1.7 and 1.10 indicates the errors in the spline approximations are reduced for x near the right end $x = 1$, where e^{ax} changes more rapidly but are increased at the left end $x = 0$ due to the larger grid spacing. Thus, some experimentation and evaluation is required when using a variable grid. For example, smaller variations in the grid spacing at both ends could be used to reduce end effects.

The preceding discussion of uniform and variable grid spline differentiation can now be used to develop SCMOL for PDEs. This is done subsequently through example applications.

1.3   Stagewise Differentiation

An alternative approach to computing numerical derivatives is by successive differentiation, termed *stagewise differentiation*. This is illustrated by the following code applied to e^{ax}:

```
#
# Previous workspaces are cleared
  rm(list=ls(all=TRUE))
#
# Define uniform grid
  xl=0;xu=1;n=11;
  x=seq(from=xl,to=xu,by=(xu-xl)/(n-1));
#
# Define function to be approximated, and its
# derivatives
    a=1;
    u=exp(a*x);
  uxx=a^2*exp(a*x);
#
# Set up spline tables for u, ux; compute uxx
# by stagewise differentiation
  utable=splinefun(x,u);
  usx=utable(x,deriv=1);
  uxtable=splinefun(x,usx);
  usxx=uxtable(x,deriv=1);
#
# Display comparison of exact and spline derivatives
#
# uxx
  cat(sprintf("\n    x          uxx        usxx
              diff"));
```

```
for(i in 1:n){
  cat(sprintf("\n%6.4f%10.5f%10.5f%12.7f",
    x[i],uxx[i],usxx[i],usxx[i]-uxx[i]));
}
```

Listing 1.5 Stagewise differentiation of e^{ax}.

We can note the following details about Listing 1.5:

- Previous workspaces are removed and a uniform grid in x with 11 points is defined.

```
#
# Previous workspaces are cleared
  rm(list=ls(all=TRUE))
#
# Define uniform grid
  xl=0;xu=1;n=11;
  x=seq(from=xl,to=xu,by=(xu-xl)/(n-1));
```

- e^{ax} and its second derivative are computed and placed in two n-vectors, u, uxx.

```
#
# Define function to be approximated, and its
# derivatives
    a=1;
    u=exp(a*x);
  uxx=a^2*exp(a*x);
```

- splinefun operates on u to produce a spline table, utable, which is then used to calculate the first derivative, usx. splinefun then operates on usx to produce a spline table, uxtable, which is then used to calculate the second derivative, usxx. usx.

```
#
# Set up spline tables for u, ux; compute uxx
# by stagewise differentiation
  utable=splinefun(x,u);
  usx=utable(x,deriv=1);
  uxtable=splinefun(x,usx);
  usxx=uxtable(x,deriv=1);
```

Two successive calls to splinefun demonstrate the stagewise differentiation (note the use of deriv=1 for each stage). In principle, the successive differentiation can be continued to calculate higher-order derivatives, but in

practice, the additional error at each stage of the differentiation will eventually give inaccurate derivatives.
- The exact and numerical second derivatives, and their difference, are displayed.

```
#
# Display comparison of exact and spline derivatives
#
# uxx
  cat(sprintf("\n     x          uxx          usxx
              diff"));
  for(i in 1:n){
    cat(sprintf("\n%6.4f%10.5f%10.5f%12.7f",
      x[i],uxx[i],usxx[i],usxx[i]-uxx[i]));
  }
```

The output from Listing 1.5 is in Table 1.11, along with the second derivative by direct differentiation from Table 1.6.

Table 1.11 Output for the function $y = e^{ax}$, $a = 1$, $n = 11$, usxx by direct and stagewise differentiation.

Stagewise differentiation

x	uxx	usxx	diff
0.00	1.00000	0.99301	-0.0069892
0.10	1.10517	1.10498	-0.0001877
0.20	1.22140	1.22201	0.0006029
0.30	1.34986	1.34954	-0.0003202
0.40	1.49182	1.49198	0.0001514
0.50	1.64872	1.64858	-0.0001371
0.60	1.82212	1.82239	0.0002701
0.70	2.01375	2.01310	-0.0006504
0.80	2.22554	2.22675	0.0012137
0.90	2.45960	2.45950	-0.0001065
1.00	2.71828	2.70213	-0.0161501

Direct differentiation (from Table 1.6)

x	uxx	usxx	diff
0.00	1.00000	0.99030	-0.0097019

Table 1.11 (Continued)

0.10	1.10517	1.10663	0.0014559
0.20	1.22140	1.21975	-0.0016548
0.30	1.34986	1.34891	-0.0009520
0.40	1.49182	1.49053	-0.0012956
0.50	1.64872	1.64739	-0.0013349
0.60	1.82212	1.82050	-0.0016196
0.70	2.01375	2.01244	-0.0013094
0.80	2.22554	2.22232	-0.0032249
0.90	2.45960	2.46267	0.0030664
1.00	2.71828	2.69693	-0.0213550

The comparison in Table 1.11 indicates that stagewise differentiation gives smaller errors than direct differentiation (this conclusion is for e^{ax} only and does not constitute a general result or proof). This validation is important since stagewise differentiation can be used to implement several types of PDE boundary conditions (BCs), as explained in examples to follow.

The preceding discussion of spline differentiation will now be used as the basis for the SCMOL integration of PDEs.

Appendix A1 – Online Documentation for `splinefun`

The technical details and options of `splinefun` are provided in the following online documentation produced at the R prompt by entering `help(splinefun)`.

```
splinefun {stats} R Documentation

Interpolating Splines

Description

Perform cubic (or Hermite) spline interpolation of
given data points, returning either a list of points
obtained by the interpolation or a function
performing the interpolation.
```

Usage

```
splinefun(x, y = NULL,
  method = c("fmm", "periodic", "natural",
   "monoH.FC"),
  ties = mean)

spline(x, y = NULL, n = 3*length(x), method = "fmm",
  xmin = min(x), xmax = max(x), xout, ties = mean)

splinefunH(x, y, m)
```
Arguments

x,y
vectors giving the coordinates of the points to be
inter-polated.   Alternatively a single plotting
structure can be specified: see xy.coords.

m
(for splinefunH()): vector of slopes m[i] at the
points (x[i],y[i]); these together determine the
Hermite ?spline? which is piecewise cubic, (only) once
differentiable continuously.

method
specifies the type of spline to be used. Possible values are "fmm","natural", "periodic" and "monoH.FC".

n
if xout is left unspecified, interpolation takes place
at n equally spaced points spanning the interval
[xmin, xmax].

xmin, xmax
left-hand and right-hand endpoint of the interpolation
interval (when xout is unspecified).

xout
an optional set of values specifying where interpolation is to take place.

ties

Handling of tied x values. Either a function with a
single vector argument returning a single number
result or the string "ordered".

Details

The inputs can contain missing values that are
deleted, so at least one complete (x, y) pair is
required. If method = "fmm", the spline used is that
of Forsythe, Malcolm and Moler (an exact cubic is fitted through the four points at each end of the data,
and this is used to determine the end conditions).
Natural splines are used when method = "natural", and
periodic splines when method = "periodic".

The new (R 2.8.0) method "monoH.FC" computes a monotone Hermite spline according to the method of Fritsch
an Carlson. It does so by determining slopes such that
the Hermite spline, determined by (x[i],y[i],m[i]), is
monotone (increasing or decreasing) iff the data are.

These interpolation splines can also be used for
extrapolation, that is prediction at points outside
the range of x. Extrapolation makes little sense for
method = "fmm"; for natural splines it is linear using
the slope of the interpolating curve at the nearest
data point.

Value

spline returns a list containing components x and y
which give the ordinates where interpolation took
place and the interpolated values.

splinefun returns a function with formal arguments x
and deriv, the latter defaulting to zero. This function can be used to evaluate the interpolating cubic
spline (deriv=0), or its derivatives (deriv=1,2,3) at

```
the points x, where the spline function interpolates
the data points originally specified. This is often
more useful than spline.
```

```
References
Becker, R. A., Chambers, J. M. and Wilks, A. R. (1988)
The New S Language. Wadsworth & Brooks/Cole.
```

```
Forsythe, G. E., Malcolm, M. A. and Moler, C. B.
(1977) Computer Methods for Mathematical Computations.
Prentice-Hall
```

```
Fritsch, F. N. and Carlson, R. E. (1980) Monotone
piecewise cubic interpolation, SIAM Journal on Numeri-
cal Analysis 17, 238-246.
```

Reference

1 Forsythe, G.E., M.A. Malcolm, and C.B. Moler (1977), *Computer Methods for Mathematical Computations*, Prentice-Hall, Englewood Cliffs, NJ, pp. 70–79.

2

One-Dimensional PDEs

In this chapter, spline collocation method of lines (SCMOL) is applied to a series of one-dimensional (1D) PDEs. The intent is to demonstrate how various PDE forms can be accommodated within the SCMOL framework.

2.1 Constant Coefficient

The discussion of SCMOL starts with one-dimensional (1D), constant coeffi-cient PDEs, specifically, the diffusion equation (Fick's second law).

$$\frac{\partial u}{\partial t} = D\frac{\partial^2 u}{\partial x^2} \tag{2.1}$$

where

u dependent variable
t initial-value independent variable
x boundary-value independent variable
D diffusivity.

The solution of Eq. (2.1) is $u(x, t)$ primarily in numerical form, with analytical solutions used in some cases to verify the numerical solutions.

Equation (2.1) is first order in t and second order in x. It therefore requires one initial condition (IC) and two boundary conditions (BCs). The IC is

$$u(x, t = 0) = g_1(x) \tag{2.2}$$

where $g_1(x)$ is a function to be specified.

The essential features of SCMOL are explained by application to Eqs. (2.1) and (2.2):

- Equation (2.1) is approximated as

$$\frac{du_i}{dt} = D\frac{d^2 u}{dx_i^2} \tag{2.3a}$$

Spline Collocation Methods for Partial Differential Equations: With Applications in R, First Edition.
William E. Schiesser.
© 2017 John Wiley & Sons, Inc. Published 2017 by John Wiley & Sons, Inc.
Companion website: www.wiley.com/go/Spline_Collocation

where u_i is the approximation of u in Eq. (2.1) on a grid of n points in x, that is, $du_i/dt, i = 1, 2, \ldots, n$. In other words, Eq. (2.3a) are a system of n initial-value ODEs in t.

- $\dfrac{d^2 u}{dx_i^2}$ is the spline approximation of $\partial^2 u / \partial x^2$ in Eq. (2.1) at point i. $\dfrac{d^2 u}{dx_i^2}$ is computed by calls to splinefun as discussed in Chapter 1.
- The ICs for Eq. (2.3a) are from Eq. (2.2)

$$u(x_i, t = 0) = g_1(x_i) \tag{2.3b}$$

Equations (2.3) constitute the SCMOL approximation of Eq. (2.1). BCs for Eq. (2.1) are included in Eq. (2.3a) as explained in a subsequent series of examples.

Equation (2.3a) can be written in the alternative form

$$\frac{du_i}{dt} - D\frac{d^2 u}{dx_i^2} = R(x_i, t) = 0 \tag{2.3c}$$

where $R(x_i, t)$ is the residual of the numerical (SCMOL) solution at point $x = x_i$. This residual is set to zero at each of the grid points which is termed *collocation*. This procedure can be considered as a special case of *weighted residuals* in which the weighting function is a delta function with the property

$$R(x_i, t) = \int_{x_l}^{x_u} \delta(\lambda - x_i) R(\lambda, t) d\lambda$$

Since the approximation of the derivative in x in Eq. (2.1) is by splines, the procedure illustrated by the collocation of Eq. (2.3c) is termed *spline collocation* (SC).

To complete the specification of the PDE problem, BCs for Eq. (2.1) must also be included. Various BCs are considered next.

2.1.1 Dirichlet BCs

If the dependent variable is specified at the boundaries, the BCs are termed *Dirichlet*. They can be stated generally as

$$u(x = x_l, t) = g_2(t); \quad u(x = x_u, t) = g_3(t) \tag{2.4a,b}$$

where $g_2(t), g_3(t)$ are functions to be specified and x_l, x_u are boundary values of x to be specified.

For the PDE application discussed next,

$$g_1(x) = \sin(\pi x)(\text{ncase} = 1); \quad g_1(x) = 1(\text{case} = 2) \tag{2.5a,b}$$

$$g_2(t) = g_3(t) = 0 \ (homogeneous \text{ (zero) Dirichlet BCs)} \tag{2.6a,b}$$

2.1.1.1 Main Program

An R main program for Eqs. (2.1) and (2.2) for the special case of Eqs. (2.4)–(2.6) is listed next.

```
#
# Delete previous workspaces
  rm(list=ls(all=TRUE))
#
# Access ODE integrator
  library("deSolve");
#
# Access functions for numerical solution
  setwd("f:/collocation/test2");
  source("pde_1a.R");
#
# Step through ncase
  for(ncase in 1:2){
#
# Grid (in x)
  n=21;xl=0;xu=1;
  x=seq(from=xl,to=xu,by=(xu-xl)/(n-1));
#
# Parameters
  D=1;
#
# Independent variable for ODE integration
  nout=6;t0=0;tf=0.25;
  tout=seq(from=t0,to=tf,by=(tf-t0)/(nout-1));
#
# Initial condition
  if(ncase==1)u0=sin(pi*x);
  if(ncase==2)u0=rep(1,n);
  ncall=0;
#
# ODE integration
  out=ode(func=pde_1a,y=u0,times=tout);
#
# Arrays for display
   u=matrix(0,nrow=n,ncol=nout);
  ua=matrix(0,nrow=n,ncol=nout);
  for(it in 1:nout){
  for(i   in 1:n){
    u[i,it]=out[it,i+1];
```

```
    if(ncase==1){ua[i,it]=sin(pi*x[i])*exp(-D*(pi^2)*
                 tout[it]);}
  }
  }
  cat(sprintf("\n ncase = %2d",ncase));
#
# Numerical, analytical solutions, ncase = 1
  if(ncase==1){
    cat(sprintf("\n        t       x      u(x,t)    ua(x,t)
                 diff"));
    for(it in 1:nout){
    for(i  in 1:n){
      cat(sprintf("\n %6.2f%6.2f%10.4f%10.4f
                   %12.6f",tout[it],x[i],u[i,it],
                   ua[i,it],u[i,it]-ua[i,it]));
    }
    }
    matplot(x,u,type="l",lwd=2,col="black",lty=1,
            xlab="x",ylab="u(x,t)");
    matpoints(x,ua,pch="o",col="black");
#
# End ncase=1
  }
#
# Numerical solution, ncase = 2
  if(ncase==2){
  cat(sprintf("\n        t       x     u(x,t)"));
    for(it in 1:nout){
    for(i  in 1:n){
      cat(sprintf("\n %6.2f%6.2f%10.4f",
        tout[it],x[i],u[i,it]));

    }
    }
    matplot(x[2:(n-1)],u[2:(n-1),],type="l",
            lwd=2,col="black",lty=1,xlab="x",
            ylab="u(x,t)");
#
# End ncase=2
  }
#
# Calls to ODE routine
```

```
  cat(sprintf("\n\n  ncall = %3d\n",ncall));
#
# Next case
  }
```

Listing 2.1 Main program for Eqs. (2.1) and (2.2) for the special case of Eqs. (2.4)–(2.6).

We can note the following details about Listing 2.1:

- Previous workspaces are removed. Then the ODE integrator library deSolve[1] is accessed. Note the use of / rather than the usual \ in setwd, and the required revision for the local computer (to define the working folder or directory for the SCMOL ODE routine pde_1a, discussed subsequently).

```
#
# Delete previous workspaces
  rm(list=ls(all=TRUE))
#
# Access ODE integrator
  library("deSolve");
#
# Access functions for numerical solution
  setwd("f:/collocation/test2");
  source("pde_1a.R");
```

- Two cases are executed corresponding to Eqs. (2.5).

```
#
# Step through ncase
  for(ncase in 1:2){
```

- A uniform grid in x of 21 points is defined with the seq utility for the interval $x_l \le x \le x_u$. Therefore, the vector x has the values $x = 0, 0.05, \ldots, 1$.

```
#
# Grid (in x)
  n=21;xl=0;xu=1;
  x=seq(from=xl,to=xu,by=(xu-xl)/(n-1));
```

---

1 deSolve is open source and is available from the R site. It contains approximately 20 quality library ODE integrators, including the LSODE (Livermore Solvers for Ordinary Differential Equations) series and daspk, radau for differential-algebraic Equations (DAE), plus several integrators for nonstiff ODEs, for example, ode45. In the main program of Listing 2.1, the generic ode integrator is used, which has as a default lsoda that automatically switches between stiff and nonstiff methods. Additional information is available by entering ??deSolve at the R prompt.

- The model parameters are defined numerically. In this case, there is only one, the constant diffusivity D in Eq. (2.1). This parameter is available to the ODE/SCMOL subordinate routine pde_1a without any special designation, a feature of R. The converse is not true. Parameters set in subroutines are not available to the superior (calling) routine without a special designation, as will be illustrated when pde_1a is considered subsequently.

```
#
# Parameters
  D=1;
```

- A uniform grid in t of 6 output points is defined with the seq utility for the interval $0 \le t \le t_f = 0.25$. Therefore, the vector tout has the values $t = 0, 0.05, \dots, 0.25$.

```
#
# Independent variable for ODE integration
  nout=6;t0=0;tf=0.25;
  tout=seq(from=t0,to=tf,by=(tf-t0)/(nout-1));
```

At this point, the intervals of x and t in Eq. (2.1) are defined.
- The IC function $g_1(x)$ in Eqs. (2.2) and (2.5) is defined for the two cases executed within the previous for.

```
#
# Initial condition
  if(ncase==1)u0=sin(pi*x);
  if(ncase==2)u0=rep(1,n);
  ncall=0;
```

Again, the vectorization available within R is used (u0 is an n-vector). For ncase=2, the IC $u(x, t = 0) = 1$ is implemented with the rep utility. Also, the counter for the calls to pde_1a is initialized (and passed to pde_1a without a special designation).
- The system of 21 SCMOL ODEs is integrated by the library integrator ode (available in deSolve). As expected, the inputs to ode are the ODE function, pde_1a, the IC vector u0, and the vector of output values of t, tout. The length of u0 (e.g., 21) informs ODE how many ODEs are to be integrated. func, y, times are reserved names.

```
#
# ODE integration
  out=ode(func=pde_1a,y=u0,times=tout);
```

The numerical solution to the ODEs is returned in matrix out. In this case, out has the dimensions $nout \times (n + 1) = 6 \times 22$. The offset $n + 1$ is required since the first element of each column has the output t (also in tout), and the $2, \ldots, n + 1 = 2, \ldots, 22$ column elements have the 21 ODE solutions. This indexing of out is used next.

- The ODE solution is placed in a 21 × 6 matrix, u, for subsequent plotting (by stepping through the solution with respect to x and t within a pair of fors). The analytical solution for ncase=1,

$$u_a(x, t) = \sin(\pi x)e^{-D\pi^2 t} \tag{2.7}$$

is also placed in array ua

```
#
# Arrays for display
   u=matrix(0,nrow=n,ncol=nout);
   ua=matrix(0,nrow=n,ncol=nout);
   for(it in 1:nout){
   for(i   in 1:n){
     u[i,it]=out[it,i+1];
     if(ncase==1){ua[i,it]=sin(pi*x[i])*
                  exp(-D*(pi^2)*tout[it]);}
   }
   }
   cat(sprintf("\n ncase = %2d",ncase));
```

The value of ncase identifies the two solutions (from the for with index ncase).

- The numerical and analytical solutions and the difference for ncase=1 are displayed, then plotted with the matplot, matpoints utilities.

```
#
# Numerical, analytical solutions, ncase = 1
   if(ncase==1){
     cat(sprintf("\n       t      x      u(x,t)     ua(x,t)
                  diff"));
     for(it in 1:nout){
     for(i   in 1:n){
       cat(sprintf("\n %6.2f%6.2f%10.4f%10.4f
                  %12.6f",tout[it],x[i],u[i,it],
                  ua[i,it],u[i,it]-ua[i,it]));
     }
     }
     matplot(x,u,type="l",lwd=2,col="black",lty=1,
             xlab="x",ylab="u(x,t)");
     matpoints(x,ua,pch="o",col="black");
```

The numerical solution in array u is plotted as a solid line, and the analytical solution in ua is plotted as points against x. This formatting appears in

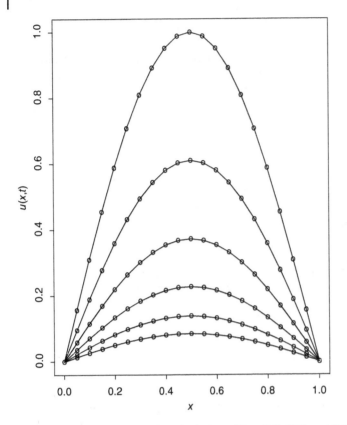

Figure 2.1 Numerical and analytical solutions of Eqs. (2.1), (2.5), and (2.6), ncase=1 solid – num, points – anal.

Figure 2.1 (discussed subsequently). matplot will plot u against x with *t* as a parameter since the row dimension of u (21) matches the length of x (also 21). The same is true in plotting ua with matpoints.

- The display of the numerical solution for ncase=2 is similar. The analytical solution is not included because of its complexity (an infinite series). The numerical solution appears as a solid line in Figure 2.2.

```
#
# Numerical solution, ncase = 2
  if(ncase==2){
  cat(sprintf("\n        t      x      u(x,t)"));
    for(it in 1:nout){
    for(i in 1:n){
      cat(sprintf("\n %6.2f%6.2f%10.4f",
        tout[it],x[i],u[i,it]));
```

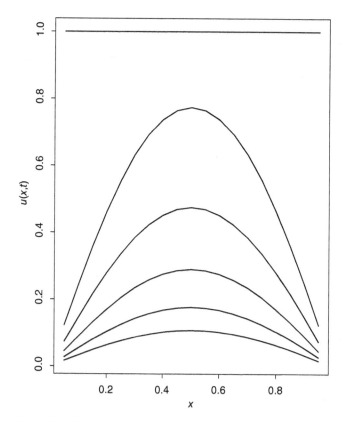

Figure 2.2 Numerical solution of Eqs. (2.1), (2.5), and (2.6), `ncase=2`.

```
    }
  }
  matplot (x[2:(n-1)],u[2:(n-1),],type="l",
           lwd=2,col="black",lty=1,xlab="x",
           ylab="u(x,t)");
#
# End ncase=2
  }
```

The discontinuity between the IC $u(x, t = 0) = 1$ of Eq. (2.5b) and the BCs of Eqs. (2.6) $u(x = 0, t) = u(x = 1, t) = 0$ is not plotted (it involves a solid line connecting the two values, from 0 to 1), so the x and u arrays are subscripted as x[2:(n−1)], u[2:(n−1),] in the call to matplot. This coding illustrates two important programming features for indices: (i) an interval in an index can be designated with : and (ii) all values of an index can be designated with , (e.g., in u). Note also that (n−1) is used rather than just n−1 since in R, these are different.

- The number of calls to pde_1a is displayed as an indication of the computational effort required to compute the solution.

```
#
# Calls to ODE routine
  cat(sprintf("\n\n   ncall = %3d\n",ncall));
#
# Next case
  }
```

The final } concludes the for in ncase.

The ODE routine called by ode is considered next.

2.1.1.2   ODE Routine

pde_1a called by the main program of Listing 2.1 is in Listing 2.2.

```
  pde_1a=function(t,u,parms) {
#
# Function pde_1a computes the t derivative
# vector of u(x,t)
#
# BCs
  u[1]=0;u[n]=0;
#
# uxx
  table=splinefun(x,u);
  uxx=table(x,deriv=2);
#
# PDE
  ut=D*uxx;
  ut[1]=0;ut[n]=0;
#
# Increment calls to pde_1
  ncall <<- ncall+1;
#
# Return derivative vector
  return(list(c(ut)));
  }
```

Listing 2.2 ODE/SCMOL routine pde_1a for Eqs. (2.1), (2.5), and (2.6).

We can note the following details about pde_1a:

- The function is defined.

  ```
  pde_1a=function(t,u,parms){
  #
  # Function pde_1a computes the t derivative
  # vector of u(x,t)
  ```

  t is the current value of t in Eq. (2.1). u is the 21-vector of ODE-dependent variables (of Eq. (2.3a)). parm is an argument to pass parameters to pde_1a (unused, but required in the argument list). The arguments must be listed in the order stated to properly interface with ode called in the main program. The derivative vector of the LHS of Eq. (2.3a) is calculated next and returned to ode.
- The homogeneous Dirichlet BCs of Eqs. (2.6) are implemented. Note the use of the subscripts 1, n corresponding to $x = x_l = 0, x = x_u = 1$.

  ```
  #
  # BCs
    u[1]=0;u[n]=0;
  ```

- splinefun calculates $\dfrac{d^2u}{dx_i^2}$ in Eq. (2.3a) as uxx.

  ```
  #
  # uxx
    table=splinefun(x,u);
    uxx=table(x,deriv=2);
  ```

  A second derivative is specified with deriv=2.
- Equation (2.1) is programmed with vectorization. The derivatives at the boundaries are set to zero so that the ODE integrator does not move them away from the prescribed values of Eqs. (2.6).

  ```
  #
  # PDE
    ut=D*uxx;
    ut[1]=0;ut[n]=0;
  ```

- The counter for the calls to pde_1a is incremented and its value is returned to the calling program (of Listing 2.1) with the <<- operator.

  ```
  #
  # Increment calls to pde_1a
    ncall <<- ncall+1;
  ```

- The derivative vector (LHS of Eq. (2.3a)) is returned to ode, which requires a list. c is the R vector utility. The combination of return, list, c gives ode (the ODE integrator called in the main program of Listing 2.1) the required derivative vector for the next step along the solution.

```
#
# Return derivative vector
  return(list(c(ut)));
  }
```

The final } concludes pde_1a.

Execution of the preceding main program and subordinate ODE routine produces the following abbreviated numerical output in Table 2.1 for ncase=1 (with the output for $t = 0, 0.05, \ldots, 0.20$ deleted to conserve space).

Table 2.1 Abbreviated output for Eqs. (2.1), (2.5), and (2.6).

```
ncase =   1
```

t	x	u(x,t)	ua(x,t)	diff
0.25	0.00	0.0000	0.0000	0.000000
0.25	0.05	0.0132	0.0133	-0.000065
0.25	0.10	0.0261	0.0262	-0.000130
0.25	0.15	0.0383	0.0385	-0.000192
0.25	0.20	0.0496	0.0498	-0.000249
0.25	0.25	0.0597	0.0600	-0.000299
0.25	0.30	0.0683	0.0686	-0.000343
0.25	0.35	0.0752	0.0756	-0.000378
0.25	0.40	0.0803	0.0807	-0.000404
0.25	0.45	0.0833	0.0838	-0.000419
0.25	0.50	0.0844	0.0848	-0.000424
0.25	0.55	0.0833	0.0838	-0.000419
0.25	0.60	0.0803	0.0807	-0.000404
0.25	0.65	0.0752	0.0756	-0.000378
0.25	0.70	0.0683	0.0686	-0.000343
0.25	0.75	0.0597	0.0600	-0.000299
0.25	0.80	0.0496	0.0498	-0.000249
0.25	0.85	0.0383	0.0385	-0.000192
0.25	0.90	0.0261	0.0262	-0.000130
0.25	0.95	0.0132	0.0133	-0.000065
0.25	1.00	0.0000	0.0000	0.000000

```
ncall = 206
```

Note in particular the symmetry around x=0.5 corresponding to a maximum (zero slope). Also, the homogeneous Dirichlet BCs at x=0,1 are clear. Table 2.1 indicates the agreement between the numerical and analytical solutions is acceptable considering that $n = 21$ points were used (the errors could be reduced by increasing the number of points). Also, $t = 0.25$ is at the end of the solution and there is no indication that integration errors have accumulated significantly. Finally, the numerical solution was computed with modest effort (ncall = 206).

The numerical output for ncase=2 is not presented, but the solutions for ncase=1,2 are plotted in Figures 2.1 and 2.2.

Figure 2.1 indicates the agreement between the numerical and analytical solutions for $t = 0, 0.05, \ldots, 0.25$. As expected from the homogeneous Dirichlet BCs (Eqs. (2.6)), the solution appears to be $u(x, t \to \infty) = 0$.

Figure 2.2 indicates the numerical solution for $t = 0, 0.05, \ldots, 0.25$ without the boundary points $x = x_l = 0, x = x_u = 1$ as discussed previously. Again, as expected from the homogeneous Dirichlet BCs (Eqs. (2.6)), the solution appears to be $u(x, t \to \infty) = 0$.

An important conclusion from Figure 2.2 is that the discontinuity between the IC and BCs did not cause a computational problem and could be accommodated within the SCMOL framework. This is due to the fact that Eq. (2.1) is parabolic, that is, it models diffusion that smooths discontinuities. This would not be so for hyperbolic PDEs that propagate discontinuities and are therefore relatively difficult to accommodate numerically. This point is discussed later when first- and second-order hyperbolic PDEs are considered.

In summary, the example of Section (2.1) indicates that the SCMOL programming of Eqs. (2.1), (2.5), and (2.6) is straightforward. The numerical solution of Eq. (2.1) for homogeneous Neumann BCs is considered next.

2.1.2 Neumann BCs

If the derivative of the dependent variable with respect to x is specified at the boundaries, the BCs are termed *Neumann*. They can be stated generally as

$$\frac{\partial u(x = x_l, t)}{\partial x} = g_2(t); \qquad \frac{\partial u(x = x_u, t)}{\partial x} = g_3(t) \qquad (2.8a,b)$$

where $g_2(t), g_3(t)$ are functions to be specified and x_l, x_u are boundary values of x to be specified.

For the PDE application discussed next,

$$g_1(x) = \cos(\pi x)(\text{ncase} = 1); \qquad g_1(x) = 1(\text{ncase} = 2) \qquad (2.9a,b)$$

$$g_2(t) = g_3(t) = 0(\text{homogeneous BCs}) \qquad (2.10a,b)$$

2.1.2.1 Main Program

An R main program for Eqs. (2.1), (2.9), and (2.10) is listed next.

```
#
# Delete previous workspaces
  rm(list=ls(all=TRUE))
#
# Access ODE integrator
  library("deSolve");
#
# Access functions for numerical solution
  setwd("f:/collocation/test3");
  source("pde_1a.R");
#
# Step through ncase
  for(ncase in 1:2){
#
# Grid (in x)
  n=21;xl=0;xu=1;
  x=seq(from=xl,to=xu,by=(xu-xl)/(n-1));
#
# Parameters
  D=1;
#
# Independent variable for ODE integration
  nout=6;t0=0;tf=0.25;
  tout=seq(from=t0,to=tf,by=(tf-t0)/(nout-1));
#
# Initial condition
  if(ncase==1)u0=cos(pi*x);
  if(ncase==2)u0=rep(1,n);
  ncall=0;
#
# ODE integration
  out=ode(func=pde_1a,y=u0,times=tout);
#
# Arrays for display
   u=matrix(0,nrow=n,ncol=nout);
  ua=matrix(0,nrow=n,ncol=nout);
  for(it in 1:nout){
  for(i in 1:n){
```

```
    u[i,it]=out[it,i+1];
   if(ncase==1){ua[i,it]=cos(pi*x[i])*exp(-D*(pi^2)*
              tout[it]);}
    }
   }
   cat(sprintf("\n ncase = %2d",ncase));
#
# Numerical, analytical solutions, ncase = 1
   if(ncase==1){
   cat(sprintf("\n        t      x      u(x,t)    ua(x,t)
              diff"));
   for(it in 1:nout){
   for(i  in 1:n){
     cat(sprintf("\n %6.2f%6.2f%10.4f%10.4f
                 %12.6f",tout[it],x[i],u[i,it],
                 ua[i,it],u[i,it]-ua[i,it]));
   }
   }
   matplot(x,u,type="l",lwd=2,col="black",lty=1,
           xlab="x",ylab="u(x,t)");
   matpoints(x,ua,pch="o",col="black");
   }
#
# End ncase=1
#
# Numerical solution, ncase = 2
   if(ncase==2){
   cat(sprintf("\n        t      x      u(x,t)"));
   for(it in 1:nout){
   for(i  in 1:n){
     cat(sprintf("\n %6.2f%6.2f%10.4f",
       tout[it],x[i],u[i,it]));
   }
   }
   matplot(x[2:(n-1)],u[2:(n-1),],type="l",
           lwd=2,col="black",lty=1,xlab="x",
           ylab="u(x,t)");
#
# End ncase=2
   }
```

```
#
# Calls to ODE routine
  cat (sprintf ("\n\n  ncall = %3d\n", ncall));
#
# Next case
  }
```

Listing 2.3 Main program for Eqs. (2.1), (2.9), and (2.10).

Listing 2.3 is similar to Listing 2.1. Therefore, only the differences are discussed next.

- The IC is changed from sin to cos for ncase=1 in accordance with Eqs. (2.9).

```
#
# Initial condition
  if (ncase==1) u0=cos (pi*x);
  if (ncase==2) u0=rep (1, n);
  ncall=0;
```

- The analytical solution has cos in place of sin.

```
#
# Arrays for display
  u=matrix (0, nrow=n, ncol=nout);
  ua=matrix (0, nrow=n, ncol=nout);
  for (it in 1:nout) {
  for (i  in 1:n) {
    u[i, it] =out [it, i+1];
    if (ncase==1) {ua [i, it] =cos (pi*x [i]) * exp (-D* (pi^2) *
                  tout [it]); }
  }
  }
  cat (sprintf ("\n ncase = %2d", ncase));
```

The ODE routine called by ode is considered next.

2.1.2.2 ODE Routine
pde_1a called by the main program of Listing 2.3 is in Listing 2.4.

```
  pde_1a=function (t, u, parms) {
#
# Function pde_1a computes the t derivative vectors of
# u (x, t)
#
# ux
```

```
      tablex=splinefun(x,u);
      ux=tablex(x,deriv=1);
#
#  BCs
      ux[1]=0;ux[n]=0;
#
#  uxx
      tablexx=splinefun(x,ux);
      uxx=tablexx(x,deriv=1);
#
#  PDE
      ut=D*uxx;
#
#  Increment calls to pde_1a
      ncall <<- ncall+1;
#
#  Return derivative vector
      return(list(c(ut)));
      }
```

Listing 2.4 ODE/SCMOL routine pde_1a for Eqs. (2.1), (2.9), and (2.10).

Listing 2.4 is similar to Listing 2.2, so only the differences are discussed next.

- The first derivative in x is computed by splinefun.

```
#
#  ux
      tablex=splinefun(x,u);
      ux=tablex(x,deriv=1);
```

- The homogeneous Neumann BCs of Eqs. (2.10) are implemented. Note the use of the subscripts 1 , n corresponding to $x = x_l = 0, x = x_u = 1$.

```
#
#  BCs
      ux[1]=0;ux[n]=0;
```

- splinefun calculates $\dfrac{d^2u}{dx_i^2}$ in Eq. (2.3a) as uxx.

```
#
#  uxx
      tablexx=splinefun(x,ux);
      uxx=tablexx(x,deriv=1);
```

This code is an application of stagewise differentiation since uxx is calculated from ux (this indicates the essential use of stagewise differentiation when derivatives are specified in BCs).

- Equation (2.1) is programmed with vectorization.

```
#
# PDE
  ut=D*uxx;
```

- The counter for the calls to pde_1a is incremented and its value returned to the calling program (of Listing 2.1) with the <<- operator.

```
#
# Increment calls to pde_1a
  ncall <<- ncall+1;
```

- The derivative vector (LHS of Eq. (2.3a)) is returned to ode with a combination of return, list, c.

```
#
# Return derivative vector
  return(list(c(ut)));
  }
```

The final } concludes pde_1a.

Execution of the preceding main program and subordinate ODE routine produces the following abbreviated numerical output in Table 2.2 for ncase=1 (with the output for $t = 0, 0.05, \ldots, 0.20$ deleted to conserve space).

Note in particular the antisymmetry (symmetry with a sign change) around x=0.5 corresponding to a zero value at $x = 0.5$. Also, the homogeneous Neumann BCs (zero slope) at x=0, 1 are clear. Table 2.2 indicates the agreement between the numerical and analytical solutions is better than six figures except at the boundaries $x = x_l, x_u$. The numerical solution was computed with modest effort (ncall = 247).

The numerical output for ncase=2 is not presented, but the solutions for ncase=1, 2 are plotted in Figures 2.3 and 2.4.

Figure 2.3 indicates the agreement between the numerical and analytical solutions for $t = 0, 0.05, \ldots, 0.25$. As expected from the homogeneous Neumann BCs (Eqs. (2.10)), the solution appears to be $u(x, t \to \infty) = 0$.

In Figure 2.4, the solution remains unchanged from the IC as expected from the homogeneous Neumann BCs (Eqs. (2.10)), that is, $u(x, t \to \infty) = 1$.

In summary, the solution of Eqs. (2.1), (2.9), and (2.10) indicates that Neumann BCs can be accommodated within the SCMOL framework by using stagewise differentiation. The following example demonstrates the solution of Eq. (2.1) with Robin BCs.

Table 2.2 Abbreviated output for Eqs. (2.1), (2.9), and (2.10).

```
ncase =   1
    t       x      u(x,t)      ua(x,t)         diff
   0.25   0.00    0.0848       0.0848      -0.000012
   0.25   0.05    0.0838       0.0838      -0.000000
   0.25   0.10    0.0807       0.0807      -0.000001
   0.25   0.15    0.0756       0.0756      -0.000000
   0.25   0.20    0.0686       0.0686      -0.000000
   0.25   0.25    0.0600       0.0600      -0.000000
   0.25   0.30    0.0498       0.0498      -0.000000
   0.25   0.35    0.0385       0.0385      -0.000000
   0.25   0.40    0.0262       0.0262      -0.000000
   0.25   0.45    0.0133       0.0133       0.000000
   0.25   0.50    0.0000       0.0000       0.000000
   0.25   0.55   -0.0133      -0.0133      -0.000000
   0.25   0.60   -0.0262      -0.0262       0.000000
   0.25   0.65   -0.0385      -0.0385       0.000000
   0.25   0.70   -0.0498      -0.0498       0.000000
   0.25   0.75   -0.0600      -0.0600       0.000000
   0.25   0.80   -0.0686      -0.0686       0.000000
   0.25   0.85   -0.0756      -0.0756       0.000000
   0.25   0.90   -0.0807      -0.0807       0.000001
   0.25   0.95   -0.0838      -0.0838       0.000000
   0.25   1.00   -0.0848      -0.0848       0.000012

ncall = 247
```

2.1.3 Robin BCs

If the derivative of the dependent variable with respect to x and the dependent variable are expressed as a linear combination at the boundaries, the BCs are termed *Robin*. They can be stated generally as

$$\frac{\partial u(x = x_l, t)}{\partial x} + k_1 u(x = x_l, t) = g_2(t);$$

$$\frac{\partial u(x = x_u, t)}{\partial x} + k_2 u(x = x_l, t) = g_3(t) \qquad (2.11a,b)$$

where $g_2(t), g_3(t)$ are functions to be specified and k_1, k_2 are constants to be specified.

Although an analytical solution for Eq. (2.1) with IC Eq. (2.2) and BCs (2.11) is available (to verify the numerical solution), it is not considered here since

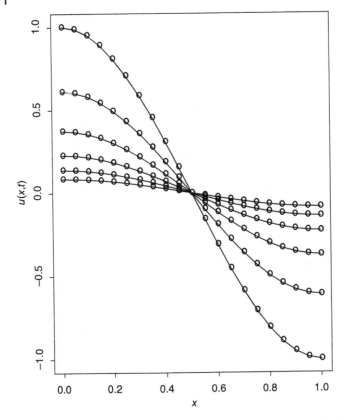

Figure 2.3 Numerical and analytical solutions of Eqs. (2.1), (2.9), and (2.10), `ncase=1` solid – num, points – anal.

it is relatively complicated (an infinite series of special eigenfunctions and associated eigenvalues). Therefore, an alternative approach is used to evaluate the numerical solution, that is, a second numerical solution based on finite differences (FDs). The details for computing the second solution are given in the routines that follow.

2.1.3.1 Main Program
An R main program for Eqs. (2.1), (2.2), and (2.11) is listed next.

```
#
# Delete previous workspaces
  rm(list=ls(all=TRUE))
#
# Access ODE integrator
  library("deSolve");
```

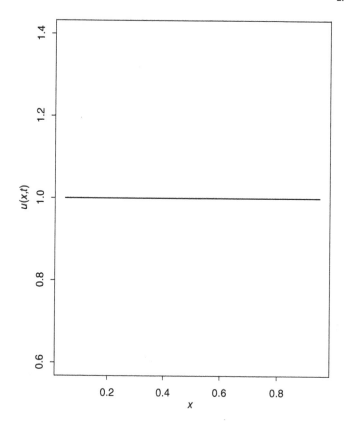

Figure 2.4 Numerical solution of Eqs. (2.1), (2.9), and (2.10), ncase=2.

```
#
# Access functions for numerical solutions
  setwd("f:/collocation/test4");
  source("pde_1a.R");
  source("dss044.R");
#
# Grid (in x)
  n=21;xl=0;xu=1;
  x=seq(from=xl,to=xu,by=(xu-xl)/(n-1));
#
# Parameters
  D=1;ua=0;
#
# Independent variable for ODE integration
  nout=6;t0=0;tf=1;
```

```
  tout=seq(from=t0,to=tf,by=(tf-t0)/(nout-1));
#
# Initial condition
  u0=rep(0,2*n);
  for(i in 1:n){
    u0[i  ]=sin(pi*x[i]);
    u0[i+n]=u0[i];
  }
  ncall=0;
#
# ODE integration
  out=ode(func=pde_1a,y=u0,times=tout);
  nrow(out)
  ncol(out)
#
# Arrays for display
  u1=matrix(0,nrow=n,ncol=nout);
  u2=matrix(0,nrow=n,ncol=nout);
  for(it in 1:nout){
  for(i  in 1:n){
    u1[i,it]=out[it,i+1];
    u2[i,it]=out[it,i+1+n];
  }
  }
#
# Numerical solution
  cat(sprintf("\n      t      x     u1(x,t)     u2(x,t)"));
  for(it in 1:nout){
  for(i  in 1:n){
    cat(sprintf("\n %6.2f%6.2f%10.4f%10.4f",
      tout[it],x[i],u1[i,it],u2[i,it]));
  }
  }
  matplot(x,u1,type="l",lwd=2,col="black",lty=1,
          xlab="x",ylab="u1(x,t),u2(x,t)");
  matpoints(x,u2,pch="o",col="black");
#
# Calls to ODE routine
  cat(sprintf("\n\n  ncall = %3d\n",ncall));
```

Listing 2.5 Main program for Eqs. (2.1), (2.2), and (2.11).

We can note the following details about Listing 2.5:

- Previous workspaces are cleared, the ODE integrator library deSolve is accessed, and routines used in the SCMOL solution are accessed. dss044 is a library routine for the FDMOL solution.

```
#
# Delete previous workspaces
  rm(list=ls(all=TRUE))
#
# Access ODE integrator
  library("deSolve");
#
# Access functions for numerical solutions
  setwd("f:/collocation/test4");
  source("pde_1a.R");
  source("dss044.R");
```

- A grid in x of 21 points is defined for the interval $x = x_l = 0 \le x \le x = x_u = 1$. Thus, $x = 0, 0.05, \ldots, 1$.

```
#
# Grid (in x)
  n=21;xl=0;xu=1;
  x=seq(from=xl,to=xu,by=(xu-xl)/(n-1));
```

- The PDE parameters are defined.

```
#
# Parameters
  D=1;ua=0;
```

These parameters are passed to subordinate routines without any special designation (a feature of R).

- A vector of output values of t of 6 points is defined for the interval $t = t_0 = 0 \le t \le t = t_f = 1$. Thus, $t = 0, 0.2, \ldots, 1$.

```
#
# Independent variable for ODE integration
  nout=6;t0=0;tf=1;
  tout=seq(from=t0,to=tf,by=(tf-t0)/(nout-1));
```

- An IC vector, u0, is defined numerically in two parts: (i) $1 \le i \le n$ is the IC for Eq. (2.1) for the SCMOL solution and (ii) $n + 1 \le i \le 2n$ is the IC for the FDMOL solution. This in effect is a system of two PDEs integrated simultaneously, although with somewhat unusual details since both PDEs are Eq. (2.1) and they are not coupled in any way. The comparison of the two solutions will indicate if the numerical solutions are essentially the same for the two numerical methods of solution.

```
#
# Initial condition
  u0=rep(0,2*n);
  for(i in 1:n){
    u0[i  ]=sin(pi*x[i]);
    u0[i+n]=u0[i];
  }
  ncall=0;
```

The counter for the calls to the ODE routine (considered next) is initialized.
- The $2n$ ODEs are integrated by ode. As expected, the input arguments to ODE are the ODE function, pde_1a, the IC vector, u0, and the vector of output values of t, tout. func,y,times are reserved names.

```
#
# ODE integration
  out=ode(func=pde_1a,y=u0,times=tout);
  nrow(out)
  ncol(out)
```

The dimensions of the solution vector out are displayed with the nrow,ncol utilities. We would expect these to be 6 and $2n + 1 = 2(21) + 1 = 43$, respectively, which is confirmed in the numerical output that follows.
- The two numerical solutions are placed in arrays u1,u2. The offset 1 in i+1,i+1+n accounts for the use of the first column position in out for t.

```
#
# Arrays for display
  u1=matrix(0,nrow=n,ncol=nout);
  u2=matrix(0,nrow=n,ncol=nout);
  for(it in 1:nout){
  for(i  in 1:n){
    u1[i,it]=out[it,i+1];
    u2[i,it]=out[it,i+1+n];
  }
  }
```

- The two numerical solutions are displayed for comparison. They are also plotted via the matplot for the first solution (to give solid lines) and matpoints for the second solution (to give points). The resulting plots are Figures 2.5 and 2.6.

```
#
# Numerical solution
  cat(sprintf("\n      t      x     u1(x,t)
                u2(x,t)"));
  for(it in 1:nout){
  for(i  in 1:n){
    cat(sprintf("\n %6.2f%6.2f%10.4f%10.4f",
      tout[it],x[i],u1[i,it],u2[i,it]));
  }
  }
  matplot(x,u1,type="l",lwd=2,col="black",lty=1,
          xlab="x",ylab="u1(x,t),u2(x,t)");
  matpoints(x,u2,pch="o",col="black");
```

- The number of calls to pde_1a is displayed as a measure of the computational effort required to produce the numerical solution.

```
#
# Calls to ODE routine
  cat(sprintf("\n\n  ncall = %3d\n",ncall));
```

The ODE routine called by ode (Listing 2.5) is considered next.

2.1.3.2 ODE Routine
pde_1a called by the main program of Listing 2.5 is in Listing 2.6.

```
  pde_1a=function(t,u,parms){
#
# Function pde_1a computes the t derivative
# vectors of u(x,t)
#
# One vector to two vectors
  u1=rep(0,n);u2=rep(0,n);
  for(i in 1:n){
    u1[i]=u[i];
    u2[i]=u[i+n];
  }
#
# u1x
  tablex=splinefun(x,u1);
  u1x=tablex(x,deriv=1);
#
```

```
# BCs
  u2x=rep(0,n);
  u1x[1]=u1[1]-ua;u1x[n]=ua-u1[n];
  u2x[1]=u2[1]-ua;u2x[n]=ua-u2[n];
#
# u1xx
  tablexx=splinefun(x,u1x);
  u1xx=tablexx(x,deriv=1);
#
# u2xx
  nl=2;nu=2;
  u2xx=dss044(xl,xu,n,u2,u2x,nl,nu);
#
# PDE
  u1t=D*u1xx;
  u2t=D*u2xx;
#
# Two vectors to one vector
  ut=rep(0,2*n);
  for(i in 1:n){
      ut[i]=u1t[i];
    ut[i+n]=u2t[i];
    }
#
# Increment calls to pde_1a
  ncall <<- ncall+1;
#
# Return derivative vector
  return(list(c(ut)));
    }
```

Listing 2.6 ODE/SCMOL routine pde_1a for Eqs. (2.1), (2.2), and (2.11).

We can note the following details about pde_1a:

- The function is defined.

```
  pde_1a=function(t,u,parms){
#
# Function pde_1a computes the t derivative
# vectors of u(x,t)
```

t is the current value of t. u is a $2(21) = 42$-vector with 21 elements for the SCMOL solution of Eq. (2.1) followed by the 21 elements for the FDMOL solution. parms for passing parameters to pde_1a is unused.

- u is placed in two 21-vectors for the two solutions of Eq. (2.1).

```
#
# One vector to two vectors
  u1=rep(0,n);u2=rep(0,n);
  for(i in 1:n){
    u1[i]=u[i];
    u2[i]=u[i+n];
  }
```

This coding demonstrates how simultaneous PDEs can be accommodated within the MOL framework.

- The first derivative of u1 in x is computed by splinefun.

```
#
# u1x
  tablex=splinefun(x,u1);
  u1x=tablex(x,deriv=1);
```

- BCs (2.11) are implemented. Note that BCs for both u1 and u2 are included. Points 1, n correspond to $x = x_l = 0, x = x_u = 1$.

```
#
# BCs
  u2x=rep(0,n);
  u1x[1]=u1[1]-ua;u1x[n]=ua-u1[n];
  u2x[1]=u2[1]-ua;u2x[n]=ua-u2[n];
```

$k_1 = -1$, $k_2 = 1$, $g_2(t) = -u_a$, $g_3 = u_a$ in Eqs. (2.11).
- The second derivative of u1 is computed as the derivative of the first derivative (stagewise differentiation).

```
#
# u1xx
  tablexx=splinefun(x,u1x);
  u1xx=tablexx(x,deriv=1);
```

splinefun calculates $\dfrac{\partial^2 u_1}{\partial x^2}$ in Eq. (2.1) as u1xx.
- The second derivative of u2 is computed by the library routine dss044. Neumann BCs, nl=nu=2, are specified since the first derivatives are defined by Eqs. (2.11) (and the BCs for u2x[1], u2x[n] were computed previously).

```
#
# u2xx
  nl=2;nu=2;
  u2xx=dss044(xl,xu,n,u2,u2x,nl,nu);
```

- The repeated use of Eq. (2.1) for u1, u2 is programmed with vectorization.

```
#
# PDE
  u1t=D*u1xx;
  u2t=D*u2xx;
```

- The two derivative vectors in t are placed in one vector, ut, for return to the ODE integrator, ode.

```
#
# Two vectors to one vector
  ut=rep(0,2*n);
  for(i in 1:n){
      ut[i]=u1t[i];
    ut[i+n]=u2t[i];
    }
```

- The counter for the calls to pde_1a is incremented and its value returned to the calling program (of Listing 2.1) with the <<- operator.

```
#
# Increment calls to pde_1a
  ncall <<- ncall+1;
```

- The derivative vector (LHS of Eq. (2.1)) is returned to ode with a combination of return, list, c.

```
#
# Return derivative vector
  return(list(c(ut)));
  }
```

The final } concludes pde_1a.

Execution of the preceding main program and subordinate ODE routine produces the following abbreviated numerical output in Table 2.3 (with the output for $t = 0, 0.2, \dots, 0.8$ deleted to conserve space).

The two solutions, u1, u2, are in close agreement, which implies that the SCMOL solution is correct (this conclusion is only for the comparison of the two numerical solutions of Eqs. (2.1), (2.2), and (2.11) and does not constitute a proof that the solutions are correct). The dimensions of the solution matrix

Table 2.3 Abbreviated output for Eqs. (2.1), (2.2), and (2.11).

```
[1] 6

[1] 43
```

t	x	u1(x,t)	u2(x,t)
1.00	0.00	0.1013	0.1013
1.00	0.05	0.1059	0.1061
1.00	0.10	0.1103	0.1105
1.00	0.15	0.1142	0.1144
1.00	0.20	0.1177	0.1179
1.00	0.25	0.1206	0.1208
1.00	0.30	0.1230	0.1232
1.00	0.35	0.1249	0.1251
1.00	0.40	0.1263	0.1265
1.00	0.45	0.1270	0.1273
1.00	0.50	0.1274	0.1275
1.00	0.55	0.1270	0.1273
1.00	0.60	0.1263	0.1265
1.00	0.65	0.1249	0.1251
1.00	0.70	0.1230	0.1232
1.00	0.75	0.1206	0.1208
1.00	0.80	0.1177	0.1179
1.00	0.85	0.1142	0.1144
1.00	0.90	0.1103	0.1105
1.00	0.95	0.1059	0.1061
1.00	1.00	0.1013	0.1013

```
ncall = 735
```

out are 6×43 as expected $(2(21) + 1 = 43$ since t is also included in out). The computational effort is modest with ncall $= 735$.

The graphical output is shown in Figure 2.5. The close agreement of u1, u2 of Table 2.3 is clear.

As expected from Eqs. (2.11) with $g_1(t) = g_2(t) = 0$ $(u_a = 0)$, the solutions appear to be $u_1(x, t \to \infty) = u_2(x, t \to \infty) = 0$.

In summary, the solution of Eqs. (2.1), (2.2), and (2.11) indicates that Robin BCs can be accommodated within the SCMOL framework by using stagewise differentiation. The following example demonstrates the solution of Eq. (2.1) with nonlinear BCs.

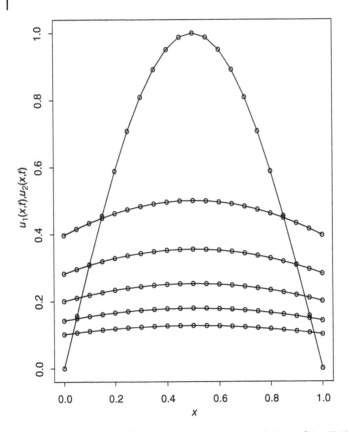

Figure 2.5 SCMOL (solid lines) and FDMOL (points) solutions of Eqs. (2.1), (2.2), and (2.11).

2.1.4 Nonlinear BCs

The following variation of BCs (2.11) can be accommodated within the MOL framework.

$$\frac{\partial u(x = x_l, t)}{\partial x} = 0; \quad \frac{\partial u(x = x_u, t)}{\partial x} = \sigma(u_a^4 - u^4(x = x_u, t)) \qquad \text{(2.12a,b)}$$

where u_a is a specified constant, for example, an ambient temperature. The fourth degree nonlinearity in Eq. (2.12b), $u^4(x = x_u, t)$, is the Stefan–Boltzmann law for radiation from a black body. Equation (2.12a) is a homogeneous Neumann BC that could, for example, reflect symmetry in x or zero heat flux at $x = x_l$.

The implementation of Eqs. (2.12) is similar to the preceding example of Robin BCs. The main program is similar to that of Listing 2.5.

2.1.4.1   Main Program

```
#
# Delete previous workspaces
  rm(list=ls(all=TRUE))
#
# Access ODE integrator
  library("deSolve");
#
# Access functions for numerical solutions
  setwd("f:/collocation/test12");
  source("pde_1a.R");
  source("dss044.R");
#
# Grid (in x)
  n=21;xl=0;xu=1;
  x=seq(from=xl,to=xu,by=(xu-xl)/(n-1));
#
# Parameters
  D=0.001;ua=298;sig=5.670373e-08;
#
# Independent variable for ODE integration
  nout=6;t0=0;tf=1000;
  tout=seq(from=t0,to=tf,by=(tf-t0)/(nout-1));
#
# Initial condition
  u0=rep(0,2*n);
  for(i in 1:n){
    u0[i  ]=1000;
    u0[i+n]=u0[i];
  }
  ncall=0;
#
# ODE integration
  out=ode(func=pde_1a,y=u0,times=tout);
  nrow(out)
  ncol(out)
#
# Arrays for display
  u1=matrix(0,nrow=n,ncol=nout);
  u2=matrix(0,nrow=n,ncol=nout);
  for(it in 1:nout){
  for(i  in 1:n){
```

```
      u1[i,it]=out[it,i+1];
      u2[i,it]=out[it,i+1+n];
    }
  }
#
# Numerical solution
    cat(sprintf("\n        t      x     u1(x,t)     u2(x,t)"));
    for(it in 1:nout){
    for(i  in 1:n){
      cat(sprintf("\n %6.2f%6.2f%10.2f%10.2f",
        tout[it],x[i],u1[i,it],u2[i,it]));
    }
    }
    matplot(x,u1,type="l",lwd=2,col="black",lty=1,
            xlab="x",ylab="u1(x,t),u2(x,t)");
    matpoints(x,u2,pch="o",col="black");
#
# Calls to ODE routine
    cat(sprintf("\n\n  ncall = %3d\n",ncall));
```

Listing 2.7 Main program for Eqs. (2.1), (2.2), and (2.12).

We can note the following differences between Listings 2.5 and 2.7:

- The parameters have the values

```
#
# Parameters
  D=1;ua=298;sig=5.670373e-08;
```

σ is the Stefan–Boltzmann constant with the SI units $W\text{-}m^{-2}\text{-}{}^{\circ}K^{-4}$. The ambient temperature u_a is therefore in ${}^{\circ}K$. D, the thermal diffusivity, has a value in $m^2\ s^{-1}$ that might reflect a physical application (cooling from an elevated temperature).

- The IC is a constant (which might be, e.g., the temperature of hot steel).

```
#
# Initial condition
  u0=rep(0,2*n);
  for(i in 1:n){
    u0[i  ]=1000;
    u0[i+n]=u0[i];
  }
  ncall=0;
```

2.1.4.2 ODE Routine

The ODE routine called by ODE in Listing 2.7 follows.

```
  pde_1a=function(t,u,parms){
#
# Function pde_1a computes the t derivative vectors of
# u(x,t)
#
# One vector to two vectors
  u1=rep(0,n);u2=rep(0,n);
  for(i in 1:n){
    u1[i]=u[i];
    u2[i]=u[i+n];
  }
#
# u1x
  tablex=splinefun(x,u1);
  u1x=tablex(x,deriv=1);
#
# BCs
  u2x=rep(0,n);
  u1x[1]=0;u1x[n]=sig*(ua^4-u1[n]^4);
  u2x[1]=0;u2x[n]=sig*(ua^4-u2[n]^4);
#
# u1xx
  tablexx=splinefun(x,u1x);
  u1xx=tablexx(x,deriv=1);
#
# u2xx
  nl=2;nu=2;
  u2xx=dss044(xl,xu,n,u2,u2x,nl,nu);
#
# PDE
  u1t=D*u1xx;
  u2t=D*u2xx;
#
# Two vectors to one vector
  ut=rep(0,2*n);
  for(i in 1:n){
      ut[i]=u1t[i];
    ut[i+n]=u2t[i];
  }
#
```

```
# Increment calls to pde_1a
  ncall <<- ncall+1;
#
# Return derivative vector
  return(list(c(ut)));
  }
```

Listing 2.8 ODE/SCMOL routine pde_1a for Eqs. (2.1), (2.2), and (2.12).

Listing 2.8 is similar to Listing 2.6, and the only difference is the coding of the nonlinear BCs (2.12) in place of the linear BCs of Eqs. (2.11).

```
#
# BCs
  u2x=rep(0,n);
  u1x[1]=0;u1x[n]=sig*(ua^4-u1[n]^4);
  u2x[1]=0;u2x[n]=sig*(ua^4-u2[n]^4);
```

In particular, the nonlinear terms

```
  -u1[n]^4;  -u2[n]^4
```

illustrate the ease of using the nonlinear BCs within the MOL format.

Abbreviated numerical output is in Table 2.4.

The two solutions, u1, u2, are in close agreement, which again implies that the SCMOL solution is correct. The dimensions of the solution matrix out are 6×43 as expected. The computational effort is quite acceptable with ncall = 1211.

The graphical output is shown in Figure 2.6. The close agreement of u1, u2 of Table 2.4 is clear.

As expected from Eqs. (2.12) with $u_a = 298$, the solutions appear to be $u_1(x, t \to \infty) = u_2(x, t \to \infty) = 298$. Thus, the effect of BCs (2.12) is clear.

In summary, the solution of Eqs. (2.1), (2.2), and (2.12) indicates that nonlinear BCs can be accommodated within the SCMOL framework by using stage-wise differentiation. The following example demonstrates the solution of PDEs with variable coefficients.

2.2 Variable Coefficient

Equation (2.1) is in 1D Cartesian coordinates and has the constant coefficient D. However, PDEs with variable coefficients, for example, a function of the spatial coordinate, can be accommodated within the SCMOL framework. This is

Table 2.4 Abbreviated output for Eqs. (2.1), (2.2), and (2.12).

```
[1]  6

[1]  43
```

t	x	u1(x,t)	u2(x,t)
1000.00	0.00	419.15	419.16
1000.00	0.05	418.86	418.88
1000.00	0.10	418.01	418.02
1000.00	0.15	416.58	416.59
1000.00	0.20	414.59	414.61
1000.00	0.25	412.06	412.07
1000.00	0.30	408.98	409.00
1000.00	0.35	405.38	405.40
1000.00	0.40	401.28	401.29
1000.00	0.45	396.69	396.70
1000.00	0.50	391.64	391.65
1000.00	0.55	386.14	386.15
1000.00	0.60	380.24	380.25
1000.00	0.65	373.95	373.96
1000.00	0.70	367.31	367.32
1000.00	0.75	360.35	360.35
1000.00	0.80	353.10	353.10
1000.00	0.85	345.59	345.60
1000.00	0.90	337.87	337.88
1000.00	0.95	329.98	329.98
1000.00	1.00	321.94	321.94

```
ncall = 1211
```

---

illustrated with the following 1D diffusion equation in cylindrical coordinates.

$$\frac{\partial u}{\partial t} = D \frac{1}{r}\left(r\frac{\partial u}{\partial r}\right) = D\left(\frac{\partial^2 u}{\partial r^2} + \frac{1}{r}\frac{\partial u}{\partial r}\right) \qquad (2.13a)$$

where

u dependent variable
t initial-value independent variable

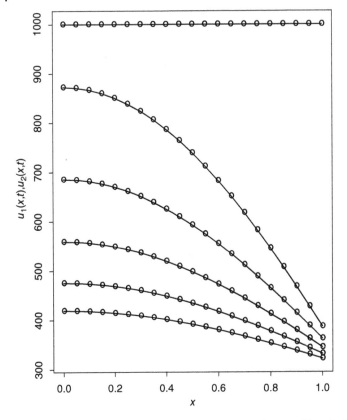

Figure 2.6 SCMOL (solid lines) and FDMOL (points) solutions of Eqs. (2.1), (2.2), and (2.12).

r radial, boundary-value dependent variable

D diffusivity.

Equation (2.13a) is a 1D special case of Fick's (Fourier's) second law in the cylindrical coordinates (r, θ, z) for which angular (in θ) and axial (in z) variations have been neglected.[2]

Equation (2.13a) is first order in t and second order in r, so that one IC and two BCs are required.

$$u(r, t = 0) = g_1(r) \tag{2.13b}$$

---

2 Equation (2.13a) follows from the differential operators listed in Appendix A1. The 3D equation in cylindrical coordinates from Appendix A1 is

$$\frac{\partial u}{\partial t} = D\left(\frac{\partial^2 u}{\partial r^2} + \frac{1}{r}\frac{\partial u}{\partial r} + \frac{1}{r^2}\frac{\partial^2 u}{\partial \theta^2} + \frac{\partial^2 u}{\partial z^2}\right)$$

$$\frac{u(r=0,t)}{\partial r} = \frac{u(r=r_0,t)}{\partial r} = 0 \qquad (2.13\text{c,d})$$

Equations (2.13c,d) are homogeneous Neumann BCs. r_0 is the outer radius of r, that is, $0 \leq r \leq r_0$.

2.2.1 Main Program

The main program for Eqs. (2.13) is listed next.

```
#
# Delete previous workspaces
  rm(list=ls(all=TRUE))
#
# Access ODE integrator
  library("deSolve");
#
# Access functions for numerical solution
  setwd("f:/collocation/test9");
  source("pde_1a.R");
  source("dss004.R");
  source("dss044.R");
#
# Grid (in r)
  n=21;rl=0;ru=1;
  r=seq(from=rl,to=ru,by=(ru-rl)/(n-1));
#
# Parameters
  D=1;
#
# Independent variable for ODE integration
  nout=6;t0=0;tf=0.2;
  tout=seq(from=t0,to=tf,by=(tf-t0)/(nout-1));
#
# Initial condition
  u0=rep(0,2*n);
  for(i in 1:n){
    u0[i]=exp(-10*r[i]^2);
    u0[i+n]=u0[i];
  }
  ncall=0;
#
# ODE integration
  out=ode(func=pde_1a,y=u0,times=tout);
  nrow(out)
```

```
   ncol(out)
#
# Arrays for display
    u1=matrix(0,nrow=n,ncol=nout);
    u2=matrix(0,nrow=n,ncol=nout);
    for(it in 1:nout){
    for(i  in 1:n){
      u1[i,it]=out[it,i+1];
      u2[i,it]=out[it,i+1+n];
    }
    }
#
# Numerical solutions
    cat(sprintf("\n          t          r     u1(r,t)    u2(r,t)
                   diff"));
    for(it in 1:nout){
    for(i  in 1:n){
      cat(sprintf("\n %6.2f%6.2f%10.4f%10.4f
                   %12.6f",tout[it],r[i],u1[i,it],
                   u2[i,it],u1[i,it]-u2[i,it]));
    }
    }
    matplot(r,u1,type="l",lwd=2,col="black",lty=1,
            xlab="x",ylab="u(r,t)");
    matpoints(r,u2,pch="o",col="black");
#
# Calls to ODE routine
    cat(sprintf("\n\n  ncall = %3d\n",ncall));
```

Listing 2.9 Main program for Eqs. (2.13).

We can note the following points about Listing 2.9:

- Previous workspaces are removed. The library of ODE integrators, deSolve, is accessed, followed by the routines for Eqs. (2.13). Note the use of / rather than the usual \ in the setwd (set working directory), and required revision for the local computer (to define the working folder or directory for the routines pde_1a, dss004, dss044).

```
#
# Delete previous workspaces
  rm(list=ls(all=TRUE))
#
# Access ODE integrator
```

```
library("deSolve");
#
# Access functions for numerical solution
setwd("f:/collocation/test9");
source("pde_1a.R");
source("dss004.R");
source("dss044.R");
```

- A grid of 21 points in r is defined for $r_l = 0 \le r \le r_u = 1$ so that $r = 0, 0.05, \ldots, 1$. The seq utility is used to produce the sequence of values in r.

```
#
# Grid (in r)
n=21;rl=0;ru=1;
r=seq(from=rl,to=ru,by=(ru-rl)/(n-1));
```

- The parameter, D in Eq. (2.13a), is defined followed by the interval in t, which has 6 values corresponding to $t = 0, 0.04, \ldots, 0.2$.

```
#
# Parameters
D=1;
#
# Independent variable for ODE integration
nout=6;t0=0;tf=0.2;
tout=seq(from=t0,to=tf,by=(tf-t0)/(nout-1));
```

- IC (2.13b) is a Gaussian function centered at $r = 0$. This function is placed in u0[i], i=1,...,n for the SCMOL solution and in u0[i], i=n+1,...,2n for the FDMOL solution (within the for).

```
#
# Initial condition
u0=rep(0,2*n);
for(i in 1:n){
   u0[i]=exp(-10*r[i]^2);
   u0[i+n]=u0[i];
}
ncall=0;
```

The counter for the calls to the ODE routine is initialized.
- The $2(21) = 42$ ODEs are integrated numerically by ode. The input arguments to ode are the MOL/ODE routine pde_1a, the IC vector u0 and the output values of t, tout. The number of ODEs to be integrated is defined by the length of u0 (42). func, y, times are reserved names.

```
#
# ODE integration
  out=ode(func=pde_1a,y=u0,times=tout);
  nrow(out)
  ncol(out)
```

The ODE solution is returned in out, a $6 \times 2(21) + 1 = 43$ element array. These dimensions are confirmed by the nrow, ncol utilities (in the numerical output considered subsequently).

- The numerical solution is placed in two 6×42 arrays, u1, u2, for subsequent numerical and graphical output. The offset of 1 in ir+1, ir+1+n is required since the 6 values of t are included in out. Two nested fors step through the nout=6 values of t and the n=21 values of r.

```
#
# Arrays for display
  u1=matrix(0,nrow=n,ncol=nout);
  u2=matrix(0,nrow=n,ncol=nout);
  for(it in 1:nout){
  for(ir in 1:n){
    u1[ir,it]=out[it,ir+1];
    u2[ir,it]=out[it,ir+1+n];
  }
  }
```

- The numerical solutions are displayed for the nout=6 values of t and n=21 values of r. The difference in the SCMOL and FDMOL solutions is also included in the output.

```
#
# Numerical solutions
  cat(sprintf("\n        t        r    u1(r,t)      u2(r,t)
              diff"));
  for(it in 1:nout){
  for(ir in 1:n){
    cat(sprintf("\n %6.2f%6.2f%10.4f%10.4f
                %12.6f",tout[it],r[i],u1[i,it],
                u2[i,it],u1[i,it]-u2[i,it]));
  }
  }
```

- The SCMOL solution is plotted with matplot and the FD solution is plotted with matpoints. The two plots are superimposed as indicated in Figure 2.6.

```
matplot(r,u1,type="l",lwd=2,col="black",lty=1,
        xlab="x",ylab="u(r,t)");
matpoints(r,u2,pch="o",col="black");
```

- The number of calls to pde_1a is displayed at the end of the solution as an indication of the computational effort required to compute the solution.

```
#
# Calls to ODE routine
  cat(sprintf("\n\n  ncall = %3d\n",ncall));
```

The ODE routine, pde_1a, called by ode is listed next.

2.2.2   ODE Routine

pde_1 has the coding for both the SCMOL and FDMOL solutions.

```
  pde_1a=function(t,u,parms){
#
# Function pde_1a computes the t derivative vectors of
# u1(r,t), u2(r,t)
#
# One vector to two vectors
  u1=rep(0,n);u2=rep(0,n);
  for(i in 1:n){
    u1[i]=u[i];
    u2[i]=u[i+n];
  }
#
# u1r
  tabler=splinefun(r,u1);
  u1r=tabler(r,deriv=1);
#
# BCs, u1
  u1r[1]=0;u1r[n]=0;
#
# u1rr
  tablerr=splinefun(r,u1r);
  u1rr=tablerr(r,deriv=1);
#
# u2r
  u2r=dss004(rl,ru,n,u2);
#
# BCs, u2
```

```
  u2r[1]=0;u2r[n]=0;
#
# u2rr
  nl=2;nu=2;
  u2rr=dss044(rl,ru,n,u2,u2r,nl,nu);
#
# PDEs
  u1t=rep(0,n);u2t=rep(0,n);
  for(i in 1:n){
    if(i==1){
      u1t[i]=2*D*u1rr[i];
      u2t[i]=2*D*u2rr[i];
    }else{
      u1t[i]=D*(u1rr[i]+(1/r[i])*u1r[i]);
      u2t[i]=D*(u2rr[i]+(1/r[i])*u2r[i]);
    }
  }
#
# Two vectors to one vector
  ut=rep(0,2*n);
  for(i in 1:n){
    ut[i]   =u1t[i];
    ut[i+n] =u2t[i];
  }
#
# Increment calls to pde_1a
  ncall <<- ncall+1;
#
# Return derivative vector
  return(list(c(ut)));
  }
```

Listing 2.10 ODE/SCMOL routine pde_1a for Eqs. (2.13).

We can note the following details about pde_1a:

- The function is defined. The input arguments are (i) the current value of t; (ii) the vector of ODE-dependent variables, a 42-vector; and (iii) input parameters, not used in this case but still required as an argument. The output of pde_1a is the vector of ODE derivatives in t, as explained at the end of pde_1a.

  ```
  pde_1a=function(t,u,parms){
  #
  ```

```
#
# Function pde_1a computes the t derivative
# vectors of u1(r,t), u2(r,t)
```

- u is placed in two 21-vectors for the SCMOL and FD solutions of Eq. (2.13a).

```
#
# One vector to two vectors
  u1=rep(0,n);u2=rep(0,n);
  for(i in 1:n){
    u1[i]=u[i];
    u2[i]=u[i+n];
  }
```

- The first derivative of u1 in r is computed by splinefun.

```
#
# u1r
  tabler=splinefun(r,u1);
  u1r=tabler(r,deriv=1);
```

- BCs (2.13c,d) are implemented for u1. Points 1,n correspond to $r = r_l = 0, r = r_u = 1$.

```
#
# BCs, u1
  u1r[1]=0;u1r[n]=0;
```

- The second derivative of u1 is computed as the derivative of the first derivative (stagewise differentiation).

```
#
# u1rr
  tablerr=splinefun(r,u1r);
  u1rr=tablerr(r,deriv=1);
```

- The first derivative of u2 in r is computed by dss004.

```
#
# u2r
  u2r=dss004(rl,ru,n,u2);
```

- BCs (2.13c,d) are implemented for u2. Points 1,n correspond to $r = r_l = 0, r = r_u = 1$.

```
#
# BCs, u2
  u2r[1]=0;u2r[n]=0;
```

- The second derivative of u2 is computed by the library routine dss044. Neumann BCs, nl=nu=2, are specified since the first derivatives are defined by Eqs. (2.13c,d) (and the BCs for u2r[1], u2r[n] were defined previously).

```
#
# u2rr
  nl=2;nu=2;
  u2rr=dss044(rl,ru,n,u2,u2r,nl,nu);
```

- Equation (2.13a) for u1, u2 is programmed.

```
#
# PDEs
  u1t=rep(0,n);u2t=rep(0,n);
  for(i in 1:n){
    if(i==1){
      u1t[i]=2*D*u1rr[i];
      u2t[i]=2*D*u2rr[i];
    }else{
      u1t[i]=D*(u1rr[i]+(1/r[i])*u1r[i]);
      u2t[i]=D*(u2rr[i]+(1/r[i])*u2r[i]);
    }
  }
```

Some additional explanation is required.
- The derivatives in t are declared (preallocated) for u1, u2.

```
      u1t=rep(0,n);u2t=rep(0,n);
```

- A for steps through the values of r. For $r = 0$ (i=1), Eq. (13.1a) has an indeterminate form which, for u_1, is $\dfrac{1}{r}\dfrac{\partial u_1}{\partial r}\big|_{r=0}$. From BC (2.13c), $\dfrac{\partial u_1(r=0,t)}{\partial r} = 0$ so the indeterminate form is 0/0. Application of l'Hospital's rule gives

$$\lim_{r\to 0} \frac{1}{r}\frac{\partial u_1}{\partial r} = \frac{\partial^2 u_1}{\partial r^2}$$

and the radial group in Eq. (2.13a) becomes

$$\frac{\partial^2 u_1}{\partial r^2} + \frac{1}{r}\frac{\partial u_1}{\partial r} = 2\frac{\partial^2 u_1}{\partial r^2}$$

This result is programmed for u_1, u_2 as (including D)

```
      if(i==1){
        u1t[i]=2*D*u1rr[i];
```

```
   u2t[i]=2*D*u2rr[i];
```

– For $r \neq 0$ ($i > 1$), Eq. (2.13a) is programmed as

```
      }else{
        u1t[i]=D*(u1rr[i]+(1/r[i])*u1r[i]);
        u2t[i]=D*(u2rr[i]+(1/r[i])*u2r[i]);
      }
```

Note in particular the variable coefficient $1/r[i]$.

- The two derivative vectors in t are placed in one vector, ut, for return to the ODE integrator, ode (called in Listing 2.9).

```
#
# Two vectors to one vector
  ut=rep(0,2*n);
  for(i in 1:n){
    ut[i]   =u1t[i];
    ut[i+n]=u2t[i];
  }
```

- The counter for the calls to pde_1a is incremented and its value returned to the calling program (of Listing 2.9) with the <<- operator.

```
#
# Increment calls to pde_1
  ncall <<- ncall+1;
```

- The derivative vector (LHS of Eq. (2.13a)) is returned to ode with a combination of return, list, c.

```
#
# Return derivative vector
  return(list(c(ut)));
  }
```

The final } concludes pde_1a.

Abbreviated numerical output for $t = 0.2$ is displayed in Table 2.5 (the output for $t = 0, 0.04, \ldots, 0.16$ is removed to conserve space).

The dimensions of out are 6 x 43 as explained previously. The two solutions, u1, u2, are in close agreement, which implies that the SCMOL solution is correct. The computational effort is modest ncall = 470.

The graphical output is shown in Figure 2.7, which confirms the agreement between u_1 and u_2 in Table 2.5.

In conclusion, this example application demonstrates the ease of programming a PDE variable coefficient within the SCMOL framework.

Table 2.5 Abbreviated output for Eqs. (2.13).

```
[1]  6

[1]  43

        0.20   0.00    0.1222     0.1227    -0.000472
        0.20   0.05    0.1225     0.1225     0.000077
        0.20   0.10    0.1218     0.1218    -0.000053
        0.20   0.15    0.1209     0.1208     0.000077
        0.20   0.20    0.1194     0.1195    -0.000053
        0.20   0.25    0.1178     0.1178     0.000078
        0.20   0.30    0.1157     0.1158    -0.000052
        0.20   0.35    0.1136     0.1136     0.000078
        0.20   0.40    0.1111     0.1112    -0.000051
        0.20   0.45    0.1088     0.1087     0.000079
        0.20   0.50    0.1061     0.1062    -0.000050
        0.20   0.55    0.1037     0.1037     0.000080
        0.20   0.60    0.1012     0.1013    -0.000049
        0.20   0.65    0.0991     0.0990     0.000081
        0.20   0.70    0.0969     0.0969    -0.000048
        0.20   0.75    0.0952     0.0951     0.000082
        0.20   0.80    0.0936     0.0936    -0.000048
        0.20   0.85    0.0925     0.0924     0.000082
        0.20   0.90    0.0915     0.0916    -0.000047
        0.20   0.95    0.0911     0.0910     0.000083
        0.20   1.00    0.0904     0.0909    -0.000461

    ncall = 470
```

2.3 Inhomogeneous, Simultaneous, Nonlinear

We now consider an example that demonstrates the application of SCMOL to a system of inhomogeneous, simultaneous, nonlinear PDEs, stated as

$$\frac{\partial u_1}{\partial t} = D\frac{\partial}{\partial x}\left[\frac{\partial u_1}{\partial x} - 2\frac{u_1}{u_2}\frac{\partial u_2}{\partial x}\right] \tag{2.14a}$$

$$\frac{\partial u_2}{\partial t} = -ku_1 \tag{2.14b}$$

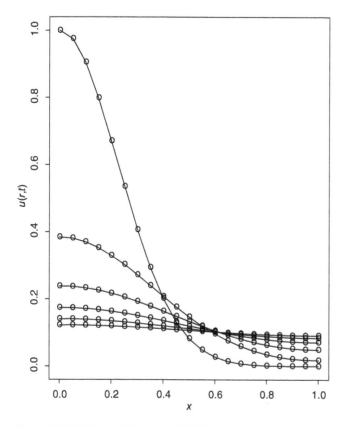

Figure 2.7 SCMOL (solid lines) and FDMOL (points) solutions of Eqs. (2.13).

Equations (2.14) are an example of a *chemotaxis* system (diffusion under the influence of a chemical gradient). In particular, Eq. (2.14a) has the nonlinear diffusion term

$$\frac{\partial}{\partial x}\left[-2\frac{u_1}{u_2}\frac{\partial u_2}{\partial x}\right]$$

in which the diffusion flux is

$$-2\frac{u_1}{u_2}\frac{\partial u_2}{\partial x}$$

Thus, the rate of diffusion is proportional to $\frac{\partial u_2}{\partial x}$, that is, the gradient of a second component u_2, which is termed an *attractant* (from Eq. (12.4b), with a linear depletion $-ku_1$).

Note the minus that gives diffusion of u_1 in the direction of increasing u_2. Also, the effective diffusivity is $\frac{u_1}{u_2}$, a nonlinear term that increases with increasing u_1 and decreases with increasing u_2. These unusual properties, and associated term in Eq. (2.14a), are included in the following SCMOL routines. An analytical solution is also included to verify the numerical solution.

Equations (2.14) are first order in t and each requires one IC.

$$u_1(x, t = 0) = f_1(x); \quad u_2(x, t = 0) = f_2(x) \tag{2.15a,b}$$

where f_1, f_2 are functions to be specified.

Equations (2.14) are second order in x (u_2 is second order in x through Eq. (2.14a)). Thus, each dependent variable, u_1, u_2, requires two BCs. These are taken as homogeneous Neumann BCs.

$$\frac{\partial u_1(x = x_l, t)}{\partial x} = \frac{\partial u_1(x = x_u, t)}{\partial x} = 0 \tag{2.16a}$$

$$\frac{\partial u_2(x = x_l, t)}{\partial x} = \frac{\partial u_2(x = x_u, t)}{\partial x} = 0 \tag{2.16b}$$

Physically, BCs (2.16) indicate that the diffusing (chemical) components do not leave the system, that is, they are no-flux BCs.

An analytical solution to Eqs. (2.14) is available for verification of the numerical solution ([2], p68)

$$u_1(z) = [1 + e^{-cz/D}]^{-1} \tag{2.17a}$$

$$u_2(z) = \frac{c^2}{kD} e^{-cz/D} [1 + e^{-cz/D}]^{-2} \tag{2.17b}$$

where $z = x - ct$; c is a constant to be specified (a velocity). Note that $u_1(z), u_2(z)$ are a function of the single Lagrangian variable z. In other words, these solutions are invariant for a constant value of z, regardless of how x and t may vary. A solution with this property is termed a traveling wave [1]. We will discuss this property further when the numerical solution to Eqs. (2.14) is considered subsequently.

Equations (2.17) are used for ICs (2.15) with $t = 0, z = x$.

$$f_1(x) = u_1(x, t = 0) = [1 + e^{-cx/D}]^{-1} \tag{2.18a}$$

$$f_2(x) = u_2(x, t = 0) = \frac{c^2}{kD} e^{-cx/D} [1 + e^{-cx/D}]^{-2} \tag{2.18b}$$

Equations (2.14)–(2.18) constitute the two-PDE model to be studied numerically. The analytical solutions, Eqs. (2.17), are used to verify the numerical solution.

2.3.1 Main Program

The main program for Eqs. (2.14)–(2.18) is listed next.

```
#
# Remove previous workspaces
```

```
  rm(list=ls(all= TRUE));
#
# Access ODE integrator
  library("deSolve");
#
# Access functions for numerical, analytical solutions
  setwd("f:/collocation/test44");
  source("pde_1a.R");
  source("u1_anal.R");
  source("u2_anal.R");
#
# Level of output
#
#   ip = 1 - graphical (plotted) solutions
#             (u1(x,t), u2(x,t)) only
#
#   ip = 2 - numerical and graphical solutions
#
  ip=2;
#
# Grid (in x)
  n=101;xl=-10;xu=15
  x=seq(from=xl,to=xu,by=0.25);
#
# Parameters
  k=1;D=1;c=1;
  cat(sprintf("\n\n k = %5.2f    D = %5.2f
                c = %5.2f\n",k,D,c));
#
# Independent variable for ODE integration
  nout=6;t0=0;tf=5;
  tout=seq(from=0,to=5,by=(tf-t0)/(nout-1));
#
# Initial condition from analytical solutions,t=0)
  u0=rep(0,2*n);
  for(i in 1:n){
    u0[i]   =u1_anal(x[i],tout[1],k,D,c);
    u0[i+n] =u2_anal(x[i],tout[1],k,D,c);
  }
  ncall=0;
#
# ODE integration
  out=lsodes(y=u0,times=tout,func=pde_1a,
```

```
        sparsetype="sparseint",rtol=1e-6,
        atol=1e-6,maxord=5);
  nrow(out)
  ncol(out)
#
# Arrays for plotting numerical, analytical solutions
  u1=matrix(0,nrow=n,ncol=nout);
  u2=matrix(0,nrow=n,ncol=nout);
 u1a=matrix(0,nrow=n,ncol=nout);
 u2a=matrix(0,nrow=n,ncol=nout);
  for(it in 1:nout){
  for(i  in 1:n){
    u1[i,it]=out[it,i+1];
    u2[i,it]=out[it,i+1+n];
    u1a[i,it]=u1_anal(x[i],tout[it],k,D,c);
    u2a[i,it]=u2_anal(x[i],tout[it],k,D,c);
  }
  }
#
# Display numerical solution
  if(ip==2){
    for(it in 1:nout){
      if((it==1)|(it==nout)){
  cat(sprintf("\n      t          x    u1(x,t)
              u1_ex(x,t)   u1_err(x,t)"));
  cat(sprintf("\n                         u2(x,t)
              u2_ex(x,t)   u2_err(x,t)\n"));
    for(i in 1:n){
      cat(sprintf("%5.1f%8.2f%10.5f%12.5f
                  %13.6f\n",tout[it],x[i],
                  u1[i,it],u1a[i,it],
                  u1[i,it]-u1a[i,it]));
      cat(sprintf("               %10.5f%12.5f
                  %13.6f\n",u2[i,it],u2a[i,it],
                  u2[i,it]-u2a[i,it]));
        }
      }
    }
  }
#
# Calls to ODE routine
  cat(sprintf("\n\n ncall = %5d\n\n",ncall));
#
```

```
# Plot u1 numerical, analytical
  par(mfrow=c(1,1));
  matplot(x=x,y=u1,type="l",xlab="x",ylab="u1(x,t)",
          xlim=c(xl,xu),lty=1,col="black",lwd=2);
  matpoints(x=x,y=u1a,xlim=c(xl,xu),col="black",lwd=2)
#
# Plot u2 numerical, analytical
  par(mfrow=c(1,1));
  matplot(x=x,y=u2,type="l",xlab="x",ylab="u2(x,t)",
          xlim=c(xl,xu),lty=1,col="black",lwd=2);
  matpoints(x=x,y=u2a,xlim=c(xl,xu),col="black",lwd=2)
```

Listing 2.11 Main program for Eqs. (2.14).

We can note the following details about Listing 2.11:

- Previous workspaces are removed. The library of ODE integrators, deSolve, is accessed, followed by the routines for Eqs. (2.14). Note the use of / rather than the usual \ in the setwd (set working directory). Also, setwd will require revision for the local computer (to define the working folder or directory for the routines pde_1a, u1_anal.R, u2_anal.R).

```
#
# Remove previous workspaces
  rm(list=ls(all= TRUE));
#
# Access ODE integrator
  library("deSolve");
#
# Access functions for numerical,
# analytical solutions
  setwd("f:/collocation/test44");
  source("pde_1a.R");
  source("u1_anal.R");
  source("u2_anal.R");
```

- The level of output is specified.

```
#
# Level of output
#
#   ip = 1 - graphical (plotted) solutions
#             (u1(x,t), u2(x,t)) only
#
#   ip = 2 - numerical and graphical solutions
```

```
#
  ip=2;
```

- A grid of 101 points in x is defined for $x_l = -10 \leq x \leq x_u = 15$ so that $x = -10, -9.75, \ldots, 15$. The seq utility is used to produce the sequence of values in x.

```
#
# Grid (in x)
  n=101;xl=-10;xu=15
  x=seq(from=xl,to=xu,by=0.25);
```

- The parameters in Eqs. (2.14) are defined.

```
#
# Parameters
  k=1;D=1;c=1;
  cat(sprintf("\n\n  k = %5.2f    D = %5.2f
               c = %5.2f\n",k,D,c));
```

- The interval in t has 6 values corresponding to $t = 0, 1, \ldots, 5$.

```
#
# Independent variable for ODE integration
  nout=6;t0=0;tf=5;
  tout=seq(from=0,to=5,by=(tf-t0)/(nout-1));
```

- ICs (2.15) are defined by two functions, u1_anal, u2_anal, that also implement analytical solution (2.17) as discussed subsequently. These functions are placed in a vector u0[i] of length 2*n with $t = t_0 = 0$ or tout[1]=0.

```
#
# Initial condition (from analytical solutions,t=0)
  u0=rep(0,2*n);
  for(i in 1:n){
     u0[i]   =u1_anal(x[i],tout[1],k,D,c);
     u0[i+n] =u2_anal(x[i],tout[1],k,D,c);
  }
  ncall=0;
```

The counter for the calls to the ODE routine is initialized.

- The $2(101) = 202$ ODEs are integrated numerically by lsodes. The input arguments to lsodes are the MOL/ODE routine pde_1a, the IC vector u0 and the output values of t, tout. The number of ODEs to be integrated is defined by the length of u0 (202). func, y, times are reserved names.

```
#
# ODE integration
  out=lsodes(y=u0,times=tout,func=pde_1a,
      sparsetype="sparseint",rtol=1e-6,
      atol=1e-6,maxord=5);
  nrow(out)
  ncol(out)
```

The ODE solution is returned in out, a $6 \times 2(101) + 1 = (6)(203) = 1218$ element array. These dimensions are confirmed by the nrow, ncol utilities (in the numerical output considered subsequently).

- The numerical solution is placed in two 6×101 arrays, u1, u2, for subsequent numerical and graphical output. The offset of 1 in i+1, i+1+n is required since the 6 values of t are included in out. Two nested fors step through the nout=6 values of t and the n=101 values of x. The analytical solution is also placed in two arrays, u1a, u2a.

```
#
# Arrays for plotting numerical,
# analytical solutions
  u1=matrix(0,nrow=n,ncol=nout);
  u2=matrix(0,nrow=n,ncol=nout);
 u1a=matrix(0,nrow=n,ncol=nout);
 u2a=matrix(0,nrow=n,ncol=nout);
  for(it in 1:nout){
  for(i  in 1:n){
    u1[i,it]=out[it,i+1];
    u2[i,it]=out[it,i+1+n];
   u1a[i,it]=u1_anal(x[i],tout[it],k,D,c);
   u2a[i,it]=u2_anal(x[i],tout[it],k,D,c);
   }
   }
```

- For ip=2, the numerical solution is displayed for the nout=6 values of t and n=101 values of x. The difference in the numerical and analytical solutions is also included in the output.

```
#
# Display numerical solution
  if(ip==2){
    for(it in 1:nout){
      if((it==1)|(it==nout)){
  cat(sprintf("\n     t         x      u1(x,t)
              u1_ex(x,t)   u1_err(x,t)"));
  cat(sprintf("\n                       u2(x,t)
```

```
                    u2_ex(x,t)   u2_err(x,t)\n"));
   for(i in 1:n){
     cat(sprintf("%5.1f%8.2f%10.5f%12.5f
                    %13.6f\n",tout[it],x[i],
                    u1[i,it],u1a[i,it],
                    u1[i,it]-u1a[i,it]));
       cat(sprintf("            %10.5f%12.5f
                    %13.6f\n",u2[i,it],u2a[i,it],
                    u2[i,it]-u2a[i,it]));
      }
    }
   }
 }
```

if((it==1)|(it==nout)) is used to limit the volume of the numerical output corresponding to $t = 0, 5$.

- The number of calls to pde_1a is displayed as an indication of the computational effort required to compute the solution.

```
#
# Calls to ODE routine
   cat(sprintf("\n\n ncall = %5d\n\n",ncall));
```

- The numerical solution is plotted with matplot and the analytical solution is plotted with matpoints. The two plots are superimposed as indicated in Figures 2.8 and 2.9.

```
#
# Plot u1 numerical, analytical
   par(mfrow=c(1,1));
   matplot(x=x,y=u1,type="l",xlab="x",ylab="u1(x,t)",
           xlim=c(xl,xu),lty=1,col="black",lwd=2);
   matpoints(x=x,y=u1a,xlim=c(xl,xu),col="black",
             lwd=2)
#
# Plot u2 numerical, analytical
   par(mfrow=c(1,1));
   matplot(x=x,y=u2,type="l",xlab="x",ylab="u2(x,t)",
           xlim=c(xl,xu),lty=1,col="black",lwd=2);
   matpoints(x=x,y=u2a,xlim=c(xl,xu),col="black",
             lwd=2)
```

The ODE routine pde_1a, called by lsodes, is listed next.

2.3.2 ODE routine

The routine for the SCMOL ODEs follows.

```
  pde_1a=function(t,u,parms){
#
# Function pde_1a computes the t derivative vectors of
# u1(x,t),u2(x,t)
#
# One vector to two vectors
  u1=rep(0,n);u2=rep(0,n);
  for(i in 1:n){
    u1[i]=u[i];
    u2[i]=u[i+n];
  }
#
# u1x,u2x
  table=splinefun(x,u1);
  u1x=table(x,deriv=1);
  table=splinefun(x,u2);
  u2x=table(x,deriv=1);
#
# BCs
  u1x[1]=0;  u1x[n]=0;
  u2x[1]=0;  u2x[n]=0;
#
# Nonlinear term
  u1u2x=rep(0,n);
  for(i in 1:n){
    u1u2x[i]=2*u1[i]/u2[i]*u2x[i];
  }
#
# u1xx,u1u2xx
  table=splinefun(x,u1x);
  u1xx=table(x,deriv=1);
  table=splinefun(x,u1u2x);
  u1u2xx=table(x,deriv=1);
#
# PDEs
  u1t=rep(0,n);u2t=rep(0,n);
  for(i in 1:n){
    u1t[i]=D*(u1xx[i]-u1u2xx[i]);
    u2t[i]=-k*u1[i];
  }
```

```
#
# Two vectors to one vector
  ut=rep(0,2*n);
  for(i in 1:n){
    ut[i]   =u1t[i];
    ut[i+n] =u2t[i];
  }
#
# Increment calls to pde_1a
  ncall <<- ncall+1;
#
# Return derivative vector
  return(list(c(ut)));
  }
```

Listing 2.12 ODE/MOL routine pde_1a for Eqs. (2.14).

We can note the following details about pde_1a:

- The function is defined. The input arguments are (i) the current value of t; (ii) the vector of ODE-dependent variables, a 202-vector; and (iii) input parameters, not used in this case but still required as an argument. The output of pde_1a is the vector of ODE derivatives in t, as explained at the end of pde_1a.

```
  pde_1a=function(t,u,parms){
#
# Function pde_1a computes the t derivative
# vectors of u1(x,t),u2(x,t)
```

- u is placed in two 101-vectors for the dependent variables u_1, u_2 of Eqs. (2.14).

```
#
# One vector to two vectors
  u1=rep(0,n);u2=rep(0,n);
  for(i in 1:n){
    u1[i]=u[i];
    u2[i]=u[i+n];
  }
```

- The first derivatives of u_1, u_2 are computed by splinefun.

```
#
# u1x,u2x
  table=splinefun(x,u1);
  u1x=table(x,deriv=1);
```

```
table=splinefun (x,u2) ;
u2x=table (x,deriv=1) ;
```

- BCs (2.16) are implemented for u_1, u_2 of Eqs. (2.14). Points 1, n correspond to $x = x_l = -10, x = x_u = 15$.

```
#
# BCs
  u1x[1]=0;  u1x[n]=0;
  u2x[1]=0;  u2x[n]=0;
```

- The nonlinear term of Eq. (2.14a), $2\dfrac{u_1}{u_2}\dfrac{\partial u_2}{\partial x}$, is computed.

```
#
# Nonlinear term
  u1u2x=rep (0,n) ;
  for(i in 1:n){
    u1u2x[i]=2*u1[i]/u2[i]*u2x[i] ;
  }
```

- The second derivatives of Eq. (2.14a) are computed as the derivative of the first derivative (stagewise differentiation).

```
#
# u1xx,u1u2xx
  table=splinefun (x,u1x) ;
  u1xx=table (x,deriv=1) ;
  table=splinefun (x,u1u2x) ;
  u1u2xx=table (x,deriv=1) ;
```

- Equations (2.14) for u_1, u_2 are programmed.

```
#
# PDEs
  u1t=rep (0,n) ;u2t=rep (0,n) ;
  for(i in 1:n){
    u1t[i]=D* (u1xx[i]-u1u2xx[i]) ;
    u2t[i]=-k*u1[i] ;
  }
```

The close correspondence of this coding with Eqs. (2.14) demonstrates an important feature of SCMOL.

- The two derivative vectors in t are placed in one vector, ut, for return to the ODE integrator, lsodes (called in Listing 2.11).

```
#
# Two vectors to one vector
```

```
ut=rep(0,2*n);
for(i in 1:n){
   ut[i]   =u1t[i];
   ut[i+n]=u2t[i];
   }
```

- The counter for the calls to pde_1a is incremented and its value returned to the calling program (of Listing 2.11) with the <<- operator.

```
#
# Increment calls to pde_1a
  ncall <<- ncall+1;
```

- The derivative vector (LHSs of Eqs. (2.14)) is returned to lsodes with a combination of return, list, c.

```
#
# Return derivative vector
  return(list(c(ut)));
  }
```

The final } concludes pde_1a.

2.3.3 Subordinate Routines

Functions u1_anal, u2_anal for ICs (2.18) are listed next.

```
u1_anal=function(x,t,k,D,c){
#
# Function u1_anal computes the analytical
# solution for u1(x,t)
#
  z=x-c*t;
  u1a=(c^2/(k*D))*exp(-c*z/D)/(1+exp(-c*z/D))^2;
#
# Return solution
  return(c(u1a));
  }
```

Listing 2.13a Function u1_anal for Eq. (2.17a).

```
u2_anal=function(x,t,k,D,c){
#
# Function u2_anal computes the analytical
```

```
# solution for u2(x,t)
#
  z=x-c*t;
  u2a=1/(1+exp(-c*z/D));
#
# Return solution
  return(c(u2a));
  }
```

Listing 2.13b Function u2_anal for Eq. (2.17b).

The coding in Listings 2.13 follows directly from Eqs. (2.17).

This completes the programming of Eqs. (2.14)–(2.18). Abbreviated numerical output is displayed in Table 2.6. The graphical output is shown in Figures 2.8 and 2.9. The dimensions of out are 6 × 203 as expected (from the preceding discussion). The close agreement of the numerical and analytical values of u_1, u_2 in Table 2.6 is clear. The computational effort was modest, ncall = 389.

Table 2.6 Abbreviated output for Eqs. (2.14)–(2.18).

```
[1] 6

[1] 203
```

t	x	u1(x,t) u2(x,t)	u1_ex(x,t) u2_ex(x,t)	u1_err(x,t) u2_err(x,t)
0.0	-10.00	0.00005	0.00005	0.000000
		0.00005	0.00005	0.000000
0.0	-9.75	0.00006	0.00006	0.000000
		0.00006	0.00006	0.000000
0.0	-9.50	0.00007	0.00007	0.000000
		0.00007	0.00007	0.000000
0.0	-9.25	0.00010	0.00010	0.000000
		0.00010	0.00010	0.000000
0.0	-9.00	0.00012	0.00012	
		.		.

(Continued)

Table 2.6 (Continued)

```
                   .                        .
                   .                        .
                   .                        .
   0.0   14.00   0.00000       0.00000      0.000000
                 1.00000       1.00000      0.000000
   0.0   14.25   0.00000       0.00000      0.000000
                 1.00000       1.00000      0.000000
   0.0   14.50   0.00000       0.00000      0.000000
                 1.00000       1.00000      0.000000
   0.0   14.75   0.00000       0.00000      0.000000
                 1.00000       1.00000      0.000000
   0.0   15.00   0.00000       0.00000      0.000000
                 1.00000       1.00000      0.000000
                   .                        .
                   .                        .
                   .                        .
           Output for t = 1 to 4 removed
                   .                        .
                   .                        .
                   .                        .

     t        x    u1(x,t)   u1_ex(x,t)   u1_err(x,t)
                    u2(x,t)   u2_ex(x,t)   u2_err(x,t)
   5.0  -10.00   0.00000       0.00000      0.000003
                 0.00000       0.00000      0.000002
   5.0   -9.75   0.00000       0.00000      0.000003
                 0.00000       0.00000      0.000001
   5.0   -9.50   0.00000       0.00000      0.000003
                 0.00000       0.00000      0.000001
   5.0   -9.25   0.00000       0.00000      0.000003
                 0.00000       0.00000      0.000001
   5.0   -9.00   0.00000       0.00000      0.000003
                   .                        .
                   .                        .
                   .                        .
```

(Continued)

Table 2.6 (Continued)

```
       Output for x = -8.75 to 3.75 removed
                  .                              .
                  .                              .
                  .                              .
 5.0      4.00      0.19658        0.19661      -0.000030
                    0.26896        0.26894       0.000017
 5.0      4.25      0.21787        0.21789      -0.000028
                    0.32084        0.32082       0.000014
 5.0      4.50      0.23498        0.23500      -0.000024
                    0.37755        0.37754       0.000011
 5.0      4.75      0.24611        0.24613      -0.000019
                    0.43783        0.43782       0.000009
 5.0      5.00      0.24999        0.25000      -0.000015
                    0.50001        0.50000       0.000008
 5.0      5.25      0.24612        0.24613      -0.000012
                    0.56218        0.56218       0.000007
 5.0      5.50      0.23499        0.23500      -0.000011
                    0.62247        0.62246       0.000007
 5.0      5.75      0.21788        0.21789      -0.000010
                    0.67918        0.67918       0.000006
 5.0      6.00      0.19660        0.19661      -0.000009
                  .                              .
                  .                              .
                  .                              .
       Output for x = 6.25 to 13.75 removed
                  .                              .
                  .                              .
                  .                              .
 5.0     14.00      0.00014        0.00012       0.000017
                    0.99986        0.99988      -0.000016
 5.0     14.25      0.00012        0.00010       0.000021
                    0.99988        0.99990      -0.000021
 5.0     14.50      0.00010        0.00007       0.000027
                    0.99990        0.99993      -0.000027
 5.0     14.75      0.00009        0.00006       0.000035
                    0.99991        0.99994      -0.000035
 5.0     15.00      0.00009        0.00005       0.000045

ncall =    389
```

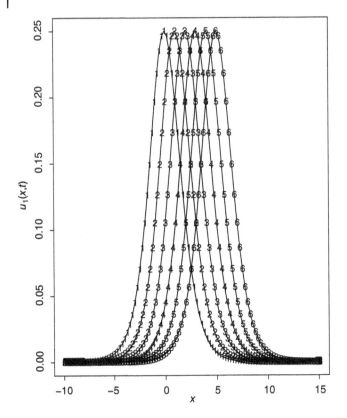

Figure 2.8 Analytical and numerical solutions $u_1(x, t)$ for Eq. (2.14a) lines – num, points – anal.

The preceding discussion demonstrates the ease of programming of the chemotaxis derivative

$$\frac{\partial}{\partial x}\left[-2\frac{u_1}{u_2}\frac{\partial u_2}{\partial x}\right]$$

in Eq. (2.14a).

As a variant of this problem, consideration of the chemotaxis derivative

$$\frac{\partial}{\partial x}\left[-2u_1\frac{\partial u_2}{\partial x}\right]$$

would be of interest if the effective diffusivity is proportional to only u_1 rather than $\frac{u_1}{u_2}$, that is, the diffusivity is proportional to the concentration of the diffusing component. This alternative can be easily investigated by changing

```
#
# Nonlinear term
```

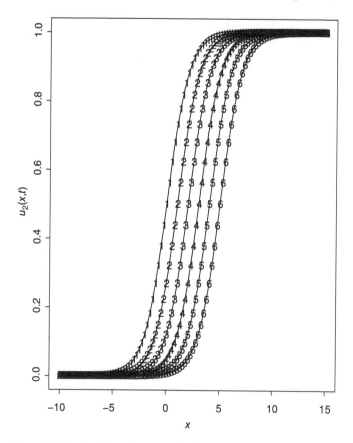

Figure 2.9 Analytical and numerical solutions $u_2(x, t)$ for Eq. (2.14b) lines – num, points – anal.

```
  u1u2x=rep(0,n);
  for(i in 1:n){
    u1u2x[i]=2*u1[i]/u2[i]*u2x[i];
  }
```

in pde_1a of Listing 2.12 to

```
#
# Nonlinear term
  u1u2x=rep(0,n);
  for(i in 1:n){
    u1u2x[i]=2*u1[i]*u2x[i];
  }
```

This change produces an unacceptable solution since u_2 assumes some negative values (a physical impossibility for a concentration). The reader can easily confirm this conclusion.

This result can be avoided by adding an *inhomogeneous source term*, $f(x, t)$, to Eq. (2.14b) that offsets the linear depletion $-ku_1$.

$$\frac{\partial u_2}{\partial t} = -ku_1 + f(x, t) \tag{2.14c}$$

For example, if $f(x, t) = e^{-cx^2}$, a Gaussian function with the programming in pde_1a of Eq. (2.14b) of Listing 2.12

```
u2t[i]=-k*u1[i]+exp(-0.1*x[i]^2);
```

$u_2(x, t)$ remains positive. This illustrates how experimentation with the model can be easily performed to investigate variants of interest.

This completes the programming and discussion of the solution of the inhomogeneous, simultaneous, nonlinear PDE system of Eqs. (2.14)–(2.18).

2.4 First Order in Space and Time

The SCMOL application to follow pertains to a PDE system first order in space and time.

$$\frac{\partial u_1}{\partial t} = -v\frac{\partial u_1}{\partial z} + k_1(u_2 - u_1) \tag{2.19a}$$

$$\frac{\partial u_2}{\partial t} = k_2(u_1 - u_2) \tag{2.19b}$$

where the dependent variables u_1 and u_2 could represent, for example, a concentration of a component in a flowing fluid stream and u_2 the concentration of the component in an adjacent stationary (nonflowing) volume. The component might be oxygen or a nutrient in a blood vessel and u_2 the concentration in the adjacent tissue. v, k_1, k_2 are parameters (constants) to be specified. Physically, v could be a fluid velocity, and k_1, k_2 mass transfer coefficients.

Equation (2.19a) is first order in t and z and therefore requires one IC and one BC.

$$u_1(z, t = 0) = f_1(z); \quad u_1(z = 0, t) = g_1(t) \tag{2.20a,b}$$

Equation (2.19b) is first order in t and requires one IC.

$$u_2(z, t = 0) = f_2(z) \tag{2.20c}$$

Since these equations are linear, an analytical solution is also available ([4], p213) to evaluate the accuracy of the numerical solution.

$$u_{1a}(z, t) = u_a(z, t) + \frac{1}{k_2}\frac{\partial u_a(z, t)}{\partial t} \tag{2.21a}$$

where

$$u_a(z, t) = e^{-(k_1/v)z} k_2 \int_0^t h(\lambda - z/v) e^{-k_2(\lambda - z/v)}$$

$$I_o \left\{ 2\sqrt{\frac{k_1 k_2}{v} z(\lambda - z/v)} \right\} d\lambda \qquad (2.21b)$$

and from *Leibniz' rule* for differentiating an integral (applied to Eq. (2.21b))

$$\frac{\partial u_a(z, t)}{\partial t} = e^{-(k_1/v)z} k_2 h(t - z/v) e^{-k_2(t-z/v)} I_o \left\{ 2\sqrt{\frac{k_1 k_2}{v} z(t - z/v)} \right\} \qquad (2.21c)$$

where I_o is the *modified Bessel function of the first kind of order zero* and $h(t - z/v)$ is the *Heaviside function* or unit step

$$h(t - z/v) = \begin{cases} 0, & (t - z/v) < 0 \\ 1, & (t - z/v) > 1 \end{cases} \qquad (2.21d)$$

At $z = z_L$, Eq. (2.21c) becomes an ODE in t

$$\frac{du_a(z = z_L, t)}{dt} = e^{-(k_1/v)z_L} k_2 h(t - z_L/v) e^{-k_2(t-z_L/v)}$$

$$I_o \left\{ 2\sqrt{\frac{k_1 k_2}{v} z(t - z_L/v)} \right\} \qquad (2.21e)$$

which is integrated numerically by adding the RHS function to the ODE routine pde_1a (discussed next). Thus, the analytical solution that is used to evaluate the numerical solution is itself partially numerical, that is, Eq. (2.21e) is integrated numerically to facilitate the analysis. Equation (2.21e) could also be integrated analytically, but that is relatively complicated and the numerical integration is straightforward.

Equation (2.19a) is a first-order *hyperbolic* PDE. A simplified version with $k_1 = 0$ is the *linear advection equation*

$$\frac{\partial u_1}{\partial t} + v \frac{\partial u_1}{\partial z} = 0 \qquad (2.21f)$$

Equation (2.22) is possibly the simplest PDE with derivatives in z and t, yet ironically, it can be one of the most difficult PDEs to integrate numerically because it can propagate discontinuities and steep fronts. We will avoid these numerical problems by selecting k_1, k_2 to smooth out discontinuities (through the transfer terms $k_1(u_2 - u_1), k_2(u_1 - u_2)$). This is necessary because splines (e.g., in splinefun) cannot accurately approximate discontinuous functions (which is also the case for other linear approximations such as FDs). Rather, nonlinear approximations are required such as flux limiters ([1], 37–43).

A main program follows for the SCMOL solution of Eqs. (2.19), and verification of the numerical solution with the analytical solution of Eqs. (2.21).

2.4.1 Main Program

```
#
# Delete previous workspaces
  rm(list=ls(all=TRUE))
#
# Access ODE integrator
  library("deSolve");
#
# Access files
  setwd("f:/collocation/test46");
  source("pde_1a.R");
  source("bessel_Io.R") ;
#
# Model parameters
   u10=0;   u20=0;
    k1=1;    k2=1;
     v=1;    zL=5;
    n=41;
   exp1=exp(-k1/v*zL);
#
# Grid in z
  z=seq(from=0,to=zL,by=(zL-0)/(n-1));
#
# Initial condition
  u0=rep(0,2*n+1);
  for(i in 1:n){
      u0[i]=u10;
    u0[i+n]=u20;
  }
  u0[2*n+1]=u10;
#
# Interval in t
  t0=0;tf=20;nout=41;
  tout=seq(from=t0,to=tf,by=(tf-t0)/(nout-1));
  ncall=0;
#
# ODE integration
  out=lsodes(y=u0,times=tout,func=pde_1a,
      sparsetype="sparseint",rtol=1e-6,
      atol=1e-6,maxord=5);
```

```
  nrow(out)
  ncol(out)
#
# Store solution
  u1=matrix(0,nrow=nout,ncol=n);
  u2=matrix(0,nrow=nout,ncol=n);
  uat=rep(0,nout);ua=rep(0,nout);
  u1a=rep(0,nout);t=rep(0,nout);
  for(it in 1:nout){
  for(i  in 1:n){
    u1[it,i]=out[it,i+1];
    u2[it,i]=out[it,i+1+n];
  }
    ua[it]=out[it,2*n+2];
     t[it]=out[it,1];
  }
#
# Display numerical solution
  cat(sprintf("\n\n      t       u1(z=zL,t)
               u1a(z=zL,t)       diff\n"));
  for(it in 1:nout){
    lam=t[it]-zL/v;
    if(lam<0){
      uat=0;
    }else{
      arg=2*sqrt((k1*k2*zL/v)*lam)
      uat=exp1*k2*exp(-k2*lam)*bessel_Io(arg)
    }
    u1a[it]=ua[it]+uat/k2;
    diff=u1[it,n]-u1a[it];
    cat(sprintf(
      "%7.2f%12.4f%12.4f%14.6f\n",
      t[it],u1[it,n],u1a[it],diff));
  }
#
# Calls to ODE routine
  cat(sprintf("\n\n ncall = %5d\n\n",ncall));
#
# Plot analytical, numerical solutions
  par(mfrow=c(1,1))
  plot(t,u1[,n],xlab="t",ylab="u1(z=zL,t)",
```

```
    col="black",lwd=2,pch="o");
lines(t,u1[,n],lty=2,lwd=2,type="l");
lines(t,ula,lty=1,lwd=2,type="l");
```

Listing 2.14 Main program for Eqs. (2.19).

We can note the following details about Listing 2.14:

- Previous workspaces are removed. The library of ODE integrators, deS-olve, is accessed, followed by the routines for Eqs. (2.19). Note the use of / rather than the usual \ in the setwd (set working directory). Also, setwd will require revision for the local computer (to define the working folder or directory for the routines pde_1a, bessel_Io.R).

```
#
# Delete previous workspaces
  rm(list=ls(all=TRUE))
#
# Access ODE integrator
  library("deSolve");
#
# Access files
  setwd("f:/collocation/test46");
  source("pde_1a.R");
  source("bessel_Io.R") ;
```

- The parameters in Eqs. (2.19) are defined.

```
#
# Model parameters
  u10=0;  u20=0;
   k1=1;   k2=1;
    v=1;   zL=5;
   n=41;
  exp1=exp(-k1/v*zL);
```

exp1 is a computed constant that is passed to the ODE routine discussed next without any special designation (a feature of R).
- A grid of 41 points in z is defined for $0 \le z \le zL = 5$ so that $z = 0, 0.125, \ldots, 5$. The seq utility is used to produce the sequence of values in z.

```
#
# Grid in z
  z=seq(from=0,to=zL,by=(zL-0)/(n-1));
```

- Homogeneous (zero) ICs for $2*n+1 = 2*41+1 = 83$ ODEs are defined.

```
#
# Initial condition
  u0=rep(0,2*n+1);
  for(i in 1:n){
      u0[i]=u10;
    u0[i+n]=u20;
  }
  u0[2*n+1]=u10;
```

The last IC, $u0[2*n+1]=u20$, is for ODE (2.21e). The homogeneous IC follows from the Heaviside function in Eq. (2.21e), $h(t - z_L/v)$, which is zero for $t = 0$ (so that $h(t - z_L/v) \leq 0$ with $z_L = 5, v = 1$).

- The interval in t is defined with $nout=41$ points for $0 \leq t \leq 20$ so that $tout = 0, 0.5, \dots, 20$.

```
#
# Interval in t
  t0=0;tf=20;nout=41;
  tout=seq(from=t0,to=tf,by=(tf-t0)/(nout-1));
  ncall=0;
```

The counter for the calls to the ODE routine is initialized.

- The $2(41) + 1 = 83$ ODEs are integrated numerically by $lsodes$. The input arguments to $lsodes$ are the MOL/ODE routine pde_1a, the IC vector u0 and the output values of t, tout. The number of ODEs to be integrated is defined by the length of u0 (83). func, y, times are reserved names.

```
#
# ODE integration
  out=lsodes(y=u0,times=tout,func=pde_1a,
      sparsetype="sparseint",rtol=1e-6,
      atol=1e-6,maxord=5);
  nrow(out)
  ncol(out)
```

The ODE solution is returned in out, a $41 \times 2(41) + 1 + 1 = (41)(84) = 1764$ element array. These dimensions are confirmed by the nrow, ncol utilities (in the numerical output considered subsequently).

- The numerical solution is placed in two 41×41 arrays, u1, u2, for subsequent numerical and graphical output. The offset of 1 in i+1, i+1+nz is required since the 41 values of t are included in out (and placed in array t). Two nested fors step through the nout=41 values of t and the n=41 values of z. The numerical portion of the analytical solution of Eq. (2.21a) (from the integration of Eq. (2.21e)) is also placed in an array, ua.

```
#
# Store solution
  u1=matrix(0,nrow=nout,ncol=n);
  u2=matrix(0,nrow=nout,ncol=n);
  uat=rep(0,nout);ua=rep(0,nout);
  u1a=rep(0,nout);t=rep(0,nout);
  for(it in 1:nout){
  for(i  in 1:n){
    u1[it,i]=out[it,i+1];
    u2[it,i]=out[it,i+1+n];
}
    ua[it]=out[it,2*n+2];
     t[it]=out[it,1];
}
```

- The numerical solution of Eq. (2.21a), is displayed for the nout=41 values of t and $z = z_L$. The difference in the numerical and analytical solutions at $z = z_L$ is also included in the output.

```
#
# Display numerical solution
  cat(sprintf("\n\n      t       u1(z=zL,t)
               u1a(z=zL,t)       diff\n"));
  for(it in 1:nout){
    lam=t[it]-zL/v;
    if(lam<0){
      uat=0;
    }else{
      arg=2*sqrt((k1*k2*zL/v)*lam)
      uat=exp1*k2*exp(-k2*lam)*bessel_Io(arg)
    }
    u1a[it]=ua[it]+uat/k2;
    diff=u1[it,n]-u1a[it];
    cat(sprintf(
      "%7.2f%12.4f%12.4f%14.6f\n",
      t[it],u1[it,n],u1a[it],diff));
}
```

The components of the analytical solution are programmed as follows:

The Lagrangian variable $\lambda - z_L/v \Rightarrow$ t[it]-zL/v.

- In Eq. (2.21b), $2\sqrt{\dfrac{k_1 k_2}{v}} z(\lambda - z/v) \Rightarrow$ 2*sqrt((k1*k2*zL/v)*lam)
  (for $\lambda - z/v > 0$).

- In Eq. (2.21e), $e^{-(k_1/v)z_L}k_2 h(t - z_L/v)e^{-k_2(t-z_L/v)}I_o\left\{2\sqrt{\dfrac{k_1 k_2}{v}}z(t - z_L/v)\right\} \Rightarrow$

  `exp1*k2*exp(-k2*lam)*bessel_Io(arg)` (for $t - z/v > 0$).

- In Eq. (2.21a), $u_a(z, t) + \dfrac{1}{k_2}\dfrac{\partial u_a(z, t)}{\partial t} \Rightarrow$ `ua[it]+uat/k2`.

Equation (2.21e) is also programmed in ODE/MOL function `pde_1a` (called by `lsodes` and discussed subsequently).

- The number of calls to `pde_1a` is displayed as an indication of the computational effort required to compute the solution.

```
#
# Calls to ODE routine
  cat(sprintf("\n\n ncall = %5d\n\n",ncall));
```

- The numerical solution is plotted against t with `plot`, `lines` for $z = z_L$ (`u1[,n]`), and the analytical solution is plotted with `lines` (`u1a`). The two plots are superimposed as indicated in Figure 2.10.

```
#
# Plot analytical, numerical solutions
  par(mfrow=c(1,1))
  plot(t,u1[,n],xlab="t",ylab="ul(z=zL,t)",
      col="black",lwd=2,pch="o");
  lines(t,u1[,n],lty=2,lwd=2,type="l");
  lines(t,u1a,lty=1,lwd=2,type="l");
```

The ODE routine `pde_1a`, called by `lsodes`, is listed next.

2.4.2 ODE Routine

The routine for the ODE/SCMOL follows.

```
  pde_1a=function(t,u,parms) {
#
# Function pde_1a computes the t derivative
# vector of the u vector
#
# One vector to two PDEs
  u1=rep(0,n);u2=rep(0,n);
  for(i in 1:n){
    u1[i]=u[i];
    u2[i]=u[i+n];
  }
#
# Boundary condition
```

```
  u1[1]=1;
#
# u1z
  table=splinefun(z,u1);
  u1z=table(z,deriv=1);
#
# u1t, u2t, u1at
  u1t=rep(0,n); u2t=rep(0,n);
  for(i in 1:n){
    if(i==1){
      u1t[i]=0;
    }else{
      u1t[i]=-v*u1z[i]+k1*(u2[i]-u1[i]);
    }
      u2t[i]=k2*(u1[i]-u2[i]);
  }
#
# Analytical ODE
  lam=t-zL/v;
  if(lam<0){
    uat=0;
  }else{
    arg=2*sqrt((k1*k2*zL/v)*lam)
    uat=exp1*k2*exp(-k2*lam)*bessel_Io(arg)
  }
#
# Two PDEs and one ODE to one derivative vector
  ut=rep(0,2*n);
  for(i in 1:n){
      ut[i]=u1t[i];
    ut[i+n]=u2t[i];
  }
  ut[2*n+1]=uat;
#
# Increment calls to pde_1a
  ncall<<-ncall+1;
#
# Return derivative vector
  return(list(c(ut)));
}
```

Listing 2.15 ODE/SCMOL routine pde_1a for Eqs. (2.19).

We can note the following details about pde_1a:

- The function is defined. The input arguments are as follows: (i) the current value of t; (ii) the vector of ODE-dependent variables, a 83-vector; and (iii) input parameters, not used in this case but still required as an argument. The output of pde_1a is the vector of ODE derivatives in t, as explained at the end of pde_1a.

```
pde_1a=function(t,u,parms){
#
# Function pde_1a computes the t derivative
# vector of the u vector
```

- u is placed in two 41-vectors for the dependent variables u_1, u_2 of Eqs. (2.19).

```
#
# One vector to two PDEs
u1=rep(0,n);u2=rep(0,n);
for(i in 1:n){
   u1[i]=u[i];
   u2[i]=u[i+n];
}
```

- BC (2.20c) is applied.

```
#
# Boundary condition
u1[1]=1;
```

- The first derivative of u_1 with respect to z is computed by splinefun.

```
#
# u1z
table=splinefun(z,u1);
u1z=table(z,deriv=1);
```

- Equations (2.19) for u_1, u_2 are programmed.

```
#
# u1t, u2t, u1at
u1t=rep(0,n); u2t=rep(0,n);
for(i in 1:n){
   if(i==1){
     u1t[i]=0;
   }else{
     u1t[i]=-v*u1z[i]+k1*(u2[i]-u1[i]);
   }
```

```
      u2t[i]=k2*(u1[i]-u2[i]);
   }
```

The close correspondence of this coding with Eqs. (2.19) demonstrates an important feature of SCMOL.

- The two derivative vectors in t are placed in one vector, ut, for return to the ODE integrator, lsodes (called in Listing 2.14).

```
#
# Two vectors to one vector
  ut=rep(0,2*n);
  for(i in 1:n){
    ut[i]   =u1t[i];
    ut[i+n] =u2t[i];
  }
```

- ODE (2.21e) is programmed as the last element in the t derivative vector.

```
#
# Analytical ODE
  lam=t-zL/v;
  if(lam<0){
    uat=0;
  }else{
    arg=2*sqrt((k1*k2*zL/v)*lam)
    uat=exp1*k2*exp(-k2*lam)*bessel_Io(arg)
  }
```

Note in particular the discontinuity at $t - z_L/v = 0$.

- The t derivatives of Eqs. (2.19) and (2.21e) are placed in a vector ut of length $2n + 1 = 2(41) + 1 = 83$.

```
#
# Two PDEs and one ODE to one derivative vector
  ut=rep(0,2*n);
  for(i in 1:n){
      ut[i]=u1t[i];
    ut[i+n] =u2t[i];
  }
  ut[2*n+1]=uat;
```

- The counter for the calls to pde_1a is incremented and its value returned to the calling program (of Listing 2.14) with the <<- operator.

```
#
# Increment calls to pde_1a
  ncall<<-ncall+1;
```

- The derivative vector ut is returned to lsodes with a combination of return, list, c.

```
#
# Return derivative vector
   return(list(c(ut)));
}
```

The final } concludes pde_1a.

2.4.3 Subordinate Routines

Function bessel_Io called in Listings 2.14 and 2.15 is listed next.

```
  bessel_Io=function(x){
#
# Function bessel_Io is a translation from Fortran
# into R of the original numerical recipes function
# BESSIO.
#
# Constants in the approximation of the Bessel
# function Io
  p1=1             ; p2=3.5156229      ; p3=3.0899424      ;
  p4=1.2067492     ; p5=0.2659732      ; p6=0.360768e-1    ;
  p7=0.45813e-2    ;
  q1=0.39894228    ; q2=0.1328592e-1 ; q3=0.225319e-2     ;
  q4=-0.157565e-2; q5=0.916281e-2    ; q6=-0.2057706e-1;
  q7=0.2635537e-1; q8=-0.1647633e-1; q9=0.392377e-2     ;
#
# Calculation of bessel_Io
  y=(x/3.75)^2;
  bessel_Io=p1+y*(p2+y*(p3+y*(p4+y*
            (p5+y*(p6+y*p7))))));
  }else{
  ax=abs(x);y=3.75/ax;
  bessel_Io=(exp(ax)/sqrt(ax))*(q1+y*
            (q2+y*(q3+y*(q4+y*(q5+y*
            (q6+y*(q7+y*(q8+y*q9)))))))));
  }
  return(c(bessel_Io));
  }
```

Listing 2.16 Function bessel_Io for the analytical solution of Eqs. (2.21a), (2.21e).

Function bessel_Io is taken from [3] with a translation into R.

This completes the programming of Eqs. (2.19)–(2.21). Abbreviated numerical output is in Table 2.7. The graphical output is shown in Figure 2.10. The dimensions of out are nout x 2*n+1+1 = 41 x 84 as expected (from the preceding discussion). The close agreement of the numerical and analytical values of $u_1(z = z_L, t)$ in Table 2.7 is clear. The computational effort was modest, ncall = 924.

Table 2.7 Abbreviated output for Eqs. (2.19)–(2.21).

```
[1]  41

[1]  84

       t        u1(z=zL,t)    u1a(z=zL,t)       diff
      0.00        0.0000        0.0000        0.000000
      0.50        0.0000        0.0000        0.000000
      1.00        0.0000        0.0000        0.000000
                    .             .
                    .             .
                    .             .
      Output for t = 1.50 to 4.50 removed
                    .             .
                    .             .
                    .             .
      5.00        0.0042        0.0068       -0.002524
      5.50        0.0304        0.0300        0.000469
      6.00        0.0658        0.0656        0.000158
      6.50        0.1126        0.1125        0.000026
      7.00        0.1686        0.1686        0.000003
      7.50        0.2313        0.2313       -0.000001
      8.00        0.2982        0.2982       -0.000003
      8.50        0.3668        0.3668       -0.000001
      9.00        0.4351        0.4351        0.000001
      9.50        0.5012        0.5012        0.000002
     10.00        0.5639        0.5639        0.000001
     10.50        0.6222        0.6222       -0.000001
     11.00        0.6756        0.6756       -0.000006
     11.50        0.7236        0.7236       -0.000012
     12.00        0.7663        0.7663       -0.000019
```

Table 2.7 (Continued)

12.50	0.8038	0.8039	-0.000025
13.00	0.8364	0.8365	-0.000030
.	.		
.	.		
.	.		

Output for t = 13.50 to 18.50 removed

.	.		
.	.		
.	.		
19.00	0.9878	0.9878	-0.000015
19.50	0.9905	0.9905	-0.000010
20.00	0.9926	0.9926	-0.000005

ncall = 924

This completes the programming and discussion of the solution of the PDE system first order in space and time, Eqs. (2.19)–(2.21). In particular, the splines from splinefun worked well (provided that the solution has a degree of smoothing to avoid discontinuities and sharp moving fronts).

2.5 Second Order in Time

The SCMOL application to follow pertains to the classical wave equation, a *second-order hyperbolic* PDE.

$$\frac{\partial^2 u}{\partial t^2} = c^2 \frac{\partial^2 u}{\partial x^2} \tag{2.22}$$

where c is a velocity.

Equation (2.22) is second order in t and x. It therefore requires two ICs and two BCs. For the ICs, we use

$$u(x, t = 0) = f(x); \quad u_t(x, t = 0) = 0 \tag{2.23a,b}$$

Thus, the PDE starts with an initial displacement $f(x)$ and 0 initial velocity.

To complete the specification of the problem, we would naturally consider the two required BCs for Eq. (2.22). However, if the PDE is analyzed over an essentially infinite domain, $-\infty \leq x \leq \infty$, and if changes in the solution occur only over a finite interval in x, then BCs at infinity have no effect; in other words, we do not have to actually specify BCs (since they have no effect).

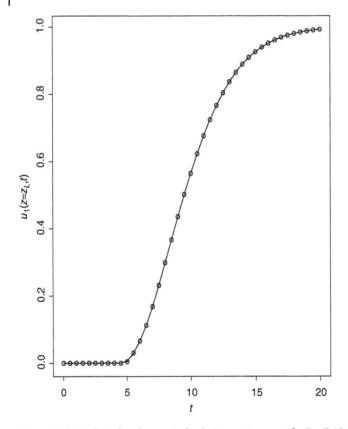

Figure 2.10 Analytical and numerical solutions $u_1 (z = z_L, t)$ for Eq. (2.19a) lines – num, points – anal.

Alternatively, two values for x can be selected that are effectively $\pm\infty$ in the sense that the solution does not depart from its initial values at the selected value of z.

Another variation is to specify a zero derivative (slope) at the finite boundaries. This second approach is illustrated subsequently and is used in the programming for Eqs. (2.22) and (2.23). In other words, the following BCs are used

$$\frac{u(x = x_L, t)}{\partial x} = \frac{u(x = x_u, t)}{\partial x} = 0 \qquad (2.24\text{a,b})$$

Equations (2.22)–(2.24) constitute the complete PDE problem.[3]

---

3 A solution to Eqs. (2.22)–(2.24) was reported by the French mathematician *Jean-le-Rond d'Alembert* (1717–1783) in 1747 in a treatise on vibrating strings. d'Alembert's remarkable solution, which used a method specific to the wave equation (based on the chain rule for

In order to apply SCMOL, Eq. (2.22) is restated as a system of two PDEs first order in t.

$$u_1(x, t) = u(x, t); \quad u_2(x, t) = \frac{\partial u}{\partial t} = \frac{\partial u_1}{\partial x} \qquad (2.26a,b)$$

The following solution is in terms of u_1, u_2.

2.5.1 Main Program

A main program follows for the SCMOL solution of Eqs. (2.22), and verification of the numerical solution with the analytical solution of Eq. (2.25).

```
#
# Delete previous workspaces
  rm(list=ls(all=TRUE))
#
# Access ODE integrator, routines
  library(deSolve);
  setwd("f:/collocation/test20");
  source("pde_1a.R");
  source("ua.R");
#
# Parameters
  c=1;cs=c^2;a=0.2;
#
# Time interval
  t0=0;tf=30;nout=4
  tout=seq(from=t0,to=tf,by=(tf-t0)/(nout-1));
  ncall=0;
#
# Spatial interval
  n=201;xl=0;xu=100;
  x=seq(from=xl,to=xu,by=(xu-xl)/(n-1));
#
# Initial conditions (ICs)
  u0=rep(0,2*n);
  u0[1:n]=ua(t0,x);
```

___

differentiation), is

$$u(x, t) = \frac{1}{2}[f(x - ct) + f(x + ct)] + \frac{1}{2c} \int_{x-ct}^{x+ct} g(\xi)d\xi \qquad (2.25)$$

It can also be obtained by a Fourier transform or by separation of variables.

Equation (2.25) is actually for a somewhat more general case than Eqs. (2.22)–(2.24). In particular, IC (2.23b) is generalized to $u_t(x, t = 0) = g(x)$. For most of the subsequent discussion, we consider the homogeneous IC $u_t(x, t = 0) = 0$ of Eq. (2.23b), that is, $g(t) = 0$.

```
  u0[(n+1):(2*n)]=0;
#
# ODE integration
  out=lsodes(y=u0,times=tout,func=pde_1a,
      sparsetype="sparseint",rtol=1e-6,
      atol=1e-6,maxord=5);
  nrow(out)
  ncol(out)
#
# Store numerical solution for plotting
   u1=matrix(0,nrow=n,ncol=nout)
  u1a=matrix(0,nrow=n,ncol=nout)
  for(it in 1:nout){
    u1a[,it]=ua(tout[it],x);
    for(i in 1:n){
    u1[i,it]=out[it,i+1];
    }
    }
#
# Display selected numerical output
    for(it in 2:nout){
      if(it==nout){
#
#    Pulse at x = 20
    cat(sprintf("\n\n\n t = %4.2f
                 \n    x %10.2f%10.2f%10.2f
                 \n   u1 %10.4f%10.4f%10.4f
                 \n u1a %10.4f%10.4f%10.4f",
                  tout[it],x[40],x[41],x[42],
                  u1[40,it], u1[41,it], u1[42,it],
                  u1a[40,it],u1a[41,it],u1a[42,it]));
#
#    Pulse at x = 80
    cat(sprintf("\n\n\n t = %4.2f
                 \n    x %10.2f%10.2f%10.2f
                 \n   u1 %10.4f%10.4f%10.4f
                 \n u1a %10.4f%10.4f%10.4f",
                  tout[it],x[160],x[161],x[162],
                  u1[160,it], u1[161,it],
                  u1[162,it],u1a[160,it],
                  u1a[161,it],u1a[162,it]));
      }
```

```
  }
#
# Calls to ODE routine
  cat(sprintf("\n\n ncall = %5d\n\n",ncall));
#
# Plot numerical, analytical solutions
  matplot(x,u1,type="l",lwd=2,col="black",
    lty=1,main="solid - num, points - anal",
    xlab="x",ylab="u(x,t)");
  matpoints(x,u1a,pch="o",col="black");
```

Listing 2.17 Main program for Eq. (2.22).

We can note the following details about Listing 2.17:

- Previous workspaces are removed. The library of ODE integrators, deS-
  olve, is accessed, followed by the routines for Eq. (2.22). Note the use of
  / rather than the usual \ in the setwd (set working directory). Also, setwd
  will require revision for the local computer (to define the working folder or
  directory for the routines pde_1a, ua.R).

```
#
# Delete previous workspaces
  rm(list=ls(all=TRUE))
#
# Access ODE integrator, routines
  library(deSolve);
  setwd("f:/collocation/test20");
  source("pde_1a.R");
  source("ua.R");
```

- The parameters in Eq. (2.22) are defined.

```
#
# Parameters
  c=1;cs=c^2;a=0.2;
```

cs is a computed constant that is passed to the ODE routine discussed next
without any special designation (a feature of R).
- The interval in t is defined with nout=4 points for $0 \leq t \leq 30$ so that tout
  $= 0,10,20,30$.

```
#
# Time interval
  t0=0;tf=30;nout=4
```

```
tout=seq(from=t0,to=tf,by=(tf-t0)/(nout-1));
ncall=0;
```

The counter for the calls to the ODE routine is initialized.

- A grid of 201 points in x is defined for $x_l = 0 \leq x \leq x_u = 100$ so that $x = 0, 0.5, \ldots, 100$. The seq utility is used to produce the sequence of values in x.

```
#
# Spatial interval
  n=201;xl=0;xu=100;
  x=seq(from=xl,to=xu,by=(xu-xl)/(n-1));
```

$x_l = 0, x_u = 100$ are effectively at $x = \pm\infty$ as discussed previously and demonstrated in the solution that follows.

- ICs (2.23a) for $2*n = 2*201 = 402$ ODEs are defined.

```
#
# Initial conditions (ICs)
  u0=rep(0,2*n);
  u0[1:n]=ua(t0,x);
  u0[(n+1):(2*n)]=0;
```

ua is a function for the analytical solution of Eq. (2.25) (discussed subsequently).

- The $2(201) + 1 = 403$ ODEs are integrated numerically by lsodes. The input arguments to lsodes are the MOL/ODE routine pde_1a (discussed subsequently), the IC vector u0 and the output values of t, tout. The number of ODEs to be integrated is defined by the length of u0 (402). func,y,times are reserved names.

```
#
# ODE integration
  out=lsodes(y=u0,times=tout,func=pde_1a,
      sparsetype="sparseint",rtol=1e-6,
      atol=1e-6,maxord=5);
  nrow(out)
  ncol(out)
```

The ODE solution is returned in out, a $4 \times 2(201) + 1 = (4)(403) = 1612$ element array. These dimensions are confirmed by the nrow, ncol utilities (in the numerical output considered subsequently).

- The numerical solution $u(x, t) = u_1(x, t)$ is placed in a 4×201 array u1 for subsequent numerical and graphical output. The offset of 1 in i+1 is required since the 4 values of t are included in out. Two nested fors step through the nout=4 values of t and the n=201 values of x. The analytical solution of Eq. (2.25) is placed in u1a (note that the second argument of u1a is the vector x, taht is, not a single value of *x*).

```
#
# Store numerical solution for plotting
   u1=matrix(0,nrow=n,ncol=nout)
   u1a=matrix(0,nrow=n,ncol=nout)
   for(it in 1:nout){
     u1a[,it]=ua(tout[it],x);
   for(i in 1:n){
     u1[i,it]=out[it,i+1];
   }
   }
```

- The numerical and analytical solutions at $x = 20, 80, t = 4$ are displayed in part.

```
#
# Display selected numerical output
   for(it in 2:nout){
     if(it==nout){
#
#     Pulse at x = 20
     cat(sprintf("\n\n\n t = %4.2f
         \n    x %10.2f%10.2f%10.2f
         \n   u1 %10.4f%10.4f%10.4f
         \n u1a %10.4f%10.4f%10.4f",
           tout[it],x[40],x[41],x[42],
           u1[40,it],u1[41,it],
           u1[42,it], u1a[40,it],
           u1a[41,it],u1a[42,it]));
#
#     Pulse at x = 80
     cat(sprintf("\n\n\n t = %4.2f
         \n    x %10.2f%10.2f%10.2f
         \n   u1 %10.4f%10.4f%10.4f
         \n u1a %10.4f%10.4f%10.4f",
           tout[it],x[160],x[161],x[162],
           u1[160,it],u1[161,it],
           u1[162,it],u1a[160,it],
           u1a[161,it],u1a[162,it]));
     }
   }
```

The selection of the values of *x* and *t* for this output is explained subsequently.

- The number of calls to pde_1a is displayed as an indication of the computational effort required to compute the solution.

```
#
# Calls to ODE routine
  cat(sprintf("\n\n ncall = %5d\n\n",ncall));
```

- The numerical solution is plotted against x with matplot with t as a parameter. The analytical solution is plotted with matpoints. The two solutions are superimposed as indicated in Figure 2.11.

```
#
# Plot numerical, analytical solutions
  matplot(x,u1,type="l",lwd=2,col="black",
    lty=1,main="solid - num, points - anal",
    xlab="x",ylab="u(x,t)");
  matpoints(x,u1a,pch="o",col="black");
```

The ODE routine pde_1a, called by lsodes, is listed next.

2.5.2 ODE Routine

The routine for the SCMOL ODEs follows.

```
  pde_1a=function(t,u,parms){
#
# Function pde_1a computes the t derivative vectors
#
# One vector to two vectors
  u1=rep(0,n);u2=rep(0,n);
  for(i in 1:n){
   u1[i]=u[i];
   u2[i]=u[i+n];
   }
#
# u1x
  table1=splinefun(x,u1);
  u1x=table1(x,deriv=1);
#
# Neumann BCs
  u1x[1]=0;u1x[n]=0;
#
# u1xx
  table2=splinefun(x,u1x);
  u1xx=table2(x,deriv=1);
#
```

```
# PDE
  u1t=rep(0,n);u2t=rep(0,n);
  for(i in 1:n){
    u1t[i]=u2[i];
    u2t[i]=cs*u1xx[i];
  }
#
# Two vectors to one vector
  ut=rep(0,2*n);
  for(i in 1:n){
      ut[i]=u1t[i];
    ut[i+n]=u2t[i];
  }
#
# Increment calls to pde_1a
  ncall<<-ncall+1;
#
# Return derivative vector
  return(list(c(ut)))
}
```

Listing 2.18 ODE/SCMOL routine pde_1a for Eq. (2.22).

We can note the following details about pde_1a:

- The function is defined. The input arguments are (i) the current value of
  t; (ii) the vector of ODE-dependent variables, a 402-vector; and (iii) input
  parameters, not used in this case but still required as an argument. The output of pde_1a is the vector of ODE derivatives in t, as explained at the end
  of pde_1a.

  ```
  pde_1a=function(t,u,parms){
  #
  # Function pde_1a computes the t derivative vectors
  ```

- u_1, u_2 of Eqs. (2.26a,b) are placed in two 201-vectors.

  ```
  #
  # One vector to two PDEs
    u1=rep(0,n);u2=rep(0,n);
    for(i in 1:n){
      u1[i]=u[i];
      u2[i]=u[i+n];
    }
  ```

- The first derivative of u_1 with respect to x is computed by splinefun.

```
#
# u1x
   table1=splinefun(x,u1);
   u1x=table1(x,deriv=1);
```

- BCs (2.24a,b) are applied.

```
#
# Neumann BCs
   u1x[1]=0;u1x[n]=0;
```

Note the subscripts 1, n corresponding to $x_l = 0, x_u = 100$.

- The second derivative of u_1 with respect to x is computed by splinefun.

```
#
# u1xx
   table2=splinefun(x,u1x);
   u1xx=table2(x,deriv=1);
```

- Equation (2.22) is programmed in terms of u_1, u_2 of Eqs. (2.26a,b).

```
#
# PDE
   u1t=rep(0,n);u2t=rep(0,n);
   for(i in 1:n){
     u1t[i]=u2[i];
     u2t[i]=cs*u1xx[i];
   }
```

- The two derivative vectors in t are placed in one vector, ut, for return to the ODE integrator, lsodes (called in Listing 2.17).

```
#
# Two vectors to one vector
   ut=rep(0,2*n);
   for(i in 1:n){
       ut[i]=u1t[i];
     ut[i+n]=u2t[i];
   }
```

- The counter for the calls to pde_1a is incremented and its value returned to the calling program (of Listing 2.17) with the <<- operator.

```
#
# Increment calls to pde_1a
   ncall<<-ncall+1;
```

- The derivative vector ut is returned to lsodes with a combination of return, list, c.

```
#
# Return derivative vector
  return(list(c(ut)))
}
```

The final } concludes pde_1a.

2.5.3 Subordinate Routine

Function ua for the analytical solution of Eq. (2.25) with $f(x)$ a pulse function is listed next.

```
  ua=function(t,x){
#
# Function ua computes the analytical solution
# to the wave equation
#
  pulse1=rep(0,n);pulse2=rep(0,n);ua=rep(0,n);
  for(i in 1:n){
    if(x[i]<(45-c*t)){pulse1[i]=0;}
    if(x[i]>(55-c*t)){pulse1[i]=0;}
    if((x[i]>=(45-c*t))&(x[i]<=(55-c*t))){
      pulse1[i]=(1/2)*exp(-a*(x[i]-(50-c*t))^2);}
    if(x[i]<(45+c*t)){pulse2[i]=0;}
    if(x[i]>(55+c*t)){pulse2[i]=0;}
    if((x[i]>=(45+c*t))&(x[i]<=(55+c*t))){
      pulse2[i]=(1/2)*exp(-a*(x[i]-(50+c*t))^2);}
    ua[i]=pulse1[i]+pulse2[i];
  }
  return(c(ua));
#
# End of ua
  }
```

Listing 2.19 Function ua for the analytical solution of Eq. (2.25).

We can note the following details of ua:

- The function is defined.

```
  ua=function(t,x){
#
```

```
# Function ua computes the analytical solution
# to the wave equation
```

t is the current value of *t* at which the analytical solution (Eq. (2.25)) is evaluated. x is the vector of n=201 spatial grid points at which the analytical solution is evaluated.

- Arrays are declared for the analytical solution, Eq. (2.25). Then the analytical solution is computed at n=201 points in *x* (with a for), that is, the computed solution ua is a 201-vector.

```
#
  pulse1=rep(0,n);pulse2=rep(0,n);ua=rep(0,n);
  for(i in 1:n){
```

- The first term in Eq. (2.25), $\frac{1}{2}[f(x-ct)]$ (with $g(\xi) = 0$), is evaluated as pulse1.

```
  if(x[i]<(45-c*t)){pulse1[i]=0;}
  if(x[i]>(55-c*t)){pulse1[i]=0;}
  if(((x[i]>=(45-c*t))&(x[i]<=(55-c*t)))){
    pulse1[i]=(1/2)*exp(-a*(x[i]-(50-c*t))^2);}
```

The term (pulse1) is zero for $x < (45 - ct)$ and $x > (55 - ct)$. For $45 \le x - ct \le 55$, the pulse is a Gaussian function (1/2)*exp(-a*(x[i]-(50-c*t))^2) centered at $ct = 50$. The Lagrangian variable $x - ct$ defines a traveling wave function as reflected in the numerical and graphical output in Table 2.8 and Figure 2.11. The parameter $a = 0.2$ is defined in the main program of Listing 2.17.

- Similarly, the second term in Eq. (2.25), $\frac{1}{2}[f(x+ct)]$ (with $g(\xi) = 0$), is evaluated as pulse2.

```
  if(x[i]<(45+c*t)){pulse2[i]=0;}
  if(x[i]>(55+c*t)){pulse2[i]=0;}
  if(((x[i]>=(45+c*t))&(x[i]<=(55+c*t)))){
    pulse2[i]=(1/2)*exp(-a*(x[i]-(50+c*t))^2);}
```

That is, it is a Gaussian function centered at $x = -ct$.

- pulse1 and pulse2 are added according to Eq. (2.25) to give the analytical solution.

```
  ua[i]=pulse1[i]+pulse2[i]
```

For $t = 0$, pulse1 = pulse2.

- ua is a 201-vector that is returned to the calling program with the return and c utilities.

```
  return(c(ua));
```

This completes the programming of Eqs. (2.22)–(2.25). Abbreviated numerical output is in Table 2.8. The graphical output is shown in Figure 2.11.

Table 2.8 Abbreviated output for Eq. (2.22).

```
[1]  4

[1]  403

t  =  30.00

   x          19.50      20.00      20.50

   u1         0.4730     0.4995     0.4776

   u1a        0.4756     0.5000     0.4756

t  =  30.00

   x          79.50      80.00      80.50

   u1         0.4776     0.4995     0.4730

   u1a        0.4756     0.5000     0.4756

ncall  =   1005
```

The dimensions of out are nout x 2*n+1+1 = 4 x 2*201+1=4 x 403 as expected (from the preceding discussion). The agreement of the numerical and analytical values of $u(x, t)$ at $x = 20, 80$ for $t = 30$ in Table 2.8 is clear. This output is from the Listing 2.17, with $x - (50 - ct) = 0$ or $x = 50 - (1)30 = 20$ and $x - (50 + ct) = 0$ or $x = 50 + (1)30 = 80$. This comparison of the numerical and analytical solutions is particularly stringent since it corresponds to two peaks (at $x = 20, 80$) where the solution is changing most rapidly, as indicated in Figure 2.11. The computational effort is modest, ncall = 1005.

This completes the programming and discussion of the solution of the wave equation (2.22). In particular, the splines from splinefun worked well for the derivative $\dfrac{\partial^2 u}{\partial x^2}$ in Eq. (2.22).

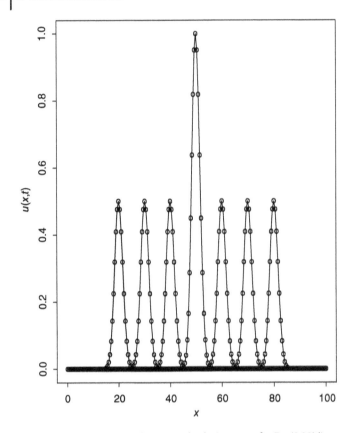

Figure 2.11 Analytical and numerical solutions $u_{(z,t)}$ for Eq. (2.22) lines – num, points – anal.

2.6 Fourth Order in Space

The following application pertains to a linear, constant coefficient PDE, which is fourth order in space, and to start, first order in time.

2.6.1 First Order in Time

The PDE is

$$\frac{\partial u}{\partial t} = c_0 u + c_2 \frac{\partial^2 u}{\partial x^2} + c_4 \frac{\partial^4 u}{\partial x^4} \tag{2.27}$$

where c_0, c_2, c_4 are constants to be specified

Equation (2.27) is fourth order in x and first order in t. It therefore requires four BCs and one IC. We start with even-order BCs.

$$u(x = 0, t) = u(x = 1, t) = 0 \tag{2.28a,b}$$

$$\frac{\partial^2 u(x = 0, t)}{\partial x^2} = \frac{\partial^2 u(x = 1, t)}{\partial x^2} = 0 \qquad (2.28c,d)$$

The IC is

$$u(x, t = 0) = \sin(\pi x) \qquad (2.29)$$

An analytical solution is used to verify the SCMOL solution computed subsequently. A trial analytical solution

$$u(x, t) = e^{at} \sin(\pi x) \qquad (2.30a)$$

when substituted in Eq. (2.27) gives

$$a e^{at} \sin(\pi x) = c_0 e^{at} \sin(\pi x) - \pi^2 c_2 e^{at} \sin(\pi x) + \pi^4 c_4 \sin(\pi x)$$

so that

$$a = c_0 - \pi^2 c_2 + \pi^4 c_4 \qquad (2.30b)$$

$u(x, t)$ of Eqs. (2.30) also satisfies BCs (2.28) and IC (2.29). It therefore constitutes a complete analytical solution to the PDE system and can be used to verify the numerical solution.

A main program for the SCMOL solution of Eqs. (2.27)–(2.29) follows.

2.6.1.1   Main Program

```
#
# Delete previous workspaces
  rm(list=ls(all=TRUE))
#
# Access ODE integrator, routines
  library(deSolve);
  setwd("f:/collocation/test21");
  source("pde_1a.R");
#
# Step through two cases
  for(ncase in 1:2){
#
# Parameters
  c0=-1.0;
  c2= 0.1;
  c4=-0.01;
  a=c0-pi^2*c2+pi^4*c4;
#
# Display parameters
  cat(sprintf("\n c0 = %4.2f   c2 = %4.2f
              c4 = %4.2f   a = %4.2f\n",
```

```
                           c0,c2,c4,a));
#
# Time interval
  t0=0;tf=0.4;nout=5
  tout=seq(from=t0,to=tf,by=(tf-t0)/(nout-1));
  ncall=0;
#
# Spatial interval
  n=31;xl=0;xu=1;
  x=seq(from=xl,to=xu,by=(xu-xl)/(n-1));
#
# Initial condition (IC)
  u0=sin(pi*x);
#
# ODE integration
  out=lsodes(y=u0,times=tout,func=pde_1a,
      sparsetype="sparseint",rtol=1e-6,
      atol=1e-6,maxord=5);
  nrow(out)
  ncol(out)
#
# Store numerical solution for plotting
   u=matrix(0,nrow=n,ncol=nout)
  ua=matrix(0,nrow=n,ncol=nout)
  for(it in 1:nout){
    ua[,it]=exp(a*tout[it])*sin(pi*x);
  for(i in 1:n){
    u[i,it]=out[it,i+1];
  }
  }
#
# Calls to ODE routine
  cat(sprintf("\n ncall = %5d\n\n",ncall));
#
# Plot numerical, analytical solutions
  matplot(x,u,type="l",lwd=2,col="black",
    lty=1,xlab="x",ylab="u(x,t)");
  matpoints(x,ua,pch="o",col="black");
#
# Next case
  }
```

Listing 2.20 Main program for Eq. (2.27), even-order BCs.

We can note the following details about Listing 2.20:

- Previous workspaces are removed. The library of ODE integrators, deS-olve, is accessed, followed by the routines for Eq. (2.27). Note the use of / rather than the usual \ in the setwd (set working directory). Also, setwd will require revision for the local computer (to define the working folder or directory for the routine pde_1a).

```
#
# Delete previous workspaces
   rm(list=ls(all=TRUE))
#
# Access ODE integrator, routines
   library(deSolve);
   setwd("f:/collocation/test21");
   source("pde_1a.R");
```

- Two cases are programmed for different calculation of the derivatives in x in the ODE/MOL routine considered next. The value of ncase is passed to the ODE/MOL routine without a special designation (a feature of R).

```
#
# Step through two cases
   for(ncase in 1:2){
```

- The parameters in Eqs. (2.27) and (2.30b) are defined.

```
#
# Parameters
   c0=-1.0;
   c2= 0.1;
   c4=-0.01;
   a=c0-pi^2*c2+pi^4*c4;
```

- The interval in t is defined with nout=5 points for $0 \le t \le 0.4$ so that tout = 0,0.1,0.2,0.3,0.4. The seq utility is used to produce the sequence of values in t.

```
#
# Time interval
   t0=0;tf=0.4;nout=5
   tout=seq(from=t0,to=tf,by=(tf-t0)/(nout-1));
   ncall=0;
```

The counter for the calls to the ODE routine is initialized.

- A grid of 31 points in x is defined for $x_l = 0 \le x \le x_u = 1$ so that $x = 0, 1/30, \ldots, 1$.

```
#
# Spatial interval
  n=31;xl=0;xu=1;
  x=seq(from=xl,to=xu,by=(xu-xl)/(n-1));
```

The value n=31 was determined by trial and error to achieve acceptable accuracy in the numerical solution by comparing it with the analytical solution.

- IC (2.29) is programmed in vectorized form (since x is a vector). With n=31, u0 is a 31-vector.

```
#
# Initial condition (IC)
  u0=sin(pi*x);
```

- The 31 ODEs are integrated numerically by lsodes. The input arguments to lsodes are the MOL/ODE routine pde_1a (discussed subsequently), the IC vector u0 and the output values of t, tout. The number of ODEs to be integrated is defined by the length of u0 (31). func, y, times are reserved names.

```
#
# ODE integration
  out=lsodes(y=u0,times=tout,func=pde_1a,
      sparsetype="sparseint",rtol=1e-6,
      atol=1e-6,maxord=5);
  nrow(out)
  ncol(out)
```

The ODE solution is returned in out, a $5 \times 31 + 1 = 156$ element array.

- The numerical solution $u(x, t)$ is placed in a 155 element array u for subsequent numerical and graphical output. The offset of 1 in i+1 is required since the 5 values of t are included in out. Two nested fors step through the nout=5 values of t and the n=31 values of x. The analytical solution of Eqs. (2.30) is placed in ua.

```
#
# Store numerical solution for plotting
    u=matrix(0,nrow=n,ncol=nout)
    ua=matrix(0,nrow=n,ncol=nout)
    for(it in 1:nout){
      ua[,it]=exp(a*tout[it])*sin(pi*x);
    for(i in 1:n){
      u[i,it]=out[it,i+1];
    }
    }
```

- The number of calls to pde_1a is displayed as an indication of the computational effort required to compute the solution.

```
#
# Calls to ODE routine
  cat(sprintf("\n ncall = %5d\n\n",ncall));
```

- The numerical solution is plotted against *x* with *t* as a parameter by matplot. The analytical solution is plotted with matpoints. The two solutions are superimposed as indicated in Figure 2.12.

```
#
# Plot numerical, analytical solutions
  matplot(x,u,type="l",lwd=2,col="black",
    lty=1,main="solid - num, points - anal",
    xlab="x",ylab="u(x,t)");
  matpoints(x,ua,pch="o",col="black");
#
# Next case
  }
```

Numerical output is not included to conserve space in this discussion. The numerical and analytical solutions are essentially identical, as reflected in Figure 2.12, but output statements can easily be added to demonstrate this agreement. The final } concludes the for in ncase.

The ODE routine pde_1a, called by lsodes, is listed next.

2.6.1.2 ODE Routine

The routine for the SCMOL ODEs follows.

```
  pde_1a=function(t,u,parms){
#
# Function pde_1a computes the t derivative vectors
#
# Four first order differentiations
  if(ncase==1){
#
# u BCs
  u[1]=0;
  u[n]=0;
#
# ux
  table1=splinefun(x,u);
  ux=table1(x,deriv=1);
#
```

```
# uxx
  table2=splinefun(x,ux);
  uxx=table2(x,deriv=1);
#
# uxx BCs
  uxx[1]=0;uxx[n]=0;
#
# uxxx
  table3=splinefun(x,uxx);
  uxxx=table3(x,deriv=1);
#
# uxxxx
  table4=splinefun(x,uxxx);
  uxxxx=table4(x,deriv=1);
  }
#
# Two second order differentiations
  if(ncase==2){
#
# u BCs
  u[1]=0;
  u[n]=0;
#
# uxx
  table2=splinefun(x,u);
  uxx=table2(x,deriv=2);
#
# uxx BCs
  uxx[1]=0;uxx[n]=0;
#
# uxxxx
  table4=splinefun(x,uxx);
  uxxxx=table4(x,deriv=2);
  }
#
# PDE
  ut=c0*u+c2*uxx+c4*uxxxx;
#
# Increment calls to pde_1a
  ncall<<-ncall+1;
#
# Return t derivative vector
  return(list(c(ut)))
```

```
#
# End of pde_1a
}
```

Listing 2.21 ODE/SCMOL routine pde_1a for Eq. (2.27), even-order BCs.

We can note the following details about pde_1a:

- The function is defined. The input arguments are (i) the current value of t; (ii) the vector of ODE-dependent variables, a 31-vector; and (iii) input parameters, not used in this case but still required as an argument. The output of pde_1a is the vector of ODE derivatives in t, as explained at the end of pde_1a.

```
    pde_1a=function(t,u,parms){
#
# Function pde_1a computes the t derivative vectors
```

- For ncase=1, four successive first-order differentiations are used to compute $\frac{\partial^4 u}{\partial x^4}$.

```
#
# Four first order differentiations
    if(ncase==1){
```

- BCs (2.28a,b) are applied.

```
#
# u BCs
    u[1]=0;
    u[n]=0;
```

The subscripts 1, n correspond to $x_l = 0, x_u = 1$.

- The second derivative $\frac{\partial^2 u}{\partial x^2}$ is computed by two consecutive first-order differentiations by splinefun.

```
#
# ux
    table1=splinefun(x,u);
    ux=table1(x,deriv=1);
#
# uxx
    table2=splinefun(x,ux);
    uxx=table2(x,deriv=1);
```

- BCs (2.28c,d) are applied.

```
#
# uxx BCs
   uxx[1]=0;uxx[n]=0;
```

The subscripts 1, n correspond to $x_l = 0, x_u = 1$.

- The fourth derivative $\dfrac{\partial^4 u}{\partial x^4}$ is computed by two consecutive first-order differentiations of the second derivative $\dfrac{\partial^2 u}{\partial x^2}$.

```
#
# uxxx
   table3=splinefun(x,uxx);
   uxxx=table3(x,deriv=1);
#
# uxxxx
   table4=splinefun(x,uxxx);
   uxxxx=table4(x,deriv=1);
   }
```

This completes ncase=1. For ncase=2, the calculation of the derivatives in x is accomplished by two successive second-order differentiations. BCs (2.28) can be applied in this process.

- BCs (2.28a,b) are applied.

```
#
# Two second order differentiations
   if(ncase==2){
#
# u BCs
   u[1]=0;
   u[n]=0;
```

- The second derivative $\dfrac{\partial^2 u}{\partial x^2}$ is computed from u.

```
#
# uxx
   table2=splinefun(x,u);
   uxx=table2(x,deriv=2);
```

- BCs (2.28c,d) are applied.

```
#
# uxx BCs
   uxx[1]=0;uxx[n]=0;
```

- The fourth derivative $\dfrac{\partial^4 u}{\partial x^4}$ is computed from the second derivative $\dfrac{\partial^2 u}{\partial x^2}$.

```
#
# uxxxx
  table4=splinefun(x,uxx);
  uxxxx=table4(x,deriv=2);
  }
```

- Equation (2.27) is programmed in vectorized form (no subscripting).

```
#
# PDE
  ut=c0*u+c2*uxx+c4*uxxxx;
```

The close resemblance of the programming and the PDE, Eq. (2.27), indicates an important feature of the MOL.

- The counter for the calls to pde_1a is incremented and its value returned to the calling program (of Listing 2.20) with the <<- operator.

```
#
# Increment calls to pde_1a
  ncall<<-ncall+1;
```

- The derivative vector ut is returned to lsodes with a combination of return, list, c.

```
#
# Return t derivative vector
  return(list(c(ut)))
#
# End of pde_1a
  }
```

The final } concludes pde_1a.

This completes the programming of Eqs. (2.27)–(2.30). Abbreviated numerical output is displayed in Table 2.9. The graphical output is shown in Figure 2.12 (for ncase=1).

Note that $a < 0$, which means that the solution of Eq. (2.30a) decays exponentially with t. This stability is clear in Figure 2.12. The computational effort is modest, ncall = 106,107.

The solutions for ncase=1,2 are essentially identical so the numerical and graphical output for ncase=2 is not presented here to conserve space. Generally, the use of four consecutive first-order differentiations in ncase=1 and two consecutive second-order differentiations in ncase=2 gives the same numerical solution.

Table 2.9 Abbreviated output for Eqs. (2.27)–(2.29).

```
c0 = -1.00   c2 = 0.10   c4 = -0.01   a = -2.96

ncall =    106

c0 = -1.00   c2 = 0.10   c4 = -0.01   a = -2.96

ncall =    170
```

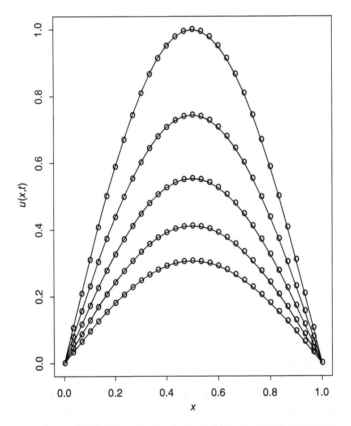

Figure 2.12 Analytical and numerical solutions $u(x, t)$ for Eq. (2.27) even-order BCs, lines – num, points – anal.

This completes the programming and discussion of the solution of Eqs. (2.27)–(2.30). In particular, the splines from splinefun worked well for the derivatives $\dfrac{\partial^2 u}{\partial x^2}$ and $\dfrac{\partial^4 u}{\partial x^4}$ in Eq. (2.27).

To conclude this example, the even-order BCs of Eqs. (2.28) (with $u, \dfrac{\partial^2 u}{\partial x^2}$ specified at the boundaries $x = 0, 1$) are replaced with the odd-order BCs.

$$\frac{u(x = 0, t)}{\partial x} = \frac{u(x = 1, t)}{\partial x} = 0 \qquad (2.31\text{a,b})$$

$$\frac{\partial^3 u(x = 0, t)}{\partial x^3} = \frac{\partial^3 u(x = 1, t)}{\partial x^3} = 0 \qquad (2.31\text{c,d})$$

The IC, that is consistent with these BCs, is taken as

$$u(x, t = 0) = \cos(\pi x) \qquad (2.32)$$

The analytical solution then becomes

$$u(x, t) = e^{at} \cos(\pi x) \qquad (2.33\text{a})$$

and when substituted in Eq. (2.27) gives

$$a e^{at} \cos(\pi x) = c_0 e^{at} \cos(\pi x) - \pi^2 c_2 e^{at} \cos(\pi x) + \pi^4 c_4 \cos(\pi x)$$

so that again

$$a = c_0 - \pi^2 c_2 + \pi^4 c_4 \qquad (2.33\text{b})$$

A main program for this problem follows closely the main program of Listing 2.21.

```
#
# Delete previous workspaces
  rm(list=ls(all=TRUE))
#
# Access ODE integrator, routines
  library(deSolve);
  setwd("f:/collocation/test21");
  source("pde_1b.R");
#
# Parameters
  c0=-1.0;
  c2= 0.1;
  c4=-0.01;
  a=c0-pi^2*c2+pi^4*c4;
#
# Display parameters
  cat(sprintf("\n c0 = %4.2f   c2 = %4.2f
              c4 = %4.2f   a = %4.2f\n",
              c0,c2,c4,a));
#
# Time interval
```

```
  t0=0;tf=0.4;nout=5
  tout=seq(from=t0,to=tf,by=(tf-t0)/(nout-1));
  ncall=0;
#
# Spatial interval
  n=31;xl=0;xu=1;
  x=seq(from=xl,to=xu,by=(xu-xl)/(n-1));
#
# Initial condition (IC)
  u0=cos(pi*x);
#
# ODE integration
  out=ode(func=pde_1b,y=u0,times=tout);
  nrow(out)
  ncol(out)
#
# Store numerical, analytical solutions
# for plotting
  u=matrix(0,nrow=n,ncol=nout)
  ua=matrix(0,nrow=n,ncol=nout)
  for(it in 1:nout){
    ua[,it]=exp(a*tout[it])*sin(pi*x);
    for(i in 1:n){
    u[i,it]=out[it,i+1];
    }
    }
#
# Calls to ODE routine
  cat(sprintf("\n ncall = %5d\n\n",ncall));
#
# Plot numerical, analytical solutions
  matplot(x,u,type="l",lwd=2,col="black",
    lty=1,xlab="x",ylab="u(x,t)");
  matpoints(x,ua,pch="o",col="black");
```

Listing 2.22 Main program for Eq. (2.27), odd-order BCs.

We can note the following differences between Listings 2.20 and 2.22:

- The ODE routine is renamed from pde_1a to pde_1b.

```
#
# Access ODE integrator, routines
  library(deSolve);
  setwd("f:/collocation/test21");
  source("pde_1b.R");
```

- In the IC, sin is changed to cos.

```
#
# Initial condition (IC)
  u0=cos(pi*x);
```

- The ODE integrator is ode in place of lsodes. This change was made because lsodes failed to complete the ODE integration and issued an error message. Although the internal problem in lsodes might have been resolved with some additional investigation (and possibly adjustment of the input parameters), this was abandoned in favor of using ode (which is based on lsoda).

```
#
# ODE integration
  out=ode(func=pde_1b,y=u0,times=tout);
  nrow(out)
  ncol(out)
```

This experience illustrates what is perhaps the most intractable difficulty in MOL analysis, that is, the failure of the ODE integrator. This problem may persist so that the use of a particular library integrator is abandoned in favor of coding an in-line integrator such as a fixed-step Euler integrator. This has the advantage that each step in the integration can be followed in detail (rather than relying on the internal operation of a library integrator). An example of an in-line integrator is given in the next chapter for 2D and 3D PDE problems.

- An output statement was added to give the numerical and analytical solutions at each *x* and *t*. The numerical output is reviewed subsequently.

```
for(it in 1:nout){
  ua[,it]=exp(a*tout[it])*cos(pi*x);
  cat(sprintf("\n\n      t    x           u           ua
            diff"));
  for(i in 1:n){
    u[i,it]=out[it,i+1];
    cat(sprintf("\n %5.2f%5.2f%10.5f%10.5f
              %10.5f",tout[it],x[i],u[i,it],
              ua[i,it],u[i,it]-ua[i,it]));
  }
}
```

Note the use of analytical solution (2.33).

ODE function pde_1b called by ode in Listing 2.22 follows.

```
  pde_1b=function(t,u,parms){
#
# Function pde_1b computes the t derivative vectors
```

```
#
# ux
  table1=splinefun(x,u);
  ux=table1(x,deriv=1);
#
# ux BCs
  ux[1]=0;
  ux[n]=0;
#
# uxx
  table2=splinefun(x,ux);
  uxx=table2(x,deriv=1);
#
# uxxx
  table3=splinefun(x,uxx);
  uxxx=table3(x,deriv=1);
#
# uxxx BCs
  uxxx[1]=0;uxxx[n]=0;
#
# uxxxx
  table4=splinefun(x,uxxx);
  uxxxx=table4(x,deriv=1);
#
# PDE
  ut=c0*u+c2*uxx+c4*uxxxx;
#
# Increment calls to pde_1b
  ncall<<-ncall+1;
#
# Return t derivative vector
  return(list(c(ut)))
#
# End of pde_1b
}
```

Listing 2.23 ODE/SCMOL routine pde_1a for Eq. (2.27), odd-order BCs.

We can note the following details about Listing 2.23:

- The function is defined. Again, the input arguments are (i) the current value of t; (ii) the vector of ODE-dependent variables, a 31-vector; and (iii) input parameters, not used in this case but still required as an argument.

```
    pde_1b=function(t,u,parms){
#
# Function pde_1b computes the t derivative vectors
```

- The first derivative $\dfrac{\partial u}{\partial x}$ is computed by splinefun.

```
#
# ux
    table1=splinefun(x,u);
    ux=table1(x,deriv=1);
```

- BCs (2.31a,b) are applied.

```
#
# ux BCs
    ux[1]=0;
    ux[n]=0;
```

The subscripts 1 , n correspond to $x_l = 0, x_u = 1$.

- The third derivative $\dfrac{\partial^3 u}{\partial x^3}$ is computed by two consecutive first-order differentiations of the first derivative $\dfrac{\partial u}{\partial x}$.

```
#
# uxx
    table2=splinefun(x,ux);
    uxx=table2(x,deriv=1);
#
# uxxx
    table3=splinefun(x,uxx);
    uxxx=table3(x,deriv=1);
```

This could also be accomplished by going from the first to the third derivative using deriv=2.

- BCs (2.31c,d) are applied.

```
#
# uxxx BCs
    uxxx[1]=0;uxxx[n]=0;
```

- The fourth derivative $\dfrac{\partial^4 u}{\partial x^4}$ is computed.

```
#
# uxxxx
    table4=splinefun(x,uxxx);
    uxxxx=table4(x,deriv=1);
```

- Equation (2.27) is programmed in vectorized form.

```
#
# PDE
  ut=c0*u+c2*uxx+c4*uxxxx;
```

- The counter for the calls to pde_1b is incremented and its value returned to the calling program (of Listing 2.22) with the <<- operator.

```
#
# Increment calls to pde_1b
  ncall<<-ncall+1;
```

- The derivative vector ut is returned to ode with a combination of return, list, c.

```
#
# Return t derivative vector
  return(list(c(ut)))
#
# End of pde_1b
```

Abbreviated numerical output is listed next.

We can note from this output that the agreement between the numerical and analytical solutions is to five figures. The graphical output in Figure 2.13 confirms this agreement. The computational effort is modest, ncall = 334.

Table 2.10 and Figure 2.13 indicate the spline approximations of spline-fun handled the odd-order BCs, Eq. (2.31), with good accuracy. However, there is an important point to consider. Both the even- and odd-order BCs are consistent with the IC (the IC and BCs have the same values at $t = 0$, $x = 0, 1$). This is an important continuity or smoothness condition. If the IC and BCs at $t = 0$ have different values, a discontinuity is introduced initially, which might cause computational problems. Equation (2.27) is parabolic (from the second derivative, $\frac{\partial^2 u}{\partial x^2}$), which tends to smooth discontinuities. The effect of the fourth derivative $\frac{\partial^4 u}{\partial x^4}$ may not have this smoothing effect. The reader can investigate the effect of discontinuities at the boundaries by modifying the IC and/or BCs.

Also, the values of c_0, c_2, c_4, that is, c0=-1.0, c2= 0.1, c4=-0.01 were given decreasing absolute values because of the multiplication of the derivatives in x of Eq. (2.27) by π^2, π^4 (from the analytical solutions, as reflected in the value of $a = c_0 - \pi^2 c_2 + \pi^4 c_4$). In other words, the choice of the coefficient values was an attempt to balance the magnitude of the three RHS terms of Eq. (2.27).

Table 2.10 Abbreviated output for Eqs. (2.27), (2.31)–(2.33).

```
c0 = -1.00   c2 = 0.10   c4 = -0.01   a = -2.96

[1] 5

[1] 32
```

t	x	u	ua	diff
0.00	0.00	1.00000	1.00000	0.00000
0.00	0.03	0.99452	0.99452	0.00000
0.00	0.07	0.97815	0.97815	0.00000
0.00	0.10	0.95106	0.95106	0.00000
0.00	0.13	0.91355	0.91355	0.00000
0.00	0.17	0.86603	0.86603	0.00000
0.00	0.20	0.80902	0.80902	0.00000
.				.
.				.
.				.

Output for x = 0.23 to 0.77 removed

.				.
.				.
.				.
0.00	0.80	-0.80902	-0.80902	0.00000
0.00	0.83	-0.86603	-0.86603	0.00000
0.00	0.87	-0.91355	-0.91355	0.00000
0.00	0.90	-0.95106	-0.95106	0.00000
0.00	0.93	-0.97815	-0.97815	0.00000
0.00	0.97	-0.99452	-0.99452	0.00000
0.00	1.00	-1.00000	-1.00000	0.00000
.				.
.				.
.				.

Output for t = 0.1 to 0.3 removed

t	x	u	ua	diff
.				.
.				.
.				.
0.40	0.00	0.30591	0.30592	-0.00001

(Continued)

Table 2.10 (Continued)

0.40	0.03	0.30425	0.30425	-0.00000
0.40	0.07	0.29924	0.29924	-0.00000
0.40	0.10	0.29095	0.29095	-0.00000
0.40	0.13	0.27947	0.27947	-0.00000
0.40	0.17	0.26494	0.26494	-0.00000
0.40	0.20	0.24750	0.24750	-0.00000

.
.
.

```
    Output for x = 0.23 to 0.77 removed
```

.
.
.

0.40	0.80	-0.24750	-0.24750	0.00000
0.40	0.83	-0.26494	-0.26494	0.00000
0.40	0.87	-0.27947	-0.27947	0.00000
0.40	0.90	-0.29095	-0.29095	0.00000
0.40	0.93	-0.29924	-0.29924	0.00000
0.40	0.97	-0.30425	-0.30425	0.00000
0.40	1.00	-0.30591	-0.30592	0.00001

```
ncall =    334
```

We next consider a PDE fourth order in space as before, but second order in time.

2.6.2 Second Order in Time

The PDE is

$$\frac{\partial^2 u}{\partial t^2} + \frac{\partial^4 u}{\partial x^4} = 0 \qquad (2.34)$$

Equation (2.34), termed the *beam equation*, is fourth order in x and second order in t. It therefore requires four BCs and two ICs.

We start with even-order BCs.

$$u(x = 0, t) = u(x = 1, t) = 0 \qquad (2.35a,b)$$

$$\frac{\partial^2 u(x = 0, t)}{\partial x^2} = \frac{\partial^2 u(x = 1, t)}{\partial x^2} = 0 \qquad (2.35c,d)$$

The ICs are

$$u(x, t = 0) = \sin(\pi x); \qquad \frac{\partial u(x, t = 0)}{\partial t} = 0 \qquad (2.36a,b)$$

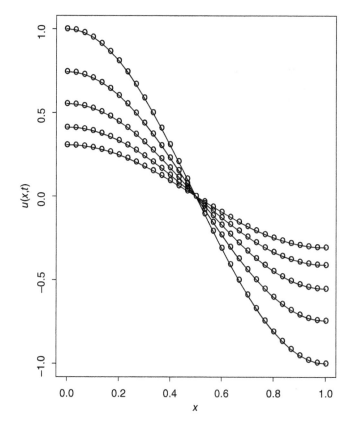

Figure 2.13 Analytical and numerical solutions $u(x, t)$ for Eq. (2.27) odd-order BCs, lines – num, points – anal.

An analytical solution is used to evaluate the SCMOL solution computed subsequently. A trial analytical solution

$$u(x, t) = \cos(at) \sin(\pi x) \tag{2.37a}$$

when substituted in Eq. (2.31) gives

$$-a^2 \cos(at) \sin(\pi x) + \cos(at)\pi^4 \sin(\pi x) = 0$$

so that

$$a = \pi^2 \tag{2.37b}$$

$u(x, t)$ of Eqs. (2.37) also satisfies BCs (2.35) and ICs (2.36). It therefore constitutes a complete analytical solution to the PDE system and can be used to evaluate the numerical solution.

A main program for the SCMOL solution of Eqs. (2.34)–(2.36) follows.

2.6.2.1 Main Program

```
#
# Delete previous workspaces
  rm(list=ls(all=TRUE))
#
# Access ODE integrator, routines
  library(deSolve);
  setwd("f:/collocation/test22");
  source("pde_1a.R");
#
# Time interval
  t0=0;tf=1;nout=11
  tout=seq(from=t0,to=tf,by=(tf-t0)/(nout-1));
  ncall=0;
#
# Spatial interval
  n=41;xl=0;xu=1;
  x=seq(from=xl,to=xu,by=(xu-xl)/(n-1));
#
# Initial conditions (ICs)
  u0=rep(0,2*n);
  for(i in 1:n){
    u0[i]  =sin(pi*x[i]);
    u0[i+n]=0;
  }
#
# ODE integration
  out=lsodes(func=pde_1a,y=u0,times=tout,
             sparsetype="sparseint",
             rtol=1e-4,atol=1e-4,maxord=5);
  nrow(out)
  ncol(out)
#
# Store numerical solution for plotting
    u1=matrix(0,nrow=n,ncol=nout)
  u1a=matrix(0,nrow=n,ncol=nout)
  for(it in 1:nout){
    u1a[,it]=cos(pi^2*tout[it])*sin(pi*x);
  for(i in 1:n){
    u1[i,it]=out[it,i+1];
  }
  }
```

```
#
# Calls to ODE routine
  cat(sprintf("\n ncall = %5d\n\n",ncall));
#
# Plot numerical, analytical solutions
  matplot(x,u1,type="l",lwd=2,col="black",
    lty=1,xlab="x",ylab="u1(x,t)");
  matpoints(x,u1a,pch="o",col="black");
```

Listing 2.24 Main program for Eq. (2.34), even-order BCs.

We can note the following details about Listing 2.24:

- Previous workspaces are removed. The library of ODE integrators, deS-olve, is accessed, followed by the routines for Eq. (2.34). Note the use of / rather than the usual \ in the setwd (set working directory). Also, setwd will require revision for the local computer (to define the working folder or directory for the routine pde_1a).

```
#
# Delete previous workspaces
  rm(list=ls(all=TRUE))
#
# Access ODE integrator, routines
  library(deSolve);
  setwd("f:/collocation/test22");
  source("pde_1a.R");
```

- Two cases are programmed for different calculation of the derivatives in x in the ODE/MOL routine considered next. The value of ncase is passed to the ODE/MOL routine without a special designation (a feature of R).
- The interval in t is defined with nout=11 points for $0 \le t \le 1$ so that tout = $0, 0.1, \ldots, 1$. The seq utility is used to produce the sequence of values in t.

```
#
# Time interval
  t0=0;tf=1;nout=11
  tout=seq(from=t0,to=tf,by=(tf-t0)/(nout-1));
  ncall=0;
```

The counter for the calls to the ODE routine is initialized.

- A grid of 41 points in x is defined for $x_l = 0 \le x \le x_u = 1$ so that $x = 0, 1/40, \ldots, 1$.

```
#
```

```
# Spatial interval
  n=41;xl=0;xu=1;
  x=seq(from=xl,to=xu,by=(xu-xl)/(n-1));
```

The value n=41 was determined by trial and error to achieve acceptable accuracy in the numerical solution by comparing it with the analytical solution.

- ICs (2.36) are programmed with n=41, u0 is a 2(41) = 82-vector.

```
#
# Initial conditions (ICs)
  u0=rep(0,2*n);
  for(i in 1:n){
    u0[i]   =sin(pi*x[i]);
    u0[i+n]=0;
  }
```

- The 82 ODEs are integrated numerically by lsodes. The input arguments to lsodes are the MOL/ODE routine pde_1a (discussed subsequently), the IC vector u0 and the output values of t, tout. The number of ODEs to be integrated is defined by the length of u0 (82). func, y, times are reserved names.

```
#
# ODE integration
  out=lsodes(func=pde_1a,y=u0,times=tout,
             sparsetype="sparseint",
             rtol=1e-4,atol=1e-4,maxord=5);
  nrow(out)
  ncol(out)
```

The ODE solution is returned in out, a 11 × 82 + 1 element array. The error tolerances, rtol=1e-4, atol=1e-4, were selected to achieve acceptable accuracy in the numerical solution with reasonable compute time (rather than the defaults rtol=1e-6, atol=1e-6).

- The numerical solution $u(x, t)$ is placed in a 11 × 82 element array u for subsequent numerical and graphical output. The offset of 1 in i+1 is required since the 11 values of t are included in out. Two nested fors step through the nout=11 values of t and the n=41 values of x. The analytical solution of Eqs. (2.37a,b) is placed in ua1.

```
#
# Store numerical solution for plotting
    u1=matrix(0,nrow=n,ncol=nout)
    u1a=matrix(0,nrow=n,ncol=nout)
```

```
   for(it in 1:nout){
     ula[,it]=cos(pi^2*tout[it])*sin(pi*x);
     for(i in 1:n){
     u1[i,it]=out[it,i+1];
     }
   }
```

- The number of calls to pde_1a is displayed as an indication of the computational effort required to compute the solution.

```
#
# Calls to ODE routine
   cat(sprintf("\n ncall = %5d\n\n",ncall));
```

- The numerical solution is plotted against *x* with matplot with *t* as a parameter. The analytical solution is plotted with matpoints. The two solutions are superimposed as indicated in Figure 2.14.

```
#
# Plot numerical, analytical solutions
   matplot(x,u1,type="l",lwd=2,col="black",
     lty=1,xlab="x",ylab="u1(x,t)");
   matpoints(x,ula,pch="o",col="black");
```

Numerical output is not included to conserve space in this discussion. The numerical and analytical solutions are essentially identical, as reflected in Figure 2.14, but output statements can easily be added to demonstrate this agreement.

The ODE routine pde_1a, called by lsodes, is listed next.

2.6.2.2 ODE Routine
The routine for the SCMOL ODEs follows.

```
   pde_1a=function(t,u,parms){
#
# Function pde_1a computes the t derivative vectors
#
# One vector to two vectors
   u1=rep(0,n);u2=rep(0,n);
   for(i in 1:n){
     u1[i]=u[i];
     u2[i]=u[i+n];
   }
#
# u1 BCs
```

```
  u1[1]=0;
  u1[n]=0;
#
# u1xx
  table2=splinefun(x,u1);
  u1xx=table2(x,deriv=2);
#
# u1xx BCs
  u1xx[1]=0;u1xx[n]=0;
#
# u1xxxx
  table4=splinefun(x,u1xx);
  u1xxxx=table4(x,deriv=2);
#
# PDE
  u1t=rep(0,n);u2t=rep(0,n);
  for(i in 1:n){
    u1t[i]=u2[i];
    u2t[i]=-u1xxxx[i];
  }
#
# Increment calls to pde_1a
  ncall<<-ncall+1;
#
# Two vectors to one vector
  ut=rep(0,2*n);
  for(i in 1:n){
      ut[i]=u1t[i];
    ut[i+n]=u2t[i];
  }
#
# Return t derivative vector
  return(list(c(ut)))
#
# End of pde_1a
}
```

Listing 2.25 ODE/SCMOL routine pde_1a for Eq. (2.34), even-order BCs.

We can note the following details about pde_1a:

- The function is defined. The input arguments are (i) the current value of t; (ii) the vector of ODE-dependent variables, a 82-vector; and (iii) input parameters, not used in this case but still required as an argument. The output of pde_1a is the vector of ODE derivatives in t, as explained at the end of pde_1a.

```
pde_1a=function(t,u,parms) {
#
# Function pde_1a computes the t derivative vectors
```

- u is placed in two 41-vectors, u1, u2, to accommodate the second derivative $\dfrac{\partial^2 u}{\partial t^2}$ of Eq. (2.34)

```
#
# One vector to two vectors
  u1=rep(0,n);u2=rep(0,n);
  for(i in 1:n){
    u1[i]=u[i];
    u2[i]=u[i+n];
  }
```

- BCs (2.35a,b) are applied.

```
#
# u1 BCs
  u1[1]=0;
  u1[n]=0;
```

The subscripts 1 , n correspond to $x_l = 0, x_u = 1$.

- The second derivative $\dfrac{\partial^2 u}{\partial x^2}$ is computed by splinefun.

```
#
# u1xx
  table2=splinefun(x,u1);
  u1xx=table2(x,deriv=2);
```

- BCs (2.35c,d) are applied.

```
#
# u1xx BCs
  u1xx[1]=0;
  u1xx[n]=0;
```

The subscripts 1 , n correspond to $x_l = 0, x_u = 1$.

- The fourth derivative $\dfrac{\partial^4 u}{\partial x^4}$ is computed from the second derivative $\dfrac{\partial^2 u}{\partial x^2}$.

```
#
# u1xxxx
  table4=splinefun(x,u1xx);
  u1xxxx=table4(x,deriv=2);
```

- Equation (2.34) is programmed as a system of first-order ODEs.

```
#
# PDE
```

```
u1t=rep(0,n);u2t=rep(0,n);
for(i in 1:n){
  u1t[i]=u2[i];
  u2t[i]=-u1xxxx[i];
}
```

This could also be done in vectorized form (no subscripting), as demonstrated subsequently with odd-order BCs, that is

```
#
# PDE
u1t=rep(0,n);u2t=rep(0,n);
u1t=u2;
u2t=-u1xxxx;
```

- The counter for the calls to pde_1a is incremented and its value returned to the calling program (of Listing 2.24) with the <<- operator.

```
#
# Increment calls to pde_1a
ncall<<-ncall+1;
```

- The two vectors u1t,u2t are placed in a single derivative vector ut to return to the ODE integrator LSODES.

```
#
# Two vectors to one vector
ut=rep(0,2*n);
for(i in 1:n){
    ut[i]=u1t[i];
  ut[i+n]=u2t[i];
}
```

- The derivative vector ut is returned to lsodes with a combination of return, list, c.

```
#
# Return t derivative vector
  return(list(c(ut)))
#
# End of pde_1a
}
```

The final } concludes pde_1a.

This concludes the programming of Eqs. (2.34)–(2.36). The graphical output is in Figure 2.14. The brief numerical output indicates a modest computational effort (ncall = 749).

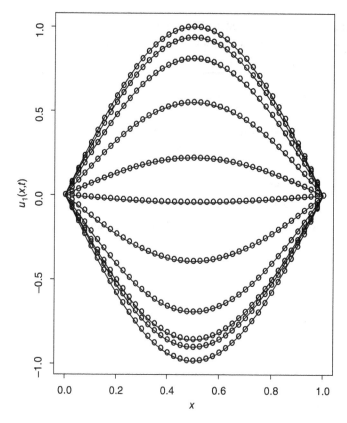

Figure 2.14 Analytical and numerical solutions $u(x, t)$ for Eq. (2.34) even-order BCs, lines – num, points – anal.

```
[1]  11

[1]  83

 ncall =    749
```

The number of values returned by lsodes, nout=11 and 2(41)+1=83, have the expected values.

The close agreement of the numerical and analytical solutions is apparent in Figure 2.14. This could be confirmed by displaying the two solutions numerically.

This completes the programming and discussion of the solution of Eqs. (2.34)–(2.36). In particular, the splines from splinefun worked well for the derivative $\dfrac{\partial^4 u}{\partial x^4}$ in Eq. (2.34).

To conclude this example, the even-order BCs of Eqs. (2.35) (with $u, \dfrac{\partial^2 u}{\partial x^2}$ specified at the boundaries $x = 0, 1$) are replaced with the odd-order BCs.

$$\frac{u(x = 0, t)}{\partial x} = \frac{u(x = 1, t)}{\partial x} = 0 \qquad (2.38a,b)$$

$$\frac{\partial^3 u(x = 0, t)}{\partial x^3} = \frac{\partial^3 u(x = 1, t)}{\partial x^3} = 0 \qquad (2.38c,d)$$

The ICs, that are consistent with these BCs, are taken as

$$u(x, t = 0) = \cos(\pi x); \qquad \frac{\partial u(x, t = 0)}{\partial t} = 0 \qquad (2.39a,b)$$

The analytical solution then becomes

$$u(x, t) = \cos(\pi^2 t) \cos(\pi x) \qquad (2.40)$$

A main program for this problem follows closely the main program of Listing 2.24.

```
#
# Delete previous workspaces
  rm(list=ls(all=TRUE))
#
# Access ODE integrator, routines
  library(deSolve);
  setwd("f:/collocation/test22");
  source("pde_1b.R");
#
# Time interval
  t0=0;tf=1;nout=11
  tout=seq(from=t0,to=tf,by=(tf-t0)/(nout-1));
  ncall=0;
#
# Spatial interval
  n=21;xl=0;xu=1;
  x=seq(from=xl,to=xu,by=(xu-xl)/(n-1));
#
# Initial conditions (ICs)
  u0=rep(0,2*n);
  for(i in 1:n){
    u0[i]   =cos(pi*x[i]);
    u0[i+n]=0;
  }
#
# ODE integration
  out=lsodes(func=pde_1b,y=u0,times=tout,
```

```
                  sparsetype="sparseint",
                  rtol=1e-4,atol=1e-4,maxord=5);
  nrow(out)
  ncol(out)
#
# Store numerical solution for plotting
  u1=matrix(0,nrow=n,ncol=nout)
  u1a=matrix(0,nrow=n,ncol=nout)
  for(it in 1:nout){
    u1a[,it]=cos(pi^2*tout[it])*cos(pi*x);
    for(i in 1:n){
    u1[i,it]=out[it,i+1];
    }
    }
#
# Calls to ODE routine
  cat(sprintf("\n ncall = %5d\n\n",ncall));
#
# Plot numerical, analytical solutions
  matplot(x,u1,type="l",lwd=2,col="black",
     lty=1,xlab="x",ylab="u1(x,t)");
  matpoints(x,u1a,pch="o",col="black");
```

Listing 2.26 Main program for Eq. (2.34), odd-order BCs.

Listing 2.26 is similar to Listing 2.24, with the following changes:

- The ODE routine is renamed from pde_1a to pde_1b.

```
  #
  # Delete previous workspaces
    rm(list=ls(all=TRUE))
  #
  # Access ODE integrator, routines
    library(deSolve);
    setwd("f:/collocation/test22");
    source("pde_1b.R");
```

- The number of grid points in x is reduced from 41 to 21.

```
  #
  # Spatial interval
    n=21;xl=0;xu=1;
    x=seq(from=xl,to=xu,by=(xu-xl)/(n-1));
```

This reduction was made to reduce the execution time.

- In the IC, `sin` is changed to `cos`, in accordance with ICs (2.39).

```
#
# Initial conditions (ICs)
  u0=rep(0,2*n);
  for(i in 1:n){
    u0[i]   =cos(pi*x[i]);
    u0[i+n]=0;
  }
```

- The ODE integrator, `lsodes`, calls `pde_1b`.

```
#
# ODE integration
  out=lsodes(func=pde_1b,y=u0,times=tout,
             sparsetype="sparseint",
             rtol=1e-4,atol=1e-4,maxord=5);
  nrow(out)
  ncol(out)
```

- The analytical solution is from Eq. (2.40) in the comparison of the numerical and analytical solutions.

```
#
# Store numerical solution for plotting
   u1=matrix(0,nrow=n,ncol=nout)
  u1a=matrix(0,nrow=n,ncol=nout)
  for(it in 1:nout){
    u1a[,it]=cos(pi^2*tout[it])*cos(pi*x);
  for(i in 1:n){
    u1[i,it]=out[it,i+1];
  }
  }
```

ODE function `pde_1b` called by `lsodes` in Listing 2.25 follows.

```
  pde_1b=function(t,u,parms){
#
# Function pde_1b computes the t derivative vectors
#
# One vector to two vectors
  u1=rep(0,n);u2=rep(0,n);
  for(i in 1:n){
    u1[i]=u[i];
    u2[i]=u[i+n];
  }
```

```
#
# u1x
  table2=splinefun(x,u1);
  u1x=table2(x,deriv=1);
#
# u1 BCs
  u1x[1]=0;
  u1x[n]=0;
#
# u1xxx
  table2=splinefun(x,u1x);
  u1xxx=table2(x,deriv=2);
#
# u1xxx BCs
  u1xxx[1]=0;
  u1xxx[n]=0;
#
# u1xxxx
  table4=splinefun(x,u1xxx);
  u1xxxx=table4(x,deriv=1);
#
# PDE
  u1t=rep(0,n);u2t=rep(0,n);
  u1t=u2;
  u2t=-u1xxxx;
#
# Increment calls to pde_1b
  ncall<<-ncall+1;
#
# Two vectors to one vector
  ut=rep(0,2*n);
  for(i in 1:n){
      ut[i]=u1t[i];
    ut[i+n]=u2t[i];
  }
#
# Return t derivative vector
  return(list(c(ut)))
#
# End of pde_1b
}
```

Listing 2.27 ODE/SCMOL routine pde_1a for Eq. (2.34), odd-order BCs.

We can note the following details about Listing 2.27:

- The function is defined. Again, the input arguments are (i) the current value of t; (ii) the vector of ODE-dependent variables, a $2(21) = 42$-vector; and (iii) input parameters, not used in this case but still required as an argument.

```
  pde_1b=function(t,u,parms){
#
# Function pde_1b computes the t derivative vectors
```

- The first derivative $\dfrac{\partial u}{\partial x}$ is computed by splinefun.

```
#
# u1x
  table2=splinefun(x,u1);
  u1x=table2(x,deriv=1);
```

- BCs (2.38a,b) are applied.

```
#
# ux BCs
  ux[1]=0;
  ux[n]=0;
```

The subscripts 1, n correspond to $x_l = 0, x_u = 1$.

- The third derivative $\dfrac{\partial^3 u}{\partial x^3}$ is computed from the first derivative $\dfrac{\partial u}{\partial x}$.

```
#
# u1xxx
  table2=splinefun(x,u1x);
  u1xxx=table2(x,deriv=2);
```

- BCs (2.38c,d) are applied.

```
#
# u1xxx BCs
  u1xxx[1]=0;
  u1xxx[n]=0;
```

- The fourth derivative $\dfrac{\partial^4 u}{\partial x^4}$ is computed.

```
#
# uxxxx
  table4=splinefun(x,uxxx);
  uxxxx=table4(x,deriv=1);
```

- Equation (2.34) is programmed in vectorized form (no subscripting).

```
#
```

```
# PDE
  u1t=rep(0,n);u2t=rep(0,n);
  u1t=u2;
  u2t=-u1xxxx;
```

- The counter for the calls to pde_1b is incremented and its value returned to the calling program (of Listing 2.25) with the <<- operator.

```
#
# Increment calls to pde_1b
  ncall<<-ncall+1;
```

- The two derivatives u1t,u2t are placed in a single vector ut of length 42 to return to lsodes.

```
#
# Two vectors to one vector
  ut=rep(0,2*n);
  for(i in 1:n){
      ut[i]=u1t[i];
    ut[i+n]=u2t[i];
  }
```

- The derivative vector ut is returned to lsodes with a combination of return, list, c.

```
#
# Return t derivative vector
  return(list(c(ut)))
#
# End of pde_1b
}
```

The brief abbreviated numerical output is listed next.

```
[1] 11

[1] 43

 ncall =   6010
```

The graphical output in Figure 2.15 confirms the agreement between the numerical and analytical solutions. The computational effort is substantial, ncall = 6010, even with the number of grid points in x reduced from 41 to 21.

Figures 2.14 and 2.15 indicate the spline approximations of splinefun handled the even- and odd-order BCs, Eqs. (2.35) and (2.38), with good accuracy.

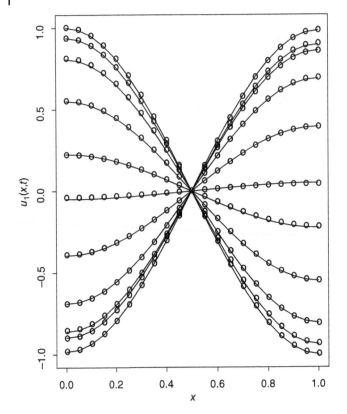

Figure 2.15 Analytical and numerical solutions $u(x, t)$ for Eq. (2.34) odd-order BCs, lines – num, points – anal.

However, we should keep in mind the possible constraints on the performance of the splines as mentioned next.

- The solution for Eq. (2.27), with derivative $\dfrac{\partial u}{\partial t}$, has the variation e^{-at}, and with $a > 0$, this factor provides a smooth decay with t. This is a general feature of parabolic PDEs such as Eq. (2.27), that is, smoothing that physically results from diffusion. In particular, parabolic PDEs will dampen or smooth steep fronts and discontinuities, which means that they are relatively easy to integrate numerically. However, Eq. (2.34), with derivative $\dfrac{\partial^2 u}{\partial t^2}$, has the variation $\sin(at)$ or $\cos(at)$, which corresponds to oscillation for all t. This is a general feature of hyperbolic PDEs such as Eq. (2.34), that is, continuing variation with no limit in t that physically results from convection or oscillation. Thus, hyperbolic PDEs can propagate steep fronts and discontinuities, which means that they are often relatively difficult to integrate numerically.

- The preceding examples are based on the use of stagewise differentiation, that is, calculation of derivatives by differentiation of lower order derivatives. For the examples considered, this worked satisfactorily, but there is no assurance in advance that it will work on a new application.
- The splines were applied in the preceding examples to generally smooth problems as reflected in the analytical solutions with exp, sin, and cos variations in space and time. However, a new application may present computational difficulties that will have to be studied, and hopefully resolved, for example, by changing the number of spatial grid points or error tolerances in the time integration. Splines are often helpful in this regard since they have inherent continuity and smoothness conditions not shared by other approaches. Also, the use of higher order splines can be explored for the resolution of computational difficulties (although this was not considered, but rather, only the splines in `splinefun` were used).
- The preceding points suggest that the development of a new SCMOL solution is generally not a mechanical procedure with a guaranteed solution in advance, that is, a solution of acceptable accuracy produced with manageable computational effort. Rather, trial-and-error experimentation may be required to achieve an acceptable solution, with careful evaluation of the numerical output to assure meaningful, useful results.
- As a corollary to the preceding point, a reported PDE numerical solution should include an error analysis to substantiate the accuracy of the solution. The presentation of the error analysis should be in sufficient detail that the reader/investigator can reproduce the analysis with reasonable effort to provide assurance of the accuracy of the solution. For example, the all too frequent statement "The solution was computed with (name of algorithm, software)" is not adequate. In particular, the details of how an error analysis was performed, the numerical results demonstrating convergence or other implied indication of accuracy, and the availability of any computer source code used to perform the analysis should be stated. This requirement is just a statement of the principle that any scientific result should be presented in enough detail that an independent investigator can reproduce and verify the reported result with reasonable effort.

In summary, the series of basic PDE applications in this chapter is complete. We now proceed to more advanced SCMOL applications in the subsequent chapters.

References

1 Griffiths, G.W., and W.E. Schiesser (2012), *Traveling Wave Analysis of Partial Differential Equations*, Elsevier, Burlington, MA.

2 Murray, J.D. (2003), *Mathematical Biology, II: Spatial Models and Biomedical Applications*, 3rd ed, Springer, New York, NY.

3 Press, W.H., et al. (2007), *Numerical Recipes: The Art of Scientific Computing*, 3rd ed., Cambridge University Press, UK

4 Schiesser, W.E.A. (1994), *Computational Mathematics in Engineering and Applied Science: ODEs, DAEs, and PDEs*, CRC Press, Boca Raton, FL.

3

Multidimensional PDEs

This chapter has PDE examples in two dimensions (2D) and three dimensions (3D). The intent is to demonstrate the application of SCMOL to multidimensional PDE problems. We start with the 2D diffusion equation in Cartesian coordinates, (x, y).

3.1 2D in Space

The problem system is

$$\frac{\partial u}{\partial t} = D\left(\frac{\partial^2 u}{\partial x^2} + \frac{\partial^2 u}{\partial y^2}\right) \tag{3.1a}$$

D is the diffusivity, a constant to be specified.

Equation (3.1a) is first order in t and second order in x, y, so that it requires one IC and two BCs for x and y.

$$u(x, y, t = 0) = \sin(\pi x)\cos(\pi y) \tag{3.1b}$$

$$u(x = x_l, y, t) = u(x = x_u, y, t) = 0 \tag{3.1c,d}$$

$$\frac{\partial u(x, y = y_l, t)}{\partial y} = \frac{\partial u(x, y = y_u, t)}{\partial y} = 0 \tag{3.1e,f}$$

Equations (3.1c,d) are homogeneous Dirichlet BCs. Equations (3.1e,f) are homogeneous Neumann BCs.

For BCs (3.1c–f), the analytical solution to Eqs. (3.1a)–(3.1f) is

$$u(x, y, t) = e^{-2D\pi^2 t}\sin(\pi x)\cos(\pi y) \tag{3.1g}$$

which is used to evaluate the numerical solution.

Spline Collocation Methods for Partial Differential Equations: With Applications in R, First Edition.
William E. Schiesser.
© 2017 John Wiley & Sons, Inc. Published 2017 by John Wiley & Sons, Inc.
Companion website: www.wiley.com/go/Spline_Collocation

3.1.1 Main Program

An R main program for Eqs. (3.1) is listed next.

```
#
# Delete previous workspaces
  rm(list=ls(all=TRUE))
#
# Access ODE integrator
  library("deSolve");
#
# Access functions for numerical ODEs
  setwd("f:/collocation/chap3");
  source("pde_2a.R");
#
# Parameters
  D=1;
#
# Grid (in x,y)
  nx=11;ny=11;xl=0;xu=1;yl=0;yu=1;
  x=seq(from=xl,to=xu,by=(xu-xl)/(nx-1));
  y=seq(from=yl,to=yu,by=(yu-yl)/(ny-1));
  out=(1-1)*nx+6;
#
# Independent variable for ODE integration
  nout=6;t0=0;tf=0.1;
  tout=seq(from=t0,to=tf,by=(tf-t0)/(nout-1));
#
# Initial condition
  u0=rep(0,nx*ny);
  for(j in 1:ny){
  for(i in 1:nx){
    u0[(j-1)*nx+i]=sin(pi*x[i])*cos(pi*y[j]);
  }
  }
  ncall=0;
#
# ODE integration
  u=u0;t=t0;h=0.0001;
  for(i1 in 1:6){
    ua=sin(pi*x[6])*cos(pi*y[1])*exp(-2*D*(pi^2)*t);
    cat(sprintf("\n t = %6.4f  u(x,y,t) = %8.4f%8.4f",
        t,u[out],ua));
    uplot=matrix(0,nrow=ny,ncol=nx);
```

```
    for(j in 1:ny){
    for(i in 1:nx){
      uplot[j,i]=u[(j-1)*nx+i];
    }
    }
    persp(x,y,uplot,theta=45,phi=45,xlim=c(xl,xu),
          ylim=c(yl,yu),zlim=c(-1,1),xlab="x",
          ylab="y",zlab="u(x,y,t)");
  for(i2 in 1:200){
    derv=pde_2a(t,u,parm);
    u=u+derv*h;
    t=t+h;
  }
  }
#
# Calls to ODE routine
  cat(sprintf("\n\n ncall = %3d\n",ncall));
```

Listing 3.1a Main program for Eqs. (3.1).

We can note the following details about Listing 3.1a:

- Previous workspaces are removed. Then the ODE/MOL routine pde_2a.R is accessed. Note that the setwd (set working directory) uses / rather than the usual \. Also, setwd requires editing for the local computer.

```
#
# Delete previous workspaces
  rm(list=ls(all=TRUE))
#
# Access functions for numerical ODEs
  setwd("f:/collocation/chap3");
  source("pde_2a.R");
```

- The model parameters are defined numerically. In this case, there is only one, the constant diffusivity D in Eq. (3.1a). This parameter is available to the ODE/MOL subordinate routine pde_2a without any special designation, a feature of R. The converse is not true. Parameters set in subordinate routines are not available to the superior (calling) routine without a special designation, as will be illustrated when pde_2a is considered subsequently.

```
#
# Parameters
  D=1;
```

- Uniform grids in x and y of 11 points each are defined with the seq utility for the intervals $x_l \leq x \leq x_u$, $y_l \leq y \leq y_u$. Therefore, with $x_l = y_l = 0$, $x_u = y_u = 1$, the vectors x, y have the values $x = 0, 0.1, \ldots, 1$, $y = 0, 0.1, \ldots, 1$.

```
#
# Grid (in x,y)
  nx=11;ny=11;xl=0;xu=1;yl=0;yu=1;
  x=seq(from=xl,to=xu,by=(xu-xl)/(nx-1));
  y=seq(from=yl,to=yu,by=(yu-yl)/(ny-1));
  out=(1-1)*nx+6;
```

One output point of the $11 \times 11 = 121$ solution points is selected, out=(1-1)*nx+6. The coding reflects a general point in the $x - y$ array with indices i, j (i for x, j for y), with a subscript (j-1)*nx+i (further explained next). Thus, the one output point corresponds to i=6,j=1 or $x = 0.5, y = 0$.

- A uniform grid in t of 6 output points is defined with the seq utility for the interval $0 \leq t \leq t_f = 0.1$. Therefore, the vector tout has the values $t = 0, 0.02, \ldots, 0.1$.

```
#
# Independent variable for ODE integration
  nout=6;t0=0;tf=0.1;
  tout=seq(from=t0,to=tf,by=(tf-t0)/(nout-1));
```

At this point, the intervals of x, y, and t in Eq. (3.1a) are defined.
- The IC vector corresponding to Eq. (3.1b) is defined numerically.

```
#
# Initial condition
  u0=rep(0,nx*ny);
  for(j in 1:ny){
  for(i in 1:nx){
    u0[(j-1)*nx+i]=sin(pi*x[i])*cos(pi*y[j]);
  }
  }
  ncall=0;
```

Also, the counter for the calls to pde_2a is initialized (and passed to pde_2a without a special designation).
- The system of 121 MOL/ODEs is integrated by an in-line Euler integrator.

```
#
# ODE integration
  u=u0;t=t0;h=0.0001;
  for(i1 in 1:nout){
```

```
ua=sin(pi*x[6])*cos(pi*y[1])*
        exp(-2*D*(pi^2)*t);
    cat(sprintf("\n t = %6.4f   u(x,y,t) =
                %8.4f%8.4f",t,u[out],ua));
uplot=matrix(0,nrow=ny,ncol=nx);
for(j in 1:ny){
for(i in 1:nx){
    uplot[j,i]=u[(j-1)*nx+i];
}
}
persp(x,y,uplot,theta=45,phi=45,xlim=c(xl,xu),
        ylim=c(yl,yu),zlim=c(-1,1),xlab="x",
        ylab="y",zlab="u(x,y,t)");
for(i2 in 1:200){
    derv=pde_2a(t,u,parm);
    u=u+derv*h;
    t=t+h;
}
}
```

Some additional explanation follows.

- The starting point for the Euler integration is the array u0.

```
u=u0;t=t0;h=0.0001;
```

The initial *t* is t0 and the Euler step is h=0.0001 (a constant throughout the numerical solution, in contrast to lsodes which adjusts the integration step in accordance with a specified error tolerance).

- The numerical solution $u(x = 0.5, y = 0, t)$ and the analytical solution of Eq. (3.1g) are displayed at the six output points $t = 0, 0.02, \ldots, 0.1$.

```
for(i1 in 1:nout){
    ua=sin(pi*x[6])*cos(pi*y[1])*
            exp(-2*D*(pi^2)*t);
        cat(sprintf("\n t = %6.4f   u(x,y,t) =
                    %8.4f%8.4f",t,u[out],ua));
```

The for with index i1 steps through these six values of *t*.

- The numerical solution is placed in array uplot for subsequent plotting in 3D perspective by persp.

```
uplot=matrix(0,nrow=ny,ncol=nx);
for(j in 1:ny){
for(i in 1:nx){
```

```
                uplot[j,i]=u[(j-1)*nx+i];
              }
            }
            persp(x,y,uplot,theta=45,phi=45,xlim=c(xl,xu),
                ylim=c(yl,yu),zlim=c(-1,1),xlab="x",
                ylab="y",zlab="u(x,y,t)");
```

persp plots $u(x, y, t)$ as a function of x, y for each of the six values of t, that is, six 3D plots are produced. The two nested fors step through the 11×11 values of $u(x, y, t)$ (with indices i, j) using the general subscript (j-1)*nx+i.

- A second (inner) for with index i2 takes 200 Euler steps (each of length h=0.0001) for the interval in t of 200(0.0001) = 0.02 (corresponding to the outer loop interval in t).

```
            for(i2 in 1:200){
              derv=pde_2a(t,u,parm);
              u=u+derv*h;
              t=t+h;
            }
          }
```

For each Euler step, the solution derivative vector in t is computed by pde_2a (discussed subsequently). The Euler step is then taken using vectorization (no subscripting). That is, this code implements the basic explicit Euler method.

$$u_{i+1} = u_i + \frac{du_i}{dt}h \qquad (3.1h)$$

For Eq. (3.1h), i is the index in t (not x). t is also advanced by h. The two }s conclude the loops in i2 and i1, respectively, for a total of 200(6) = 1200 Euler steps.

- The number of calls to pde_2a is displayed at the end of the solution.

```
#
# Calls to ODE routine
  cat(sprintf("\n\n ncall = %3d\n",ncall));
```

As will be observed, ncase=1200 as expected.

The advantage of this Euler method implementation is basically its simplicity. In other words, the details of the Euler steps are apparent (rather than automatic adjustment of the step in a library integrator such as lsodes). Each Euler step can be examined in detail, including the solution and derivative vectors, and even the individual RHS terms in the ODEs. The disadvantages of the explicit Euler method are limited accuracy and stability.

However, for the purpose of developing a new PDE application, the simplicity of the Euler method can be very helpful, particularly if difficulties (e.g., integration failures) with a library integrator are encountered.

The ODE/MOL routine pde_2a called in the Euler integration is considered next.

3.1.2 ODE Routine

pde_2a is listed next.

```
  pde_2a=function(t,u1d,parms){
#
# Function pde_2a computes the t derivative vector
# of u(x,y,t)
#
# 1D vector to 2D array
  u=matrix(0,nrow=ny,ncol=nx);
  for(j in 1:ny){
  for(i in 1:nx){
    u[j,i]=u1d[(j-1)*nx+i];
  }
  }
#
# BCs, x=0,xL
  u[, 1]=0;
  u[,nx]=0;
#
# uy
  uy=matrix(0,nrow=ny,ncol=nx);
  for(i in 1:nx){
    tabley=splinefun(y,u[,i]);
    uy[,i]=tabley(y,deriv=1);
  }
#
# BCs, y=0,yL
  uy[ 1,]=0;
  uy[ny,]=0;
#
# uxx
  uxx=matrix(0,nrow=ny,ncol=nx);
  for(j in 1:ny){
    tablexx=splinefun(x,u[j,]);
    uxx[j,]=tablexx(x,deriv=2);
  }
```

```
#
# uyy
  uyy=matrix(0,nrow=ny,ncol=nx);
  for(i in 1:nx){
    tableyy=splinefun(y,uy[,i]);
    uyy[,i]=tableyy(y,deriv=1);
  }
#
# PDE
  ut=matrix(0,nrow=ny,ncol=nx);
  ut=uxx+uyy;
#
# BCs, x=0,xL
  ut[, 1]=0;
  ut[,nx]=0;
#
# 2D array to 1D vector
  u1dt=rep(0,nx*ny);
  for(j in 1:ny){
  for(i in 1:nx){
    u1dt[(j-1)*nx+i]=ut[j,i];
  }
  }
#
# Increment calls to pde_2a
  ncall <<- ncall+1;
#
# Return derivative vector
  return(c(u1dt));
  }
```

Listing 3.1 ODE/MOL routine pde_2a for Eq. (3.1a).

We can note the following details about pde_2a:

- The function is defined.

```
  pde_2a=function(t,u1d,parms){
#
# Function pde_2a computes the t derivative vector
# of u(x,y,t)
```

t is the current value of t in Eq. (3.1a). u1d is the 121-vector of ODE-dependent variables (SCMOL approximation of Eq. (3.1a)). parm is an argument to pass parameters to pde_2a (unused, but required in the argument list). The arguments must be listed in the order stated to properly interface with the call to pde_2a in the main program of Listing 3.1a (derv=pde_2a(t,u,parm)).

- The 1D array u1d is placed in a 2D array u(i,j) to facilitate the programming of Eq. (3.1a).

```
#
# 1D vector to 2D array
  u=matrix(0,nrow=ny,ncol=nx);
  for(j in 1:ny){
  for(i in 1:nx){
    u[j,i]=u1d[(j-1)*nx+i];
  }
  }
```

The 1D subscript (j-1)*nx+i is consistent with the subscripting in the Euler integrator of Listing 3.1a.

- BCs (3.1c,d) are programmed as

```
#
# BCs, x=0,xL
  u[, 1]=0;
  u[,nx]=0;
```

All values of y are included through the [,]. $x = x_l, x_u$ are specified with indices 1, nx.

- $\dfrac{\partial u(x,y,t)}{\partial y}$ is computed from $u(x, y, t)$ by splinefun and placed in uy. All values of y are used with [,]. All values of x are used with the for in i.

```
#
# uy
  uy=matrix(0,nrow=ny,ncol=nx);
  for(i in 1:nx){
    tabley=splinefun(y,u[,i]);
    uy[,i]=tabley(y,deriv=1);
  }
```

- BCs (3.1e,f) are programmed as

```
#
# BCs, y=0,yL
  uy[ 1,]=0;
  uy[ny,]=0;
```

All values of x are included through the [,]. $y = y_l, y_u$ are specified with indices 1, ny.

- $\dfrac{\partial^2 u}{\partial x^2}$ in Eq. (3.1a) is computed from u.

```
#
#  uxx
   uxx=matrix(0,nrow=ny,ncol=nx);
   for(j in 1:ny){
     tablexx=splinefun(x,u[j,]);
     uxx[j,]=tablexx(x,deriv=2);
   }
```

All values of x are included through the [,].

- $\dfrac{\partial^2 u}{\partial y^2}$ in Eq. (3.1a) is computed from $\dfrac{\partial u}{\partial y}$.

```
#
#  uyy
   uyy=matrix(0,nrow=ny,ncol=nx);
   for(i in 1:nx){
     tableyy=splinefun(y,uy[,i]);
     uyy[,i]=tableyy(y,deriv=1);
   }
```

All values of y are included through the [,].

- Equation (3.1a) is programmed through vectorization (no subscripting). $D = 1$ is implied with this programming.

```
#
#  PDE
   ut=matrix(0,nrow=ny,ncol=nx);
   ut=uxx+uyy;
```

- Since BCs (3.1c,d) specify constants (0) at the boundaries, the derivatives in t are set to 0.

```
#
#  BCs,  x=0,xL
   ut[,  1]=0;
   ut[,nx]=0;
```

- The 2D matrix of derivatives in t is placed in a 1D vector for integration by the Euler method in Listing 3.1a.

```
#
#  2D array to 1D vector
   u1dt=rep(0,nx*ny);
   for(j in 1:ny){
```

```
for(i in 1:nx){
   u1dt[(j-1)*nx+i]=ut[j,i];
}
}
```

The subscripting (j-1)*nx+i is used again and ensures that each ODE-dependent variable derivative in u1dt is placed in the corresponding position of u1d entering pde_2a. This correspondence of positions of the dependent variables and the derivatives in *t* is essential for the correct SCMOL representation of Eq. (3.1a).

- The counter for the calls to pde_2a is incremented and returned to the main program of Listing 3.1a through the <<- operator.

```
#
# Increment calls to pde_2a
  ncall <<- ncall+1;
```

- The derivative vector (LHS of Eq. (3.1a)) is returned to the main program as a numerical vector (rather than as a list as required by the ODE integrators in deSolve such as lsodes). This return gives the Euler integrator the 11 × 11 = 121 element derivative vector for the next step along the solution.

```
#
# Return derivative vector
  return(c(u1dt));
  }
```

The final } concludes pde_2a.

This completes the programming of Eqs. (3.1). The numerical output is displayed in Table 3.1a.

Table 3.1a Output for Eqs. (3.1) from the main program of Listing of 3.1a, nx=ny=11.

```
t = 0.0000   u(x,y,t) =    1.0000   1.0000
t = 0.0200   u(x,y,t) =    0.6713   0.6738
t = 0.0400   u(x,y,t) =    0.4513   0.4540
t = 0.0600   u(x,y,t) =    0.3035   0.3059
t = 0.0800   u(x,y,t) =    0.2041   0.2062
t = 0.1000   u(x,y,t) =    0.1373   0.1389

ncall = 1200
```

The numerical (first column) and analytical (second column) solutions agree to approximately 2+ figures. This marginal agreement is not unexpected since only 11 points in x and y were used. To investigate this limited spatial convergence, the main program in Listing 3.1a was modified to

```
#
# Grid (in x,y)
   nx=21;ny=21;xl=0;xu=1;yl=0;yu=1;
   x=seq(from=xl,to=xu,by=(xu-xl)/(nx-1));
   y=seq(from=yl,to=yu,by=(yu-yl)/(ny-1));
   out=(1-1)*nx+11;
         .          .
         .          .
         .          .
   ua=sin(pi*x[11])*cos(pi*y[1])*exp(-2*D*(pi^2)*t);
```

With nx=ny=21, the numerical output is

Table 3.1b Output for Eqs. (3.1) from the main program of Listing of 3.1a, nx=ny=21.

```
t = 0.0000    u(x,y,t) =    1.0000    1.0000
t = 0.0200    u(x,y,t) =    0.6732    0.6738
t = 0.0400    u(x,y,t) =    0.4533    0.4540
t = 0.0600    u(x,y,t) =    0.3052    0.3059
t = 0.0800    u(x,y,t) =    0.2055    0.2062
t = 0.1000    u(x,y,t) =    0.1383    0.1389

ncall = 1200
```

The improvement in the spatial convergence is clear from a comparison of Tables 3.1a and 3.1b (an example of h-refinement).

The graphical output for $t = 0, 0.1$ is given in Figure 3.1a,b for nx=ny=11 and in Figure 3.1c,d for nx=ny=21. As expected, the solution is $u(x, y, t \to \infty) = 0$ (which follows from the analytical solution of Eq. (3.1g)). An animation of the solution can be constructed from the six plots for $t = 0, 0.02, \dots, 0.1$ (the animated solution can also be observed by stepping quickly through the six plots). Only two plots (for $t = 0, 0.1$) are presented here to conserve space.

We should keep in mind that Eqs. (3.1) constitute a smooth problem since $u(x, y, t)$ from Eq. (3.1g) has smooth derivatives of all orders in x, y, t. Also, IC (3.1b) and BCs (3.1c–f) are consistent, that is, the switch from the IC at $t = 0$ to the BCs for $t > 0$ does not introduce discontinuities. Thus, Eqs. (3.1) should be

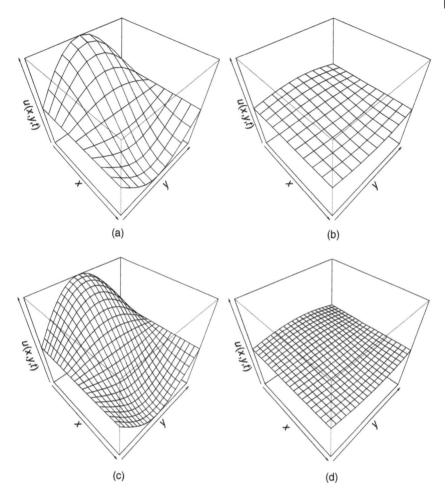

(a)

(b)

(c)

(d)

Figure 3.1 Numerical solution of Eqs. (3.1), (a) $t = 0$, nx=ny=11; (b) $t = 0.1$, nx=ny=11; (c) $t = 0$, nx=ny=21; (d) $t = 0.1$, nx=ny=21.

considered a relatively straightforward problem, and the accuracy of the preceding results may not carry over to more difficult problems without at least further study.

However, the preceding numerical approach can, in principle, be applied to more difficult problems, for example, simultaneous, inhomogeneous, nonlinear PDEs with nonlinear BCs. The application of the SCMOL to a variety of PDEs of varying complexity is considered in subsequent chapters.

We next consider the extension of the preceding analysis of a 2D problem to an analogous 3D problem.

3.2   3D in Space

The problem system is

$$\frac{\partial u}{\partial t} = D\left(\frac{\partial^2 u}{\partial x^2} + \frac{\partial^2 u}{\partial y^2} + \frac{\partial^2 u}{\partial z^2}\right) \tag{3.2a}$$

D, the diffusivity, again is taken as a constant to be specified.

Equation (3.2a) is first order in t and second order in x, y, z, so that it requires one IC and two BCs for $x, y,$ and z.

$$u(x, y, z, t = 0) = \cos(\pi x)\cos(\pi y)\cos(\pi z) \tag{3.2b}$$

$$\frac{\partial u(x = x_l, y, z, t)}{\partial x} = \frac{\partial u(x = x_u, y, z, t)}{\partial x} = 0 \tag{3.2c,d}$$

$$\frac{\partial u(x, y = y_l, z, t)}{\partial y} = \frac{\partial u(x, y = y_u, z, t)}{\partial y} = 0 \tag{3.2e,f}$$

$$\frac{\partial u(x, y, z = z_l, t)}{\partial z} = \frac{\partial u(x, y, z = z_u, t)}{\partial z} = 0 \tag{3.2g,h}$$

Equations (3.2c–h) are homogeneous Neumann BCs.

For $x_l = y_l = z_l = 0, x_u = y_u = z_u = 1$, the analytical solution to Eqs. (3.2) is

$$u(x, y, z, t) = e^{-3D\pi^2 t}\cos(\pi x)\cos(\pi y)\cos(\pi z) \tag{3.2i}$$

which is used to evaluate the numerical solution.

A main program for Eqs. (3.2) follows.

3.2.1   Main Program, Case 1

An R main program for Eqs. (3.2), which follows closely from the main program of Listing 3.1a, is listed next.

```
#
# Delete previous workspaces
  rm(list=ls(all=TRUE))
#
# Access ODE integrator
  library("deSolve");
#
# Access functions for numerical solution
  setwd("f:/collocation/chap3");
  source("pde_3a.R");
#
# Parameters
  D=1;
#
# Grid (in x,y)
```

```
  nx=11;ny=11;nz=11;
  xl=0;xu=1;yl=0;yu=1;zl=0;zu=1;
  x=seq(from=xl,to=xu,by=(xu-xl)/(nx-1));
  y=seq(from=yl,to=yu,by=(yu-yl)/(ny-1));
  z=seq(from=zl,to=zu,by=(zu-zl)/(nz-1));
  out=(1-1)*nx*ny+(1-1)*nx+1;
#
# Independent variable for ODE integration
  nout=21;t0=0;tf=0.1;
  tout=seq(from=t0,to=tf,by=(tf-t0)/(nout-1));
#
# Initial condition
  u0=rep(0,nx*ny*nz);
  for(k in 1:nz){
  for(j in 1:ny){
  for(i in 1:nx){
    u0[(k-1)*nx*ny+(j-1)*nx+i]=cos(pi*x[i])*
                               cos(pi*y[j])*
                               cos(pi*z[k]);
  }
  }
  }
  ncall=0;
#
# ODE integration
   tplot=rep(0,nout);
   uplot=rep(0,nout);
  uaplot=rep(0,nout);
  u=u0;t=t0;h=0.0001;
  for(i1 in 1:nout){
    ua=cos(pi*x[1])*cos(pi*y[1])*cos(pi*z[1])*
           exp(-3*D*(pi^2)*t);
    cat(sprintf("\n t = %6.4f  u(0,0,0,t) =
                 %8.4f%8.4f",t,u[out],ua));
    tplot[i1]=t;
    uplot[i1]=u[out];
   uaplot[i1]=ua;
  for(i2 in 1:50){
    derv=pde_3a(t,u,parm);
    u=u+derv*h;
    t=t+h;
  }
```

```
    }
#
# Plot u(x=0,y=0,z=0,t) against t
  par(mfrow=c(1,1))
  plot(tplot,uplot,xlab="t",ylab="u(0,0,0,t)",
       type="l",lwd=2);
  points(tplot,uaplot,pch="o",lwd=2);
#
# Calls to ODE routine
  cat(sprintf("\n\n ncall = %3d\n",ncall));
```

Listing 3.2a Main program for Eqs. (3.2).

We can note the following details about Listing 3.2a:

- Previous workspaces are removed. Then the ODE/MOL routine pde_3a.R is accessed.

```
#
# Delete previous workspaces
  rm(list=ls(all=TRUE))
#
# Access ODE integrator
  library("deSolve");
#
# Access functions for numerical solution
  setwd("f:/collocation/chap3");
  source("pde_3a.R");
```

- The model parameter D in Eq. (3.2a) is defined numerically.

```
#
# Parameters
  D=1;
```

- Uniform grids in x, y, and z of 11 points each are defined with the seq utility for the intervals $x_l \le x \le x_u$, $y_l \le y \le y_u$, $z_l \le z \le z_u$. Therefore, with $x_l = y_l = z_l = 0$, $x_u = y_u = z_u = 1$, the vectors x, y, z have the values $x = 0, 0.1, \ldots, 1$, $y = 0, 0.1, \ldots, 1$ $z = 0, 0.1, \ldots, 1$.

```
#
# Grid (in x,y)
  nx=11;ny=11;nz=11;
  xl=0;xu=1;yl=0;yu=1;zl=0;zu=1;
  x=seq(from=xl,to=xu,by=(xu-xl)/(nx-1));
  y=seq(from=yl,to=yu,by=(yu-yl)/(ny-1));
```

```
z=seq(from=zl,to=zu,by=(zu-zl)/(nz-1));
out=(1-1)*nx*ny+(1-1)*nx+1;
```

One output point of the $11 \times 11 \times 11 = 1331$ solution points is selected, out=(1-1)*nx*ny+(1-1)*nx+1. This coding reflects a general index (k-1)*nx*ny+(j-1)*nx+i, and for $x = y = z = 0$, $i = 1, j = 1, k = 1$.

- A uniform grid in t of 21 output points is defined with the seq utility for the interval $0 \le t \le t_f = 0.1$. Therefore, the vector tout has the values $t = 0, 0.005, \ldots, 0.1$.

```
#
# Independent variable for ODE integration
  nout=21;t0=0;tf=0.1;
  tout=seq(from=t0,to=tf,by=(tf-t0)/(nout-1));
```

At this point, the intervals of x, y, z, and t in Eq. (3.2a) are defined.

- The IC vector corresponding to Eq. (3.2b) is defined numerically.

```
#
# Initial condition
  u0=rep(0,nx*ny*nz);
  for(k in 1:nz){
  for(j in 1:ny){
  for(i in 1:nx){
    u0[(k-1)*nx*ny+(j-1)*nx+i]=cos(pi*x[i])*
                               cos(pi*y[j])*
                               cos(pi*z[k]);
  }
  }
  }
  ncall=0;
```

Note the use of the general 1D index (k-1)*nx*ny+(j-1)*nx+i. Also, the counter for the calls to pde_3a is initialized.

- The system of 1331 MOL/ODEs is integrated by an in-line Euler integrator.

```
#
# ODE integration
   tplot=rep(0,nout);
   uplot=rep(0,nout);
  uaplot=rep(0,nout);
  u=u0;t=t0;h=0.0001;
  for(i1 in 1:nout){
    ua=cos(pi*x[1])*cos(pi*y[1])*cos(pi*z[1])*
           exp(-3*D*(pi^2)*t);
```

```
    cat (sprintf ("\n t = %6.4f  u(0,0,0,t) =
                  %8.4f%8.4f",t,u[out],ua));
    tplot[i1]=t;
    uplot[i1]=u[out];
   uaplot[i1]=ua;
  for(i2 in 1:50){
     derv=pde_3a(t,u,parm);
     u=u+derv*h;
     t=t+h;
  }
  }
```

This integrator is explained in detail in the discussion of Listing 3.1a. In particular, $(21)(50) = 1050$ Euler steps, each of length h=0.0001 corresponds to the output points $t = 0, 0.005, \ldots, 0.1$ (these details are reflected in the numerical output that follows). The ODE/MOL routine pde_3a is discussed next. The analytical solutions of Eq. (3.2i), ua, are evaluated at $x = y = z = 0$ and displayed along with the numerical solution.

- After the completion of the Euler integration, $u(x = 0, y = 0, z = 0, t)$ (numerical as points, analytical as a line) is plotted against t.

```
#
# Plot u(x=0,y=0,z=0,t) against t
  par (mfrow=c(1,1))
  plot (tplot,uplot,xlab="t",ylab="u(0,0,0,t)",
        type="l",lwd=2);
  points (tplot,uaplot,pch="o",lwd=2);
```

- The number of calls to pde_3a is displayed at the end of the solution.

```
#
# Calls to ODE routine
  cat (sprintf ("\n\n ncall = %3d\n",ncall));
```

As will be observed, ncall=1050 as expected.

The ODE/MOL routine pde_3a called in the Euler integration is considered next.

3.2.2 ODE Routine

pde_3a is listed next.

```
  pde_3a=function (t,u1d,parms) {
#
# Function pde_3a computes the t derivative vector
```

```
# of u(x,y,z,t)
#
# 1D vector to 3D array
  u=array(u1d,c(nx,ny,nz));
#
# ux
  ux=array(0,c(nx,ny,nz));
  for(k in 1:nz){
  for(j in 1:ny){
    tablex=splinefun(x,u[k,j,]);
    ux[k,j,]=tablex(x,deriv=1);
  }
  }
#
# uy
  uy=array(0,c(nx,ny,nz));
  for(k in 1:nz){
  for(i in 1:nx){
    tabley=splinefun(y,u[k,,i]);
    uy[k,,i]=tabley(y,deriv=1);
  }
  }
#
# uz
  uz=array(0,c(nx,ny,nz));
  for(j in 1:ny){
  for(i in 1:nx){
    tablez=splinefun(z,u[,j,i]);
    uz[,j,i]=tablez(z,deriv=1);
  }
  }
#
# BCs, x=0,xL
  ux[,, 1]=0;
  ux[,,nx]=0;
#
# BCs, y=0,yL
  uy[, 1,]=0;
  uy[,ny,]=0;
#
# BCs, z=0,zL
  uz[ 1,,]=0;
  uz[nz,,]=0;
```

```
#
# uxx
  uxx=array(0,c(nx,ny,nz));
  for(k in 1:nz){
  for(j in 1:ny){
    tablexx=splinefun(x,ux[k,j,]);
    uxx[k,j,]=tablexx(x,deriv=1);
  }
  }
#
# uyy
  uyy=array(0,c(nx,ny,nz));
  for(k in 1:nz){
  for(i in 1:nx){
    tableyy=splinefun(y,uy[k,,i]);
    uyy[k,,i]=tableyy(y,deriv=1);
  }
  }
#
# uzz
  uzz=array(0,c(nx,ny,nz));
  for(j in 1:ny){
  for(i in 1:nx){
    tablezz=splinefun(z,uz[,j,i]);
    uzz[,j,i]=tablezz(z,deriv=1);
  }
  }
#
# PDE
  ut=array(0,c(nx,ny,nz));
  ut=uxx+uyy+uzz;
#
# 3D array to 1D vector
    u1dt=ut[,,];
#
# Increment calls to pde_3a
  ncall <<- ncall+1;
#
# Return derivative vector
  return(c(u1dt));
  }
```

Listing 3.2b ODE/MOL routine pde_3a for Eqs. (3.2).

We can note the following details about pde_3a:

- The function is defined.

```
pde_3a=function(t,u1d,parms){
#
# Function pde_3a computes the t derivative vector
# of u(x,y,z,t)
```

t is the current value of t in Eq. (3.2a). u1d is the 1331-vector of ODE-dependent variables (SCMOL approximation of Eq. (3.2a)). parm is an argument to pass parameters to pde_3a (unused but required in the argument list). The arguments must be listed in the order stated to properly interface with the call to pde_3a in the main program of Listing 3.3a (derv=pde_3a(t,u,parm)).

- The 1D array u1d is placed in a 3D array u(i,j,k) to facilitate the programming of Eq. (3.2a).

```
#
# 1D vector to 3D array
  u=array(u1d,c(nx,ny,nz));
```

The two arguments of the utility array are as follows:
1. The 1D array (vector, u1d) is to be changed to a multidimensional array (u).
2. The number of dimensions of u is defined in a vector (using the vector operator c), with the elements of the vector defining the number of points in each dimension (direction). Thus, u is defined as a 3D array with nx points in the first dimension, ny points in the second dimension, and nz in the third dimension, that is, u(nx,ny,nz). In the 2D case, array is equivalent to matrix used previously.

- The partial derivative $\dfrac{\partial u(x, y, z, t)}{\partial x}$ is computed.

```
#
# ux
  ux=array(0,c(nx,ny,nz));
  for(k in 1:nz){
  for(j in 1:ny){
    tablex=splinefun(x,u[k,j,]);
    ux[k,j,]=tablex(x,deriv=1);
  }
  }
```

The utility array is used to define the 3D array for the partial derivative (ux). All values of x are included through the [,]. All values of y and z are included using two nested fors.

- The partial derivative $\dfrac{\partial u(x,y,z,t)}{\partial y}$ is computed.

```
#
# uy
  uy=array(0,c(nx,ny,nz));
  for(k in 1:nz){
  for(i in 1:nx){
    tabley=splinefun(y,u[k,,i]);
    uy[k,,i]=tabley(y,deriv=1);
  }
  }
```

The utility array is used to define the 3D array for the partial derivative (uy). All values of y are included through the [,]. All values of x and z are included using two nested fors.

- The partial derivative $\dfrac{\partial u(x,y,z,t)}{\partial z}$ is computed.

```
#
# uz
  uz=array(0,c(nx,ny,nz));
  for(j in 1:ny){
  for(i in 1:nx){
    tablez=splinefun(z,u[,j,i]);
    uz[,j,i]=tablez(z,deriv=1);
  }
  }
```

The utility array is used to define the 3D array for the partial derivative (uz). All values of z are included through the [,]. All values of x and y are included using two nested fors.

- BCs (3.2c–h) are programmed as

```
#
# BCs, x=0,xL
  ux[,, 1]=0;
  ux[,,nx]=0;
#
# BCs, y=0,yL
  uy[, 1,]=0;
  uy[,ny,]=0;
#
# BCs, z=0,zL
  uz[ 1,,]=0;
  uz[nz,,]=0;
```

All values of particular spatial independent variables are included through the [,], for example, all values of y and z for BCs (3.2c,d) with ux[, , 1]=0, ux[, ,nx]=0). Boundaries are specified with particular values of the index, for example, for BCs (3.2c,d), (1,nx) for $x = x_l = 0$, $x = x_u = 1$.

- The second derivatives in Eq. (3.2a), $\dfrac{\partial^2 u}{\partial x^2}, \dfrac{\partial^2 u}{\partial y^2}, \dfrac{\partial^2 u}{\partial z^2}$ are computed from the corresponding first derivatives.

```
#
# uxx
  uxx=array(0,c(nx,ny,nz));
  for(k in 1:nz){
  for(j in 1:ny){
    tablexx=splinefun(x,ux[k,j,]);
    uxx[k,j,]=tablexx(x,deriv=1);
  }
  }
#
# uyy
  uyy=array(0,c(nx,ny,nz));
  for(k in 1:nz){
  for(i in 1:nx){
    tableyy=splinefun(y,uy[k,,i]);
    uyy[k,,i]=tableyy(y,deriv=1);
  }
  }
#
# uzz
  uzz=array(0,c(nx,ny,nz));
  for(j in 1:ny){
  for(i in 1:nx){
    tablezz=splinefun(z,uz[,j,i]);
    uzz[,j,i]=tablezz(z,deriv=1);
  }
  }
```

Again, a combination of array, [,] and two nested fors is used to step through the indices in 3D.
- Equation (3.2a) is programmed through vectorization (no subscripting). $D = 1$ is implied with this programming.

```
#
# PDE
  ut=array(0,c(nx,ny,nz));
```

```
ut=uxx+uyy+uzz;
```

- The 3D matrix of derivatives in t, ut, is placed in a 1D vector, u1dt, for integration by the Euler method in Listing 3.2a.

```
#
# 3D array to 1D vector
   u1dt=ut[,,];
```

The required subscripting is handled by [,,] (all values of i,j,k).
- The counter for the calls to pde_3a is incremented and then returned to the main program of Listing 3.2a through the <<- operator.

```
#
# Increment calls to pde_3a
   ncall <<- ncall+1;
```

- The derivative vector (LHS of Eq. (3.2a)) is returned to the main program as a numerical vector. This return gives the Euler integrator the 1331 element derivative vector for the next step along the solution.

```
#
# Return derivative vector
   return(c(u1dt));
   }
```

The final } concludes pde_3a.

This completes the programming of Eq. (3.2a). The numerical output is displayed in Table 3.2.

The numerical (first column) and analytical (second column) solutions agree to approximately 2+ figures. This limited agreement is not unexpected since only 11 points in x, y, and z were used. The numerical solution appears to be approaching the limiting value $u(x = 0, y = 0, z = 0, t) \rightarrow \infty) = 0$, which follows from Eq. (3.2i), $u(x = 0, y = 0, z = 0, t) = e^{-3D\pi^2 t} \cos(0) \cos(0) \cos(0)$.

Even with the coarse gridding, 1331 ODEs are integrated, which demonstrates the rapid growth in the total number of ODEs with increasing numbers of grid points in each direction, the so-called "curse of dimensionality." For example, if the spatial convergence of the numerical solution is investigated by doubling the number of grid points (as discussed in Section (3.1) for 2D), the number of ODEs would increase to $(2^3)(1331) = 10{,}648$. Thus, for higher dimensional PDEs, a balance between the number of grid points and the accuracy of the numerical solution is required.

The graphical output is shown in Figure 3.2, which demonstrates the agreement between the numerical and the analytical solutions.

Table 3.2 Output for Eqs. (3.2) from the main program of Listing of 3.2a, nx=ny=nz=11.

```
t = 0.0000   u(0,0,0,t)  =    1.0000   1.0000
t = 0.0050   u(0,0,0,t)  =    0.8601   0.8624
t = 0.0100   u(0,0,0,t)  =    0.7406   0.7437
t = 0.0150   u(0,0,0,t)  =    0.6380   0.6414
t = 0.0200   u(0,0,0,t)  =    0.5498   0.5531
t = 0.0250   u(0,0,0,t)  =    0.4738   0.4770
t = 0.0300   u(0,0,0,t)  =    0.4084   0.4114
t = 0.0350   u(0,0,0,t)  =    0.3520   0.3548
t = 0.0400   u(0,0,0,t)  =    0.3035   0.3059
t = 0.0450   u(0,0,0,t)  =    0.2616   0.2638
t = 0.0500   u(0,0,0,t)  =    0.2255   0.2275
t = 0.0550   u(0,0,0,t)  =    0.1944   0.1962
t = 0.0600   u(0,0,0,t)  =    0.1676   0.1692
t = 0.0650   u(0,0,0,t)  =    0.1445   0.1459
t = 0.0700   u(0,0,0,t)  =    0.1246   0.1259
t = 0.0750   u(0,0,0,t)  =    0.1074   0.1085
t = 0.0800   u(0,0,0,t)  =    0.0926   0.0936
t = 0.0850   u(0,0,0,t)  =    0.0798   0.0807
t = 0.0900   u(0,0,0,t)  =    0.0688   0.0696
t = 0.0950   u(0,0,0,t)  =    0.0593   0.0600
t = 0.1000   u(0,0,0,t)  =    0.0512   0.0518

ncall = 1050
```

To conclude this chapter, we consider a variation of Eqs. (3.2). Equation (3.2a) is again the PDE (renumbered to (3.3a)).

$$\frac{\partial u}{\partial t} = D\left(\frac{\partial^2 u}{\partial x^2} + \frac{\partial^2 u}{\partial y^2} + \frac{\partial^2 u}{\partial z^2}\right) \tag{3.3a}$$

D, the diffusivity, again is taken as a constant to be specified.

Equation (3.3a) is first order in t and second order in x, y, z, so that it requires one IC and two BCs for $x, y,$ and z.

$$u(x, y, z, t = 0) = 0 \tag{3.3b}$$

$$u(x = x_l, y, z, t) = u(x = x_u, y, z, t) = 1 \tag{3.3c,d}$$

$$\frac{\partial u(x, y = y_l, z, t)}{\partial y} = \frac{\partial u(x, y = y_u, z, t)}{\partial y} = 0 \tag{3.3e,f}$$

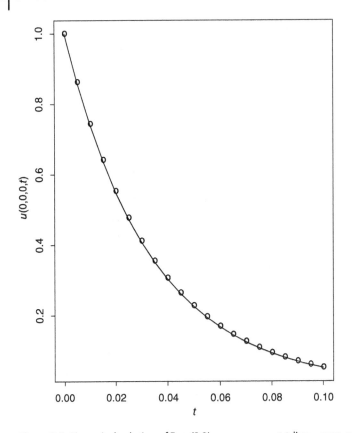

Figure 3.2 Numerical solution of Eqs. (3.2), nx=ny=nz=11 line – num, points – anal.

$$\frac{\partial u(x, y, z = z_l, t)}{\partial z} = -(u_a - u(x, y, z = z_l, t))$$

$$\frac{\partial u(x, y, z = z_u, t)}{\partial z} = (u_a - u(x, y, z = z_u, t)) \tag{3.3g,h}$$

Equations (3.3c,d) are nonhomogeneous (nonzero) Dirichlet BCs. Equations (3.3e,f) are homogeneous Neumann BCs. Equations (3.3g,h) are Robin (third type) BCs, with an ambient constant u_a. The intent with this example is to demonstrate the implementation of the three types of BCs.

Although a product analytical solution of the form of Eq. (3.2i) is available, it is relatively complicated (particularly because of BCs (3.3c,d) and (3.3g,h)). Thus, the analytical solution is not included in this example.

The numerical solution reflects the response of the system of Eqs. (3.3) to the unit value of BCs (3.3c,d), starting from the zero IC of Eq. (3.3b) (an increase in $u(x, y, z, t)$ from zero). The no flux (zero flux, impermeable, no diffusion)

BCs (3.3e,f) prevent any change in the solution from transfer across the boundaries at $y = y_l, y_u$.[1] BCs (3.3g,h) will reduce the increase in $u(x, y, z, t)$ from BCs (3.3c,d) if $u_a = 0$. These various cases are tested with the routines that follow.

A main program for Eqs. (3.3) follows.

3.2.3 Main Program, Case 2

An R main program for Eqs. (3.3), which follows closely from the main program of Listing 3.2a, is listed next.

```
#
# Delete previous workspaces
  rm(list=ls(all=TRUE))
#
# Access ODE integrator
  library("deSolve");
#
# Access functions for numerical solution
  setwd("f:/collocation/chap3");
  source("pde_3b.R");
#
# Parameters
  D=1;ua=1;
#
# Grid (in x,y)
  nx=11;ny=11;nz=11;
  xl=0;xu=1;yl=0;yu=1;zl=0;zu=1;
  x=seq(from=xl,to=xu,by=(xu-xl)/(nx-1));
  y=seq(from=yl,to=yu,by=(yu-yl)/(ny-1));
  z=seq(from=zl,to=zu,by=(zu-zl)/(nz-1));
  out=(6-1)*nx*ny+(6-1)*nx+6;
#
# Independent variable for ODE integration
  nout=21;t0=0;tf=0.5;
  tout=seq(from=t0,to=tf,by=(tf-t0)/(nout-1));
```

---

1 In a physical application, BCs (3.3g,h) model transfer or transport of mass or energy across the boundaries at $z = z_l, z_u$. If $u_a = 0$, the transfer will be out of the system to the surroundings. If $u_a > 0$, the transfer will be into or out of the system, depending on the values of $u(x, y, z = z_l, t)$ and $u(x, y, z = z_u, t)$ relative to u_a. For example, at $z = z_l$, $(u_a - u(x, y, z = z_l, t)) > 0$ in BC (3.3g) corresponds to transfer at $z = z_l$ from the surroundings into the system, thus tending to increase $u(x, y, z, t)$, while $(u_a - u(x, y, z = z_l, t)) < 0$ in BC (3.3g) corresponds to transfer from the system to the surroundings, thus tending to decrease $u(x, y, z, t)$. A similar interpretation applies to $(u_a - u(x, y, z = z_u, t)) > 0$ in BC (3.3h) (transfer into the system) and $(u_a - u(x, y, z = z_u, t)) < 0$ in BC (3.3h) (transfer out of the system) at $z = z_u$.

```
#
# Initial condition
  u0=rep(0,nx*ny*nz);
  ncall=0;
#
# ODE integration
    tplot=rep(0,nout);
    uplot=rep(0,nout);
  u=u0;t=t0;h=0.0005;
  for(i1 in 1:nout){
    cat(sprintf("\n t = %6.4f  u(0.5,0.5,0.5,t) =
                  %8.4f",t,u[out]));
    tplot[i1]=t;
    uplot[i1]=u[out];
  for(i2 in 1:50){
    derv=pde_3b(t,u,parm);
    u=u+derv*h;
    t=t+h;
  }
  }
#
# Plot u(x=0.5,y=0.5,z=0.5,t) against t
  par(mfrow=c(1,1))
  plot(tplot,uplot,
    xlab="t",ylab="u(0.5,0.5,0.5,t)",
    type="l",lwd=2);
#
# Calls to ODE routine
  cat(sprintf("\n\n ncall = %3d\n",ncall));
```

Listing 3.3a Main program for Eqs. (3.3).

We can note the following details about Listing 3.3a.

- Previous workspaces are removed. Then the ODE/MOL routine pde_3b.R is accessed.

```
#
# Delete previous workspaces
  rm(list=ls(all=TRUE))
#
# Access ODE integrator
  library("deSolve");
#
```

```
# Access functions for numerical solution
  setwd("f:/collocation/chap3");
  source("pde_3b.R");
```

- The model parameters D in Eq. (3.3a) and u_a in BCs (3.3g,h) are defined numerically.

```
#
# Parameters
  D=1;ua=1;
```

- Uniform grids in x, y and z of 11 points each are defined with the seq utility for the intervals $x_l \le x \le x_u$, $y_l \le y \le y_u$, $z_l \le z \le z_u$. Therefore, with $x_l = y_l = z_l = 0$, $x_u = y_u = z_u = 1$, the vectors x, y, z have the values $x = 0, 0.1, \ldots, 1, y = 0, 0.1, \ldots, 1\ z = 0, 0.1, \ldots, 1$.

```
#
# Grid (in x,y)
  nx=11;ny=11;nz=11;
  xl=0;xu=1;yl=0;yu=1;zl=0;zu=1;
  x=seq(from=xl,to=xu,by=(xu-xl)/(nx-1));
  y=seq(from=yl,to=yu,by=(yu-yl)/(ny-1));
  z=seq(from=zl,to=zu,by=(zu-zl)/(nz-1));
  out=(6-1)*nx*ny+(6-1)*nx+6;
```

One output point of the $11 \times 11 \times 11 = 1331$ solution points is selected, out=(6-1)*nx*ny+(6-1)*nx+6. This coding reflects a general index (k-1)*nx*ny+(j-1)*nx+i, and for $x = y = z = 0.5$, $i = 6, j = 6, k = 6$.

- A uniform grid in t of 21 output points is defined with the seq utility for the interval $0 \le t \le t_f = 0.5$. Therefore, the vector tout has the values $t = 0, 0.025, \ldots, 0.5$.

```
#
# Independent variable for ODE integration
  nout=21;t0=0;tf=0.5;
  tout=seq(from=t0,to=tf,by=(tf-t0)/(nout-1));
```

At this point, the intervals of x, y, z, and t in Eq. (3.3a) are defined.

- The IC vector corresponding to Eq. (3.3b) is defined numerically.

```
#
# Initial condition
  u0=rep(0,nx*ny*nz);
  ncall=0;
```

Note, the value of 0 from `rep(0,nx*ny*nz)` for the `nx*ny*nz = 1331` elements of the vector u0 in accordance with IC (3.3b). Also, the counter for the calls to pde_3a is initialized.

- The system of 1331 MOL/ODEs is integrated by an in-line Euler integrator.

```
#
# ODE integration
   tplot=rep(0,nout);
   uplot=rep(0,nout);
  u=u0;t=t0;h=0.0005;
  for(i1 in 1:nout){
    cat(sprintf("\n t = %6.4f   u(0.5,0.5,0.5,t) =
                   %8.4f",t,u[out]));
    tplot[i1]=t;
    uplot[i1]=u[out];
  for(i2 in 1:50){
    derv=pde_3b(t,u,parm);
    u=u+derv*h;
    t=t+h;
  }
  }
```

This integrator is explained in detail in the discussion of Listing 3.1a. In particular, (21)(50) = 1050 Euler steps, each of length h=0.0005 corresponds to the output points $t = 0, 0.025, \dots, 0.5$ (these details are reflected in the numerical output that follows). The ODE/MOL routine pde_3b is discussed next. The output at point out=`(6-1)*nx*ny+(6-1)*nx+6=` `5*11*11+5*11+6=666` is displayed (the midpoint in the vector of length 1331).

- After the completion of the Euler integration, $u(x = 0.5, y = 0.5, z = 0.5, t)$ is plotted against t.

```
#
# Plot u(x=0.5,y=0.5,z=0.5,t) against t
  par(mfrow=c(1,1))
  plot(tplot,uplot,
    xlab="t",ylab="u(0.5,0.5,0.5,t)",
    type="l",lwd=2);
```

- The number of calls to pde_3b is displayed at the end of the solution.

```
#
# Calls to ODE routine
  cat(sprintf("\n\n ncall = %3d\n",ncall));
```

As will be observed, ncase=1050 as expected.

The ODE/MOL routine pde_3b called in the Euler integration is considered next.

3.2.4 ODE Routine

pde_3b is listed next.

```
pde_3b=function(t,u1d,parms){
#
# Function pde_3b computes the t derivative vector
# of u(x,y,z,t)
#
# 1D vector to 3D array
  u=array(u1d,c(nx,ny,nz));
#
# BCs, x=0,xL
  u[,, 1]=1;
  u[,,nx]=1;
#
# ux
  ux=array(0,c(nx,ny,nz));
  for(k in 1:nz){
  for(j in 1:ny){
    tablex=splinefun(x,u[k,j,]);
    ux[k,j,]=tablex(x,deriv=1);
  }
  }
#
# uy
  uy=array(0,c(nx,ny,nz));
  for(k in 1:nz){
  for(i in 1:nx){
    tabley=splinefun(y,u[k,,i]);
    uy[k,,i]=tabley(y,deriv=1);
  }
  }
#
# uz
  uz=array(0,c(nx,ny,nz));
  for(j in 1:ny){
  for(i in 1:nx){
    tablez=splinefun(z,u[,j,i]);
```

```
    uz[,j,i]=tablez(z,deriv=1);
  }
  }
#
# BCs, y=0,yL
  uy[, 1,]=0;
  uy[,ny,]=0;
#
# BCs, z=0,zL
  uz[ 1,,]=-(ua-u[ 1,,]);
  uz[nz,,]= (ua-u[nz,,]);
#
# uxx
  uxx=array(0,c(nx,ny,nz));
  for(k in 1:nz){
  for(j in 1:ny){
    tablexx=splinefun(x,ux[k,j,]);
    uxx[k,j,]=tablexx(x,deriv=1);
  }
  }
#
# uyy
  uyy=array(0,c(nx,ny,nz));
  for(k in 1:nz){
  for(i in 1:nx){
    tableyy=splinefun(y,uy[k,,i]);
    uyy[k,,i]=tableyy(y,deriv=1);
  }
  }
#
# uzz
  uzz=array(0,c(nx,ny,nz));
  for(j in 1:ny){
  for(i in 1:nx){
    tablezz=splinefun(z,uz[,j,i]);
    uzz[,j,i]=tablezz(z,deriv=1);
  }
  }
#
# PDE
  ut=array(0,c(nx,ny,nz));
  ut=uxx+uyy+uzz;
#
```

```
# 3D array to 1D vector
  u1dt=ut[,,];
#
# Increment calls to pde_3b
  ncall <<- ncall+1;
#
# Return derivative vector
  return(c(u1dt));
  }
```

Listing 3.3b ODE/MOL routine pde_3b for Eqs. (3.3).

We can note the following details about pde_3b:

- The function is defined.

```
  pde_3b=function(t,u1d,parms){
#
# Function pde_3b computes the t derivative vector
# of u(x,y,z,t)
```

  t is the current value of t in Eq. (3.3a). u1d is the 1331-vector of ODE-dependent variables (SCMOL approximation of Eq. (3.3a)). parm is an argument to pass parameters to pde_3b (unused but required in the argument list). The arguments must be listed in the order stated to properly interface with the call to pde_3a in the main program of Listing 3.2a (derv=pde_3b(t,u,parm)).

- The 1D array u1d is placed in a 3D array u(i,j,k) to facilitate the programming of Eq. (3.3a).

```
#
# 1D vector to 3D array
  u=array(u1d,c(nx,ny,nz));
```

  The arguments of the utility array are discussed in detail after Listing 3.2b.

- BCs (3.3c,d) are applied.

```
#
# BCs, x=0,xL
  u[,, 1]=1;
  u[,,nx]=1;
```

  Subscripts 1, nx correspond to $x = x_l = 0, x = x_u = 1$. All values of y, z are included with [,,].

- The partial derivatives $\dfrac{\partial u(x,y,z,t)}{\partial x}, \dfrac{\partial u(x,y,z,t)}{\partial y}, \dfrac{\partial u(x,y,z,t)}{\partial z}$ (ux, uy, uz) are computed.

```
#
# ux
  ux=array(0,c(nx,ny,nz));
  for(k in 1:nz){
  for(j in 1:ny){
    tablex=splinefun(x,u[k,j,]);
    ux[k,j,]=tablex(x,deriv=1);
  }
  }
#
# uy
  uy=array(0,c(nx,ny,nz));
  for(k in 1:nz){
  for(i in 1:nx){
    tabley=splinefun(y,u[k,,i]);
    uy[k,,i]=tabley(y,deriv=1);
  }
  }
#
# uz
  uz=array(0,c(nx,ny,nz));
  for(j in 1:ny){
  for(i in 1:nx){
    tablez=splinefun(z,u[,j,i]);
    uz[,j,i]=tablez(z,deriv=1);
  }
  }
```

Additional discussion is given after Listing 3.2b.
- BCs (3.3e–h) are programmed as

```
#
# BCs, y=0,yL
  uy[, 1,]=0;
  uy[,ny,]=0;
#
# BCs, z=0,zL
  uz[ 1,,]=-(ua-u[ 1,,]);
  uz[nz,,]= (ua-u[nz,,]);
```

Note in particular the straightforward programming of the Robin BCs of Eqs. (3.3g,h). ua is defined numerically in the main program of Listing 3.3a.

- The second derivatives in Eq. (3.3a), $\frac{\partial^2 u}{\partial x^2}, \frac{\partial^2 u}{\partial y^2}, \frac{\partial^2 u}{\partial z^2}$ are computed from the corresponding first derivatives.

```
#
# uxx
  uxx=array(0,c(nx,ny,nz));
  for(k in 1:nz){
  for(j in 1:ny){
    tablexx=splinefun(x,ux[k,j,]);
    uxx[k,j,]=tablexx(x,deriv=1);
  }
  }
#
# uyy
  uyy=array(0,c(nx,ny,nz));
  for(k in 1:nz){
  for(i in 1:nx){
    tableyy=splinefun(y,uy[k,,i]);
    uyy[k,,i]=tableyy(y,deriv=1);
  }
  }
#
# uzz
  uzz=array(0,c(nx,ny,nz));
  for(j in 1:ny){
  for(i in 1:nx){
    tablezz=splinefun(z,uz[,j,i]);
    uzz[,j,i]=tablezz(z,deriv=1);
  }
  }
```

Again, a combination of `array`, `[,]` and two nested `for`s is used to step through the indices in 3D.

- Equation (3.3a) is programmed through vectorization (no subscripting). $D = 1$ is implied with this programming.

```
#
# PDE
  ut=array(0,c(nx,ny,nz));
  ut=uxx+uyy+uzz;
```

- The 3D matrix of derivatives in t, `ut`, is placed in a 1D vector, `u1dt`, for integration by the Euler method in Listing 3.3a.

```
#
# 3D array to 1D vector
   u1dt=ut[,,];
```

The required subscripting is handled by [,,] (all values of i, j, k).

- The counter for the calls to pde_3b is incremented, then returned to the main program of Listing 3.3a through the <<- operator.

```
#
# Increment calls to pde_3b
   ncall <<- ncall+1;
```

- The derivative vector (LHS of Eq. (3.3a)) is returned to the main program as a numerical vector. This return gives the Euler integrator the 1331 element derivative vector for the next step along the solution.

```
#
# Return derivative vector
   return(c(u1dt));
   }
```

The final } concludes pde_3a.

This completes the programming of Eq. (3.3a). The numerical output is in Table 3.3a (for ICs (3.3b), BCs (3.3c–h) with $u_a = 1$).

We can note the following details about the output in Table 3.3a:

- t covers the interval $0 \le t \le 0.5$ in steps of 0.025, as expected from the programming in Listing 3.3a.
- IC (3.3b) is confirmed ($u(0.5, 0.5, 0.5, t = 0) = 0$).
- The solution appears to approach 1 as expected from the combined effect of BCs (3.3c,d) and (3.3f,g) with $u_a = 1$.

The graphical output is shown in Figure 3.3a, which agrees with the output in Table 3.3a.

The kink from $u(0.5, 0.5, 0.5, t = 0.025) = 0.0545$ could be smoothed by using more output points in t.

If $u_a = 0$ in the main program of Listing 3.3a, the following numerical output results.

The solution appears to approach 0.86, $\cdots$, which is less than 1, as expected from the combined effect of BCs (3.3c,d) and (3.3f,g) with $u_a = 0$. The graphical output in Figure 3.3b confirms the output in Table 3.3b.

As another case of Eqs. (3.3), if the 1 in BCs (3.3c,d) is changed to 0 (zero) and $u_a = 0$ in BCs (3.3g,h), there is nothing in the system of Eqs. (3.3) to move the solution away from the IC of Eq. (3.3b), so the solution remains at 0 (for $x = y = z = 0.5$). This may seem like a trivial case, but it is worth executing since the solution moving away from IC Eq. (3.3b) would indicate an error in the code.

Table 3.3a Output for Eqs. (3.3) from the main program of Listing of 3.3a, $u_a = 1$.

```
t = 0.0000   u(0.5,0.5,0.5,t)  =    0.0000
t = 0.0250   u(0.5,0.5,0.5,t)  =    0.0545
t = 0.0500   u(0.5,0.5,0.5,t)  =    0.2497
t = 0.0750   u(0.5,0.5,0.5,t)  =    0.4320
t = 0.1000   u(0.5,0.5,0.5,t)  =    0.5739
t = 0.1250   u(0.5,0.5,0.5,t)  =    0.6810
t = 0.1500   u(0.5,0.5,0.5,t)  =    0.7613
t = 0.1750   u(0.5,0.5,0.5,t)  =    0.8215
t = 0.2000   u(0.5,0.5,0.5,t)  =    0.8664
t = 0.2250   u(0.5,0.5,0.5,t)  =    0.9001
t = 0.2500   u(0.5,0.5,0.5,t)  =    0.9253
t = 0.2750   u(0.5,0.5,0.5,t)  =    0.9441
t = 0.3000   u(0.5,0.5,0.5,t)  =    0.9582
t = 0.3250   u(0.5,0.5,0.5,t)  =    0.9687
t = 0.3500   u(0.5,0.5,0.5,t)  =    0.9766
t = 0.3750   u(0.5,0.5,0.5,t)  =    0.9825
t = 0.4000   u(0.5,0.5,0.5,t)  =    0.9869
t = 0.4250   u(0.5,0.5,0.5,t)  =    0.9902
t = 0.4500   u(0.5,0.5,0.5,t)  =    0.9927
t = 0.4750   u(0.5,0.5,0.5,t)  =    0.9945
t = 0.5000   u(0.5,0.5,0.5,t)  =    0.9959

ncall = 1050
```

3.3 Summary and Conclusions

This completes the discussion of the 2D and 3D application of SCMOL. The intent is to demonstrate the basic coding for a range of PDE applications (in Chapters 2 and 3). This discussion could continue for other types of PDE formulations such as the following:

- Nonlinear PDEs with, for example, a
  - Source term u^n, $n \neq 1$.
  - Convection (advection) term $u\dfrac{\partial u}{\partial x}$.
  - Diffusion term $\dfrac{\partial\left(D(u)\dfrac{\partial u}{\partial x}\right)}{\partial x}$.

  included in the RHS of Eq. (3.3a).

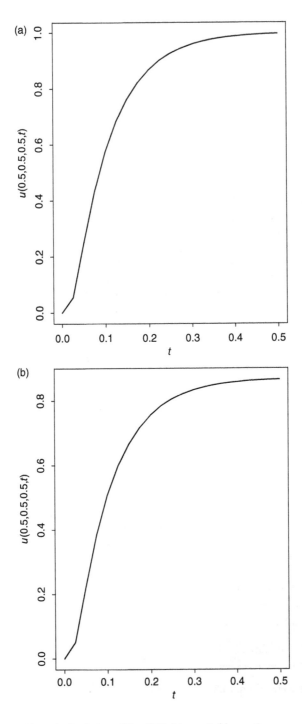

Figure 3.3 Numerical solution of Eqs. (3.3), (a)$u_a = 1$; (b) $u_a = 0$.

Table 3.3b Output for Eqs. (3.3) from the main program of Listing of 3.3a, $u_a = 0$.

```
t = 0.0000   u(0.5,0.5,0.5,t) =   0.0000
t = 0.0250   u(0.5,0.5,0.5,t) =   0.0509
t = 0.0500   u(0.5,0.5,0.5,t) =   0.2257
t = 0.0750   u(0.5,0.5,0.5,t) =   0.3844
t = 0.1000   u(0.5,0.5,0.5,t) =   0.5063
t = 0.1250   u(0.5,0.5,0.5,t) =   0.5978
t = 0.1500   u(0.5,0.5,0.5,t) =   0.6664
t = 0.1750   u(0.5,0.5,0.5,t) =   0.7176
t = 0.2000   u(0.5,0.5,0.5,t) =   0.7560
t = 0.2250   u(0.5,0.5,0.5,t) =   0.7847
t = 0.2500   u(0.5,0.5,0.5,t) =   0.8062
t = 0.2750   u(0.5,0.5,0.5,t) =   0.8222
t = 0.3000   u(0.5,0.5,0.5,t) =   0.8342
t = 0.3250   u(0.5,0.5,0.5,t) =   0.8432
t = 0.3500   u(0.5,0.5,0.5,t) =   0.8499
t = 0.3750   u(0.5,0.5,0.5,t) =   0.8550
t = 0.4000   u(0.5,0.5,0.5,t) =   0.8587
t = 0.4250   u(0.5,0.5,0.5,t) =   0.8615
t = 0.4500   u(0.5,0.5,0.5,t) =   0.8636
t = 0.4750   u(0.5,0.5,0.5,t) =   0.8652
t = 0.5000   u(0.5,0.5,0.5,t) =   0.8664

ncall = 1050
```

- Nonlinear BCs, for example, a fourth-order radiation BC in place of BCs. Equation (3.3g,h),

$$\frac{\partial u(x, y, z = z_l, t)}{\partial z} = -(u_a^4 - u^4(x, y, z = z_l, t))$$

$$\frac{\partial u(x, y, z = z_u, t)}{\partial z} = (u_a^4 - u^4(x, y, z = z_u, t))$$

- Simultaneous PDEs (for more than one dependent variable).[2]
- Various coordinate systems, for example, Cartesian, cylindrical, spherical.
- Self-contained R routines, for example, the main program of Listing 3.3a with the explicit Euler integrator replaced by the classical fourth-order

---

2 This topic is discussed briefly in Section (2.3). It is covered in more detail in the applications that follow.

Runge–Kutta method. This would substantially improve the accuracy (order) of the integration in t.

- Variable spatial grids since the splines (in `splinefun`) do not require uniform spacing. Variable grids are particularly useful for improved spatial resolution in regions of the PDE solution where the solution is changing rapidly, for example, the resolution of boundary layers. All that is required to use a variable grid is to define the grid point locations in the main program (rather than using `seq`, which gives a uniform spatial grid).

- Irregular geometries that do not fit into an established coordinate system. The basic idea is that the splines can be used to calculate derivatives at any spatial point (location) by using a small number of surrounding values of the variable to be differentiated. But the location of the point of differentiation can be wherever the analyst specifies, that is, it does not have to be located on points in a specified coordinate system such as Cartesian, cylindrical, or spherical coordinates.

- Moving boundaries where the two preceding properties can be used to adjust the location of the points for differentiation as the boundary moves.

These types of PDE variants are important but would require extended explanation. Rather than using the remaining space in this book for this purpose, a series of SCMOL applications is considered next through the use of named (legacy) PDEs. These applications illustrate additional uses of splines for approximating PDE numerical solutions.

4

Navier–Stokes, Burgers' Equations

We now consider a series of legacy (named) PDEs to demonstrate the application of the spline collocation method of lines (SCMOL) to PDEs with various forms and features, starting with the Navier–Stokes (NS) equations. The intention is to provide an introduction to the numerical integration (solution) of the PDEs, with a minimum of mathematical formality and complexity.

4.1   PDE Model

The NS equations of fluid dynamics are a system of nonlinear PDEs expressing mass conservation (continuity) and momentum conservation for a Newtonian fluid. For a 1D case in Cartesian coordinates, we start with[1]

$$\rho \frac{\partial u}{\partial t} + \rho u \frac{\partial u}{\partial x} + \frac{\partial p}{\partial x} - \mu \frac{\partial^2 u}{\partial x^2} = 0 \tag{4.1}$$

If the pressure gradient is neglected, Eq. (4.1) reduces to

$$\rho \frac{\partial u}{\partial t} + \rho u \frac{\partial u}{\partial x} - \mu \frac{\partial^2 u}{\partial x^2} = 0 \tag{4.2}$$

which is Burgers' equation. This is an unusual (and widely studied) PDE because of the following:

1. It is nonlinear, yet has a known, exact (analytical) solution.
2. The solution exhibits moving fronts that can be made arbitrarily sharp by decreasing the kinematic viscosity[2] $v = \mu/\rho$.

---

[1] Equation (4.1) defines the x component of velocity v_x. Following the usual convention in the numerical analysis literature, the dependent variable is designated as u (rather than v_x).
[2] Equation (4.2) can be divided by ρ to give the kinematic viscosity v as a coefficient of the second derivative term. Burgers' equation for $v = \mu/\rho = 0$ reduces to the inviscid Burgers' equation or nonlinear advection equation

$$\frac{\partial u}{\partial t} + u \frac{\partial u}{\partial x} = 0$$

We now consider the numerical solution of Burgers' equation (4.2) through a series of SCMOL routines. An analytical solution to Eq. (4.2), u_a [1], is used to confirm the numerical solution

$$u_a(x, t) = \frac{0.1e^{-A} + 0.5e^{-B} + e^{-C}}{e^{-A} + e^{-B} + e^{-C}} \tag{4.3a}$$

where

$$A = \frac{0.05}{v}(x - 0.5 + 4.95t) \tag{4.3b}$$

$$B = \frac{0.25}{v}(x - 0.5 + 0.75t) \tag{4.3c}$$

$$C = \frac{0.5}{v}(x - 0.375) \tag{4.3d}$$

Equation (4.2) is first order in t and second order in x. It therefore requires one initial condition (IC), taken as Eqs. (4.3) with $t = 0$ and two boundary conditions (BCs).

$$u(x, t = 0) = u_a(x, t = 0) \tag{4.4a}$$

$$\frac{\partial u(x = 0, t)}{\partial x} = \frac{\partial u(x = 1, t)}{\partial x} = 0 \tag{4.4b,c}$$

For the numerical solution, $v = 0.003$ which, as we shall observe, produces a solution with a steep moving front that sharpens with increasing t.

A main program for Eqs. (4.2)–(4.4) follows.

4.2 Main Program

```
#
# Burgers
#
# Delete previous workspaces
  rm(list=ls(all=TRUE))
#
# Access ODE integrator
  library("deSolve");
#
# Access functions for numerical solution
  setwd("f:/collocation/chap4");
  source("pde_1a.R");source("phi.R");
#
# Parameters
  xl=0;xu=1;
  vis=0.003;
```

```
#
# Spatial grid
  xl=0;xu=1;n=201;
  x=seq(from=xl,to=xu,by=(xu-xl)/(n-1));
#
# Independent variable for ODE integration
  t0=0;tf=1;nout=11;
  tout=seq(from=t0,to=tf,by=(tf-t0)/(nout-1));
  ncall=0;
#
# Initial condition
  u0=rep(0,n);
  for(i in 1:n){
    u0[i]=phi(x[i],t0);
  }
#
# ODE integration
  out=lsodes(func=pde_1a,y=u0,times=tout,
             sparsetype="sparseint")
  nrow(out)
  ncol(out)
#
# Arrays for display
   u=matrix(0,nrow=n,ncol=nout);
  ua=matrix(0,nrow=n,ncol=nout);
  for(it in 1:nout){
  for(i   in 1:n){
    u[i,it]=out[it,i+1];
   ua[i,it]=phi(x[i],tout[it]);
  }
  }
#
# Numerical, analytical solutions
  cat(sprintf("\n       t     x     u(x,t)    ua(x,t)
              diff"));
  for(it in 1:nout){
  iv=seq(from=1,to=n,by=10);
  for(i   in iv){
  cat(sprintf("\n %6.2f%6.2f%10.4f%10.4f%12.6f",
              tout[it],x[i],u[i,it],ua[i,it],
              u[i,it]-ua[i,it]));
  }
  }
```

```
matplot (x,u,type="l",lwd=2,col="black",lty=1,
   xlab="x",ylab="u(x,t)",main="Burgers");
matpoints (x,ua,pch="o",col="black");
#
# Calls to ODE routine
cat (sprintf ("\n\n  ncall = %3d\n",ncall));
```

Listing 4.1 Main program for Eqs. (4.2)–(4.4).

We can note the following details about Listing 4.1.

- Previous workspaces are removed. Then the ODE integrator library deS-olve is accessed. Note that the setwd (set working directory) uses / rather than the usual \.

```
#
# Burgers
#
# Delete previous workspaces
  rm (list=ls (all=TRUE))
#
# Access ODE integrator
  library ("deSolve");
#
# Access functions for numerical solution
  setwd ("f:/collocation/chap4");
  source ("pde_1a.R");source ("phi.R");
```

pde_1a is the routine for the SCMOL ODEs (discussed subsequently). phi.R is the routine for the analytical solution of Eqs. (4.3).

- A uniform grid in x of 201 points is defined with the seq utility for the interval $x_l = 0 \le x \le x_u = 1$. Therefore, the vector x has the values $x = 0, 0.005, \dots, 1$.

```
#
# Parameters
  xl=0;xu=1;
  vis=0.003;
#
# Spatial grid
  xl=0;xu=1;n=201;
  x=seq (from=xl,to=xu,by= (xu-xl) / (n-1));
```

The kinematic viscosity in Eq. (4.2) is also defined numerically.

- A uniform grid in t of 11 output points is defined with the seq utility for the interval $0 \le t \le t_f = 1$. Therefore, the vector tout has the values $t = 0, 0.1, \ldots, 1$.

```
#
# Independent variable for ODE integration
  t0=0;tf=1;nout=11;
  tout=seq(from=t0,to=tf,by=(tf-t0)/(nout-1));
  ncall=0;
```

At this point, the intervals of x and t in Eq. (4.2) are defined. Also, the counter for the calls to pde_1a is initialized (and passed to pde_1a without a special designation).

- The IC function in Eq. (4.4a) is defined using the analytical solution in phi for $t = 0$.

```
#
# Initial condition
  u0=rep(0,n);
  for(i in 1:n){
    u0[i]=phi(x[i],t0);
  }
```

- The system of 201 SCMOL ODEs is integrated by the library integrator lsodes (available in deSolve) with the sparse matrix option specified. As expected, the inputs to lsodes are the ODE function, pde_1a, the IC vector u0, and the vector of output values of t, tout. The length of u0 (e.g., 201) informs lsodes how many ODEs are to be integrated. func, y, times are reserved names.

```
#
# ODE integration
  out=lsodes(func=pde_1a,y=u0,times=tout,
             sparsetype="sparseint")
  nrow(out)
  ncol(out)
```

The numerical solution to the ODEs is returned in matrix out. In this case, out has the dimensions $nout \times (n+1) = 6 \times 202$. The offset $n+1$ is required since the first element of each column has the output t (also in tout), and the $2, \ldots, n+1 = 2, \ldots, 202$ column elements have the 201 ODE solutions. This indexing of out is used next.

- The ODE solution is placed in a 201×11 matrix, u, for subsequent plotting (by stepping through the solution with respect to x and t within a pair of fors). The analytical solution of Eqs. (4.3) is placed in ua.

```
#
# Arrays for display
   u=matrix(0,nrow=n,ncol=nout);
   ua=matrix(0,nrow=n,ncol=nout);
   for(it in 1:nout){
   for(i  in 1:n){
     u[i,it]=out[it,i+1];
     ua[i,it]=phi(x[i],tout[it]);
     }
     }
```

- The numerical and analytical solutions, and the difference, are then plotted with the matplot, matpoints utilities.

```
#
# Numerical, analytical solutions
   cat(sprintf("\n       t      x      u(x,t)     ua(x,t)
              diff"));
   for(it in 1:nout){
   iv=seq(from=1,to=n,by=10);
   for(i  in iv){
   cat(sprintf("\n %6.2f%6.2f%10.4f%10.4f%12.6f",
              tout[it],x[i],u[i,it],ua[i,it],
              u[i,it]-ua[i,it]));
     }
     }
   matplot(x,u,type="l",lwd=2,col="black",lty=1,
      xlab="x",ylab="u(x,t)",main="Burgers");
   matpoints(x,ua,pch="o",col="black");
```

To conserve space, only every tenth value in *x* of the solutions is displayed numerically (using the subscript iv). The numerical solution in array u is plotted as a solid line, and the analytical solution in ua is plotted as points against *x*. This formatting appears in Figure 4.1 (discussed subsequently). matplot will plot u against x with *t* as a parameter since the row dimension of u (201) matches the length of x (also 201). The same is true in plotting ua with matpoints.

- The number of calls to pde_1a is displayed as an indication of the computational effort required to compute the solution.

```
#
# Calls to ODE routine
   cat(sprintf("\n\n  ncall = %3d\n",ncall));
```

pde_1a called by the main program of Listing 4.1 is listed next.

4.3 ODE Routine

```
  pde_1a=function(t,u,parms){
#
# Function pde_1a computes the derivative
# vector in t
#
# ux
  table1=splinefun(x,u);
  ux=table1(x,deriv=1);
#
# BC at x = 0
  ux[1]=0;
#
# BC at x = 1
  ux[n]=0;
#
# uxx
  table2=splinefun(x,ux);
  uxx=table2(x,deriv=1);
#
# PDE
  ut=rep(0,n);
  for(i in 1:n){
    ut[i]=vis*uxx[i]-u[i]*ux[i];
  }
#
# Increment calls to pde_1a
  ncall << ncall+1;
#
# Return derivative vector
  return(list(c(ut)));
}
```

Listing 4.2 ODE/SCMOL routine pde_1a for Eqs. (4.2)–(4.4).

We can note the following details about pde_1a.

- The function is defined.

```
  pde_1a=function(t,u,parms){
#
# Function pde_1a computes the derivative
# vector in t
```

t is the current value of t in Eq. (4.2). u is the 201-vector of ODE/SCMOL dependent variables. parm is an argument to pass parameters to pde_1a (unused, but required in the argument list). The arguments must be listed in the order stated to properly interface with lsodes called in the main program. The derivative vector of the LHS of Eq. (4.2) is calculated next and returned to lsodes.

- $\dfrac{\partial u}{\partial x}$ in Eq. (4.2) is computed with splinefun.

```
#
# ux
   table1=splinefun(x,u);
   ux=table1(x,deriv=1);
```

- The homogeneous Neumann BCs of Eqs. (4.4b,c) are implemented. Note the use of the subscripts 1, n corresponding to $x = x_l = 0, x = x_u = 1$.

```
#
# BC at x = 0
   ux[1]=0;
#
# BC at x = 1
   ux[n]=0;
```

- splinefun calculates $\dfrac{\partial^2 u}{\partial x^2}$ in Eq. (4.2) from the first derivative (stagewise differentiation).

```
#
# uxx
   table2=splinefun(x,ux);
   uxx=table2(x,deriv=1);
```

- Equation (4.2) is programmed.

```
#
# PDE
   ut=rep(0,n);
   for(i in 1:n){
     ut[i]=vis*uxx[i]-u[i]*ux[i];
   }
```

The calculation of ut could have also been done with vectorization, that is, ut=vis*uxx-u*ux, which would be computationally more efficient (by eliminating the subscripting).

- The counter for the calls to pde_1a is incremented and its value is returned to the calling program (of Listing 4.1) with the «- operator.

```
#
# Increment calls to pde_1a
  ncall << ncall+1;
```

- The derivative vector (LHS of Eq. (4.2)) is returned to lsodes, which requires a list. c is the R vector utility. The combination of return, list, c gives lsodes (the ODE integrator called in the main program of Listing 4.1) the required derivative vector for the next step along the solution.

```
#
# Return derivative vector
  return(list(c(ut)));
}
```

The final } concludes pde_1a.

4.4 Subordinate Routine

The analytical solution of Eqs. (4.3) is programmed in function phi.

```
  phi=function(x,t){
#
# Function phi computes the exact solution of
# Burgers' equation for comparison with the
# numerical solution.   It is also used to define
# the initial and boundary conditions for the
# numerical solution.
#
# Analytical solution
  a=(0.05/vis)*(x-0.5+4.95*t);
  b=(0.25/vis)*(x-0.5+0.75*t);
  c=( 0.5/vis)*(x-0.375);
  ea=exp(-a);
  eb=exp(-b);
  ec=exp(-c);
  ua=(0.1*ea+0.5*eb+ec)/(ea+eb+ec);
  return(c(ua));
}
```

Listing 4.3 Function phi for the analytical solution of Eqs. (4.3).

The coding in Listing 4.3 follows directly from Eqs. (4.3). The numerical value of u_a is returned as ua.

4.5 Model Output

Execution of the preceding main program and subordinate routines produces the following abbreviated numerical output in Table 4.1.

Table 4.1 Abbreviated output for Eqs. (4.2)–(4.4).

[1] 11

[1] 202

t	x	u(x,t)	ua(x,t)	diff
0.00	0.00	1.0000	1.0000	0.000000
0.00	0.05	1.0000	1.0000	0.000000
0.00	0.10	1.0000	1.0000	0.000000
0.00	0.15	0.9999	0.9999	0.000000
0.00	0.20	0.9924	0.9924	0.000000
0.00	0.25	0.7500	0.7500	0.000000
0.00	0.30	0.5076	0.5076	0.000000
0.00	0.35	0.5001	0.5001	0.000000
0.00	0.40	0.4995	0.4995	0.000000
0.00	0.45	0.4862	0.4862	0.000000
0.00	0.50	0.3000	0.3000	0.000000
0.00	0.55	0.1138	0.1138	0.000000
0.00	0.60	0.1005	0.1005	0.000000
0.00	0.65	0.1000	0.1000	0.000000
0.00	0.70	0.1000	0.1000	0.000000
0.00	0.75	0.1000	0.1000	0.000000
0.00	0.80	0.1000	0.1000	0.000000
0.00	0.85	0.1000	0.1000	0.000000
0.00	0.90	0.1000	0.1000	0.000000
0.00	0.95	0.1000	0.1000	0.000000
0.00	1.00	0.1000	0.1000	0.000000

Table 4.1 (Continued)

```
                 .                        .
                 .                        .
                 .                        .
        Output for t = 0.1 to 0.4 removed
                 .                        .
                 .                        .
                 .                        .
 0.50   0.00      1.0000      1.0000      -0.000000
 0.50   0.05      1.0000      1.0000       0.000000
 0.50   0.10      1.0000      1.0000       0.000000
 0.50   0.15      1.0000      1.0000       0.000000
 0.50   0.20      1.0000      1.0000       0.000000
 0.50   0.25      1.0000      1.0000       0.000000
 0.50   0.30      1.0000      1.0000       0.000000
 0.50   0.35      1.0000      1.0000       0.000000
 0.50   0.40      1.0000      1.0000      -0.000000
 0.50   0.45      1.0000      1.0000      -0.000000
 0.50   0.50      1.0000      1.0000      -0.000000
 0.50   0.55      0.9990      0.9990      -0.000001
 0.50   0.60      0.9413      0.9413       0.000005
 0.50   0.65      0.3410      0.3410       0.000002
 0.50   0.70      0.1138      0.1138       0.000001
 0.50   0.75      0.1005      0.1005      -0.000000
 0.50   0.80      0.1000      0.1000      -0.000000
 0.50   0.85      0.1000      0.1000      -0.000000
 0.50   0.90      0.1000      0.1000      -0.000000
 0.50   0.95      0.1000      0.1000      -0.000000
 0.50   1.00      0.1000      0.1000       0.000000
                 .                        .
                 .                        .
                 .                        .
        Output for t = 0.6 to 0.9 removed
                 .                        .
                 .                        .
```

(Continued)

Table 4.1 (Continued)

1.00	0.00	1.0000	1.0000	0.000000
1.00	0.05	1.0000	1.0000	0.000000
1.00	0.10	1.0000	1.0000	0.000000
1.00	0.15	1.0000	1.0000	0.000000
1.00	0.20	1.0000	1.0000	0.000000
1.00	0.25	1.0000	1.0000	0.000001
1.00	0.30	1.0000	1.0000	0.000001
1.00	0.35	1.0000	1.0000	0.000001
1.00	0.40	1.0000	1.0000	0.000000
1.00	0.45	1.0000	1.0000	0.000000
1.00	0.50	1.0000	1.0000	0.000000
1.00	0.55	1.0000	1.0000	0.000001
1.00	0.60	1.0000	1.0000	0.000001
1.00	0.65	1.0000	1.0000	0.000000
1.00	0.70	1.0000	1.0000	-0.000001
1.00	0.75	1.0000	1.0000	-0.000001
1.00	0.80	1.0000	1.0000	-0.000003
1.00	0.85	0.9999	0.9999	0.000015
1.00	0.90	0.8571	0.8569	0.000130
1.00	0.95	0.1026	0.1026	-0.000011
1.00	1.00	0.1000	0.1000	0.000002

ncall = 1198

The computational effort is modest; ncall = 1198. The agreement of the numerical and analytical solutions is satisfactory (note the last column). This agreement is confirmed by Figure 4.1.

4.6 Summary and Conclusions

Listing 4.2 indicates that the coding of Burgers' equation (4.2) is straightforward, and the numerical solution is acceptable, even with the front sharpening with increasing t. The is particularly noteworthy since the nonlinearity of Eq. (4.2), $u\dfrac{\partial u}{\partial x}$, which produces the front sharpening, is easily included in the SCMOL solution.

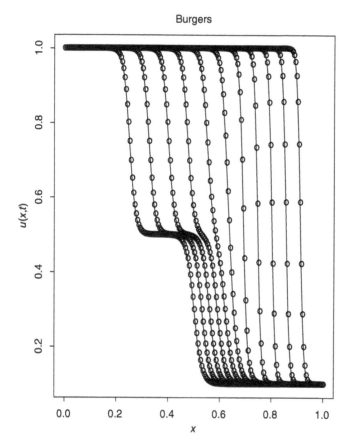

Figure 4.1 Numerical and analytical solutions of Eqs. (4.2)–(4.4) lines–num, points–anal.

Reference

1 Madsen, N.K., and R.F. Sincovec (1976), General software for partial differential equations, in *Numerical Methods for Differential Systems*, L. Lapidus and W.E. Schiesser (eds.), Academic Press, San Diego, CA, 1976.

5

Korteweg–de Vries Equation

The Korteweg–de Vries (KdV) equation is an established test problem for PDE numerical methods. It has the same general features as that of Burgers' equation of Chapter 4, that is, a nonlinear PDE with an analytical traveling wave solution that can be used to verify the numerical solution.

5.1 PDE Model

The KdV equation is another example from fluid dynamics (in addition to Burgers' equation in Chapter 4). In particular, the KdV equation demonstrates the properties of solitons. For this chapter, the following concepts are discussed:

1. A nonlinear PDE with an exact solution that can be used to assess the accuracy of a numerical method of lines (MOL) solution
2. Some of the basic properties of *solitons* that have broad applications in many areas of physics, engineering, and the biological sciences
3. Nonlinear interaction of two (or more) solitons
4. Computer analysis of a PDE over an essentially infinite spatial domain
5. Evaluation of invariants of the solution (integrals of functions of the solution that do not change with t).

The one-dimensional (1D) KdV equation is [1]

$$\frac{\partial u}{\partial t} + 6u\frac{\partial u}{\partial x} + \frac{\partial^3 u}{\partial x^3} = 0 \tag{5.1}$$

where originally $u(x, t)$ was interpreted as the height of a water wave in a channel [3].

Equation (5.1) is first order in t and third order in x, and therefore requires one initial condition (IC) and three boundary conditions (BCs). The IC is taken from the analytical solution (for a single soliton with velocity c)

$$u(x, t) = \frac{1}{2}c\,\mathrm{sech}^2\left\{\frac{1}{2}\sqrt{c}(x - ct)\right\} \tag{5.2}$$

Spline Collocation Methods for Partial Differential Equations: With Applications in R, First Edition.
William E. Schiesser.
© 2017 John Wiley & Sons, Inc. Published 2017 by John Wiley & Sons, Inc.
Companion website: www.wiley.com/go/Spline_Collocation

with $t = 0$,

$$u(x, t = 0) = \frac{1}{2}c \operatorname{sech}^2\left\{\frac{1}{2}\sqrt{c}(x)\right\} \tag{5.3}$$

To complete the specification of the problem, we would logically consider the three required BCs for Eq. (5.1). However, if the PDE is analyzed over an essentially infinite domain, $-\infty \leq x \leq \infty$, and if changes in the solution occur only over a finite interval in x, then BCs at infinity have no effect; in other words, we do not have to actually specify BCs (since they have no effect). This situation will be clarified through the PDE solution.

Consequently, Eqs. (5.1) and (5.3) constitute the complete PDE problem. The analytical solution to this problem, Eq. (5.2), is used in the subsequent programming and analysis to verify the numerical MOL solution. Note that Eq. (5.2) is a traveling wave solution since the argument of the sech function is the Lagrangian variable $x - ct$.

We now consider the numerical solution of the KdV equation (5.1) through a series of spline collocation method of lines (SCMOL) routines.

A main program for Eqs. (5.1)–(5.3) follows.

5.2 Main Program

The following main program is for two cases (ncase $= 1, 2$) corresponding to one and two solitons.

```
#
# Korteweg-de Vries
#
# Delete previous workspaces
  rm(list=ls(all=TRUE))
#
# Access ODE integrator
  library("deSolve");
#
# Access functions for numerical solution
  setwd("f:/collocation/chap5");
  source("pde_1a.R");source("inital_1.R");
  source("uxxx7c.R");source("simp.R")      ;
  source("ua.R")      ;source("dss004.R")   ;
#
# Select case; 1 - one soliton; 2 - two solitons;
  ncase=1;
  if(ncase==1){c1=1;n=201;}
  if(ncase==2){c1=2;c2=0.5;n=301;}
```

```
#
# Spatial grid
  xl=-30;xu=70;dx=(xu-xl)/(n-1);
  x=seq(from=xl,to=xu,by=dx);
#
# Independent variable for ODE integration
  t0=0;tf=30;nout=4;
  tout=seq(from=t0,to=tf,by=(tf-t0)/(nout-1));
  ncall=0;
#
# Initial condition
  u0=inital_1(x,t0);
#
# ODE integration
  if(ncase==1){
    out=lsodes(func=pde_1a,y=u0,times=tout,
               sparsetype="sparseint");}
  if(ncase==2){
    out=lsodes(func=pde_1a,y=u0,times=tout,
               sparsetype="sparseint",
               rtol=1e-10,atol=1e-10,maxord=5);}
  nrow(out)
  ncol(out)
#
# Output, ncase=1
  if(ncase==1){
#
# Store analytical solution, errors
# in numerical solution
    u=matrix(0,nrow=n,ncol=nout);
  u_a=matrix(0,nrow=n,ncol=nout);
  err=matrix(0,nrow=n,ncol=nout);
    for(it in 1:nout){
      for(i in 1:n){
          u[i,it]=out[it,i+1];
        u_a[i,it]=ua(x[i],tout[it]);
        err[i,it]=u[i,it]-u_a[i,it];
      }
    }
#
#   Display selected output
    cat(sprintf("\n       t         x              u(it,i)
                 u_a(it,i)        err(it,i)\n"));
```

```
      iv=seq(from=1,to=n,by=5);
      for(i in iv){
        cat(sprintf("%6.2f%8.3f%15.6f%15.6f%15.6f\n",
                    tout[it],x[i],u[i,it],u_a[i,it],
                    err[i,it]));
#
#       Calculate and display three invariants
        ui=u[,it];
        uint=simp(xl,xu,n,ui);
        cat(sprintf("\n Invariants at t = %5.2f",
                    tout[it]));
        cat(sprintf("\n    I1 = %10.4f    Mass
                    conservation",uint[1]));
        cat(sprintf("\n    I2 = %10.4f    Energy
                    conservation",uint[2]));
        cat(sprintf("\n    I3 = %10.4f    Whitham
                    invariant\n\n",uint[3]));
      }
      cat(sprintf("    ncall = %4d\n\n",ncall));
#
#     Plot numerical and analytical solutions
      matplot(x,u,type="l",lwd=2,col="black",lty=1,
              xlab="x",ylab="u(x,t)",
              main="Korteweg-deVries");
      matpoints(x,u_a,pch="o",col="black");
#
# End of ncase=1
  }
#
# Output, ncase=2
  if(ncase==2){
#
#   Store numerical solution
    u=matrix(0,nrow=n,ncol=nout);
    for(it in 1:nout){
      for(i in 1:n){
        u[i,it]=out[it,i+1];
      }
    }
#
#   Display selected output
    cat(sprintf("\n ncase = %2d    c1 = %5.2f    c2
                = %5.2f\n",ncase,c1,c2));
```

```
    for(it in 1:nout){
      cat(sprintf("        t           x       u(i,it)
                   \n"));
      iv=seq(from=1,to=n,by=5);
      for(i in iv){
        cat(sprintf("%6.2f%10.3f%12.6f\n",tout[it],
                     x[i],u[i,it]));
      }
      cat(sprintf("\n"));
#
#     Calculate and display three invariants
      ui=u[,it];
      uint=simp(xl,xu,n,ui);
      cat(sprintf("\n Invariants at t = %5.2f",
                   tout[it]));
      cat(sprintf("\n    I1 = %10.4f    Mass
                   conservation",uint[1]));
      cat(sprintf("\n    I2 = %10.4f    Energy
                   conservation",uint[2]));
      cat(sprintf("\n    I3 = %10.4f    Whitham
                   invariant\n\n",uint[3]));
    }
    cat(sprintf("  ncall = %4d\n\n",ncall));
#
#   Plot numerical solution
    par(mfrow=c(2,2))
    for(it in 1:nout){
      if(it==1){
        plot(x,u[,1],type="l",lwd=2,col="black",
             lty=1,xlab="x",ylab="u(x,t=0)",main=
             "u(x,t=0)");}
      if(it==2){
        plot(x,u[,2],type="l",lwd=2,col="black",
             lty=1,xlab="x",ylab="u(x,t=10)",main=
             "u(x,t=10)");}
      if(it==3){
        plot(x,u[,3],type="l",lwd=2,col="black",
             lty=1,xlab="x",ylab="u(x,t=20)",main=
             "u(x,t=20)");}
      if(it==4){
        plot(x,u[,4],type="l",lwd=2,col="black",
             lty=1,xlab="x",ylab="u(x,t=30)",main=
             "u(x,t=30)");}
```

```
       }
#
# End of ncase=2
   }
```

Listing 5.1 Main program for Eqs. (5.1)–(5.3).

We can note the following details about Listing 5.1:

- Previous workspaces are removed and the ODE integrator library `deSolve` is accessed. Note that the `setwd` (set working directory) uses / rather than the usual \. Also, `setwd` requires editing for the local computer.

```
#
# Korteweg-deVries
#
# Delete previous workspaces
  rm(list=ls(all=TRUE))
#
# Access ODE integrator
  library("deSolve");
#
# Access functions for numerical solution
  setwd("f:/collocation/chap5");
  source("pde_1a.R");source("inital_1.R");
  source("uxxx7c.R");source("simp.R")     ;
  source("ua.R")     ;source("dss004.R")   ;
```

pde_1a is the routine for the SCMOL ODEs (discussed subsequently). ua.R is the routine for the analytical solution of Eq. (5.2). The function of the other routines is explained when they are first discussed.

- Two cases are programmed for one and two solitons.

```
#
# Select case; 1 - one soliton; 2 - two solitons;
  ncase=1;
  if(ncase==1){c1=1;n=201;}
  if(ncase==2){c1=2;c2=0.5;n=301;}
```

In order to spatially resolve the two solitons (ncase=2), the number of spatial grid points is increased from 201 to 301.

- A uniform grid in x of 201 points for ncase=1 is defined with the seq utility for the interval $x_l = -30 \leq x \leq x_u = 70$ with the interval $dx = (70 - (-30))/(201 - 1) = 0.5$. Therefore, the vector x has the values $x = -30, -29.5, \cdots, 70$.

```
#
# Spatial grid
  xl=-30;xu=70;dx=(xu-xl)/(n-1);
  x=seq(from=xl,to=xu,by=dx);
```

For ncase=2, the vector x has the values $x = -30, -(30 - 1/3), \cdots, 70$.

- A uniform grid in t of 4 output points is defined with the seq utility for the interval $0 \leq t \leq t_f = 30$. Therefore, the vector tout has the values $t = 0, 10, \cdots, 30$.

```
#
# Independent variable for ODE integration
  t0=0;tf=30;nout=4;
  tout=seq(from=t0,to=tf,by=(tf-t0)/(nout-1));
  ncall=0;
```

At this point, the intervals of x and t in Eq. (5.1) are defined. Also, the counter for the calls to pde_1a is initialized (and passed to pde_1a without a special designation).

- The IC function in Eq. (5.3) is defined using the analytical solution in ua called by inital_1 for $t = 0$.

```
#
# Initial condition
  u0=inital_1(x,t0);
```

ua, inital_1 are discussed subsequently.

- The system of 201 or 301 MOL/ODEs is integrated by the library integrator lsodes (available in deSolve) with the sparse matrix option specified. As expected, the inputs to lsodes are the ODE function, pde_1a, the IC vector u0, and the vector of output values of t, tout. The length of u0 (201 or 301) informs lsodes how many ODEs are to be integrated. func,y,times are reserved names.

```
#
# ODE integration
  if(ncase==1){
    out=lsodes(func=pde_1a,y=u0,times=tout,
               sparsetype="sparseint");}
  if(ncase==2){
    out=lsodes(func=pde_1a,y=u0,times=tout,
               sparsetype="sparseint",
               rtol=1e-10,atol=1e-10,maxord=5);}
  nrow(out)
  ncol(out)
```

For ncase=2, the error tolerances were reduced to rtol= 1e-10,atol=1e-10 to improve the accuracy of the numerical solution[1] (the default used for ncase=1 is rtol=1e-6,atol=1e-6). The numerical solution to the ODEs is returned in matrix out. In this case, out has the dimensions $nout \times (n+1) = 11 \times 202{,}302$. The offset $n+1$ is required since the first element of each column has the output t (also in tout), and the $2, \cdots, n+1 = 2, \cdots, 202{,}302$ column elements have the 201,301 ODE solutions. This indexing of out is used next.

- For ncase=1, the ODE solution is placed in a 201×11 matrix, u, for subsequent plotting (by stepping through the solution with respect to x and t within a pair of fors). The analytical solution of Eqs. (5.2) is placed in u_a. The difference between the numerical and analytical solutions is placed in err.

```
#
# Output, ncase=1
  if(ncase==1){
#
# Store analytical solution, errors
# in numerical solution
    u=matrix(0,nrow=n,ncol=nout);
  u_a=matrix(0,nrow=n,ncol=nout);
  err=matrix(0,nrow=n,ncol=nout);
    for(it in 1:nout){
      for(i in 1:n){
          u[i,it]=out[it,i+1];
        u_a[i,it]=ua(x[i],tout[it]);
        err[i,it]=u[i,it]-u_a[i,it];
      }
    }
```

- The numerical and analytical solutions, and the difference, are then displayed numerically and plotted with the matplot, matpoints utilities.

```
cat(sprintf("\n     t          x           u(it,i)
            u_a(it,i)        err(it,i)\n"));
iv=seq(from=1,to=n,by=5);
for(i in iv){
  cat(sprintf("%6.2f%8.3f%15.6f%15.6f%15.6f\n",
              tout[it],x[i],u[i,it],u_a[i,it],
              err[i,it]));
          }
```

---

1 The adjustment of the error tolerances for the integration in t was based on the apparent resolution of the two solitons as they interacted. An analytical solution was not used for ncase=2.

To conserve space, only every fifth value in x of the solutions is displayed numerically (using the subscript iv). Thus, the displayed spacing in x is $5(0.5) = 2.5$.

- Three invariants for Eq. (5.1) are calculated by simp that implements Simpson's rule for numerical quadrature (integration). These invariants are subsequently discussed along with simp.

```
#
#      Calculate and display three invariants
       ui=u[,it];
       uint=simp(xl,xu,n,ui);
       cat(sprintf("\n Invariants at t = %5.2f",
                   tout[it]));
       cat(sprintf("\n     I1 = %10.4f    Mass
                   conservation",uint[1]));
       cat(sprintf("\n     I2 = %10.4f    Energy
                   conservation",uint[2]));
       cat(sprintf("\n     I3 = %10.4f    Whitham
                   invariant\n\n",uint[3]));
       }
       cat(sprintf("    ncall = %4d\n\n",ncall));
       }
```

At the end of the solution (for ncase=1), the number of calls to pde_1a is displayed as an indication of the computational effort required to compute the solution. The concluding } ends the for in it.

- The numerical solution in array u is plotted as a solid line, and the analytical solution in u_a is plotted as points against x. This formatting appears in Figure 5.1 (discussed subsequently). matplot plots u against x with t as a parameter since the row dimension of u (201) matches the length of x (also 201). The same is true in plotting u_a with matpoints.

```
#
#    Plot numerical and analytical solutions
     matplot(x,u,type="l",lwd=2,col="black",lty=1,
             xlab="x",ylab="u(x,t)",
             main="Korteweg-deVries");
     matpoints(x,u_a,pch="o",col="black");
#
# End of ncase=1
   }
```

This concludes the solution and output for ncase=1. The numerical and graphical output is displayed in Table 5.1 and Figure 5.1, discussed subsequently.

- For ncase=2, with $n = 301$ spatial points, the main program of Listing 5.1 computes and saves only the numerical solution (an analytical solution is available for the two-soliton case, but it is not used here).

```
#
# Output, ncase=2
  if(ncase==2){
#
```

Table 5.1 Abbreviated output for Eqs. (5.1)–(5.3), ncase=1.

```
[1]  4

[1]  202

ncase =   1   c1 =   1.00
```

t	x	u(it,i)	u_a(it,i)	err(it,i)
0.00	-30.000	0.000000	0.000000	0.000000
0.00	-27.500	0.000000	0.000000	0.000000
0.00	-25.000	0.000000	0.000000	0.000000
0.00	-22.500	0.000000	0.000000	0.000000
0.00	-20.000	0.000000	0.000000	0.000000
0.00	-17.500	0.000000	0.000000	0.000000
0.00	-15.000	0.000001	0.000001	0.000000
0.00	-12.500	0.000007	0.000007	0.000000
0.00	-10.000	0.000091	0.000091	0.000000
0.00	-7.500	0.001105	0.001105	0.000000
0.00	-5.000	0.013296	0.013296	0.000000
0.00	-2.500	0.140207	0.140207	0.000000
0.00	0.000	0.500000	0.500000	0.000000
0.00	2.500	0.140207	0.140207	0.000000
0.00	5.000	0.013296	0.013296	0.000000
0.00	7.500	0.001105	0.001105	0.000000
0.00	10.000	0.000091	0.000091	0.000000
0.00	12.500	0.000007	0.000007	0.000000
0.00	15.000	0.000001	0.000001	0.000000
0.00	17.500	0.000000	0.000000	0.000000
0.00	20.000	0.000000	0.000000	0.000000
.		.		
.		.		
.		.		

Table 5.1 (Continued)

```
          Output for x = 22.5 to 57.5 removed
                          .                  .
                          .                  .
                          .                  .
   0.00     60.000    0.000000      0.000000       0.000000
   0.00     62.500    0.000000      0.000000       0.000000
   0.00     65.000    0.000000      0.000000       0.000000
   0.00     67.500    0.000000      0.000000       0.000000
   0.00     70.000    0.000000      0.000000       0.000000

   Invariants at t =  0.00
      I1 =    2.0000    Mass conservation
      I2 =    0.3333    Energy conservation
      I3 =    0.4005    Whitham invariant
             .                  .
             .                  .
             .                  .
   Numerical, analytical solutions
     removed for for t = 10, 20
             .                  .
             .                  .
             .                  .
   Invariants at t = 10.00
      I1 =    2.0002    Mass conservation
      I2 =    0.3333    Energy conservation
      I3 =    0.4004    Whitham invariant

   Invariants at t = 20.00
      I1 =    2.0000    Mass conservation
      I2 =    0.3334    Energy conservation
      I3 =    0.4000    Whitham invariant

    t        x         u(it,i)      u_a(it,i)      err(it,i)
   30.00   -30.000    0.000000      0.000000       0.000000
   30.00   -27.500   -0.001854      0.000000      -0.001854
   30.00   -25.000    0.001335      0.000000       0.001335
   30.00   -22.500    0.001909      0.000000       0.001909
   30.00   -20.000   -0.000966      0.000000      -0.000966
```

(Continued)

Table 5.1 (Continued)

```
                  .              .
                  .              .
                  .              .
          Output for x = -17.5 to 17.5 removed
                  .              .
                  .              .
                  .              .
   30.00   20.000   -0.002413    0.000091   -0.002504
   30.00   22.500    0.000348    0.001105   -0.000757
   30.00   25.000    0.012597    0.013296   -0.000699
   30.00   27.500    0.133569    0.140207   -0.006638
   30.00   30.000    0.501964    0.500000    0.001964
   30.00   32.500    0.149354    0.140207    0.009146
   30.00   35.000    0.015721    0.013296    0.002425
   30.00   37.500    0.000833    0.001105   -0.000272
   30.00   40.000    0.000845    0.000091    0.000754
                  .              .
                  .              .
                  .              .
          Output for x = 42.5 to 57.5 removed
                  .              .
                  .              .
                  .              .
   30.00   60.000   -0.000359    0.000000   -0.000359
   30.00   62.500   -0.004077    0.000000   -0.004077
   30.00   65.000    0.002806    0.000000    0.002806
   30.00   67.500   -0.001477    0.000000   -0.001477
   30.00   70.000    0.000000    0.000000   -0.000000

   Invariants at t = 30.00
      I1 =    1.9990   Mass conservation
      I2 =    0.3335   Energy conservation
      I3 =    0.3982   Whitham invariant

      ncall = 6666
```

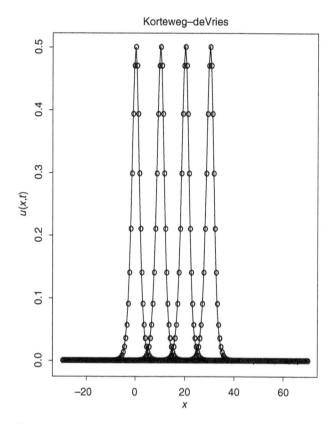

Figure 5.1 Numerical and analytical solutions of Eqs. (5.1)–(5.3) ncase=1, lines – num, points – anal.

```
#   Store numerical solution
    u=matrix(0,nrow=n,ncol=nout);
    for(it in 1:nout){
      for(i in 1:n){
        u[i,it]=out[it,i+1];
      }
    }
```

- Every fifth value of the numerical solution is displayed as a function of x and t using two `for`s.

```
#
#   Display selected output
    cat(sprintf("\n ncase = %2d    c1 = %5.2f    c2
                = %5.2f\n",ncase,c1,c2));
```

```
for(it in 1:nout){
  cat(sprintf("       t           x       u(i,it)
              \n"));
  iv=seq(from=1,to=n,by=5);
  for(i in iv){
    cat(sprintf("%6.2f%10.3f%12.6f\n",tout[it],
                x[i],u[i,it]));
  }
  cat(sprintf("\n"));
```

- The three integral invariants of Eqs. (5.4a–5.4c) are computed with simp.

```
#
#       Calculate and display three invariants
  ui=u[,it];
  uint=simp(xl,xu,n,ui);
  cat(sprintf("\n Invariants at t = %5.2f",
              tout[it]));
  cat(sprintf("\n      I1 = %10.4f    Mass
              conservation",uint[1]));
  cat(sprintf("\n      I2 = %10.4f    Energy
              conservation",uint[2]));
  cat(sprintf("\n      I3 = %10.4f    Whitham
              invariant\n\n",uint[3]));
  }
  cat(sprintf("  ncall = %4d\n\n",ncall));
```

The number of calls to pde_1a is displayed at the end of the solution.
- The solution is plotted as a 2×2 matrix (using par(mfrow=c(2,2)) to demonstrate the movement of the two solitons through $t = 0, 10, 20, 30$ (it=1,2,3,4).

```
#
#       Plot numerical solution
  par(mfrow=c(2,2))
  for(it in 1:nout){
    if(it==1){
      plot(x,u[,1],type="l",lwd=2,col="black",
           lty=1,xlab="x",ylab="u(x,t=0)",main=
           "u(x,t=0)");}
    if(it==2){
      plot(x,u[,2],type="l",lwd=2,col="black",
           lty=1,xlab="x",ylab="u(x,t=10)",main=
           "u(x,t=10)");}
    if(it==3){
      plot(x,u[,3],type="l",lwd=2,col="black",
           lty=1,xlab="x",ylab="u(x,t=20)",main=
```

```
               "u(x,t=20)");}
         if(it==4){
            plot(x,u[,4],type="1",lwd=2,col="black",
                  lty=1,xlab="x",ylab="u(x,t=30)",main=
                  "u(x,t=30)");}
      }
#
# End of ncase=2
   }
```

For each value of t, the corresponding plot has a label indicating the value of t (to clarify the composite plot of Figure 5.2). The end for ncase=2 concludes the main program.

pde_1a called by the main program of Listing 5.1 is listed next.

5.3 ODE Routine

pde_1a for Eq. (5.1) follows.

```
   pde_1a=function(t,u,parms){
#
# Function pde_1a computes the t derivative vector
# of u(x,t)
#
# ux
   table1=splinefun(x,u);
   ux=table1(x,deriv=1);
#
# uxxx
   uxxx=uxxx7c(xl,xu,n,u);
#
# PDE
   ut=rep(0,n);
   for(i in 1:n){
      ut[i]=-uxxx[i]-6*u[i]*ux[i];
   }
#
# Increment calls to pde_1a
   ncall <<- ncall+1;
#
```

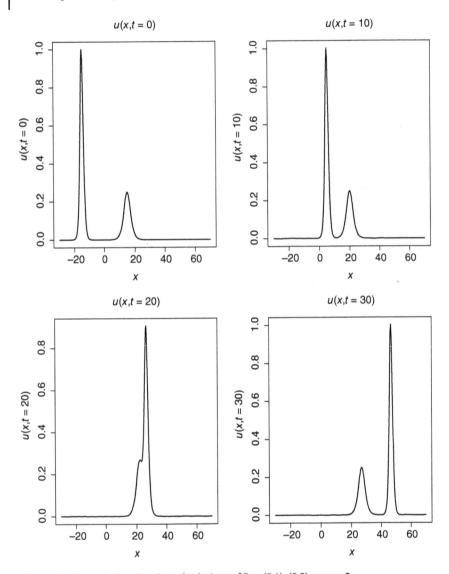

Figure 5.2 Numerical and analytical solutions of Eqs. (5.1)–(5.3) ncase=2.

```
# Return derivative vector
  return(list(c(ut)));
}
```

Listing 5.2 ODE/MOL routine pde_1a for Eq. (5.1).

We can note the following details about pde_1a.

- The function is defined.

```
pde_1a=function(t,u,parms){
#
# Function pde_1a computes the t derivative vector
# of u(x,t)
```

t is the current value of t in Eq. (5.1). u is the 201-vector for ncase=1 or 301-vector for ncase=2 of ODE/MOL dependent variables. parm is an argument to pass parameters to pde_1a (unused, but required in the argument list). The arguments must be listed in the order stated to properly interface with lsodes called in the main program of Listing 5.1. The derivative vector of the LHS of Eq. (5.1) is calculated next and returned to lsodes.

- $\dfrac{\partial u}{\partial x}$ in Eq. (5.1) is computed with splinefun.

```
#
# ux
  table1=splinefun(x,u);
  ux=table1(x,deriv=1);
```

- $\dfrac{\partial^3 u}{\partial x^3}$ in Eq. (5.1) is calculated with uxxx7c, a function especially constructed for this purpose. uxxx7c is based on a seven-point centered finite difference (FD) [2] so that the overall numerical algorithm is a combination of SC and FD^2.

```
#
# uxxx
  uxxx=uxxx7c(xl,xu,n,u);
```

---

2 The third derivative could not be computed accurately with splinefun, which is based on cubic (third-order) splines (polynomials). This order is too low to give accurate third-order numerical derivatives (using deriv=3). In other words, the third derivative from the splines in splinefun is piecewise constant over each interval in x, and this failure is not unexpected since the solution has very sharp variations in x near the traveling wave peaks. Stagewise differentiation did not work either since after the equivalent of three differentiations, the accuracy of the computed third derivative is apparently too low (the solutions in the examples in Chapter 2 were sufficiently smooth that stagewise differentiation up to fourth derivatives produced accurate solutions). The only possibility is to use higher order splines. Although this would be straightforward, we do not consider it here. But an alternative to splinefun based on higher order splines would be a good (useful) development. To test the idea that a higher order approximation is required for the third derivative, the seven-point FD of function uxxx7c was implemented, and it did produce accurate numerical solutions.

- Equation (5.1) is programmed.

```
#
# PDE
  ut=rep(0,n);
  for(i in 1:n){
    ut[i]=-uxxx[i]-6*u[i]*ux[i];
  }
```

The calculation of ut could have also been done with vectorization, that is, ut=-uxxx-6*u*ux, which would be computationally more efficient (by eliminating the subscripting). However, using the subscripting demonstrates the precise details of the calculations. This can be important for more complex PDE systems when vectorization, if it can be used, may not provide a clear picture of the details of the calculations.

- The counter for the calls to pde_1a is incremented and its value is returned to the calling program (of Listing 5.1) with the ≪- operator.

```
#
# Increment calls to pde_1a
  ncall <<- ncall+1;
```

- The derivative vector (LHS of Eq. (5.1)) is returned to lsodes, which requires a list. c is the R vector utility. The combination of return, list, c gives lsodes (the ODE integrator called in the main program of Listing 5.1) the required derivative vector for the next step along with the solution.

```
#
# Return derivative vector
  return(list(c(ut)));
}
```

The final } concludes pde_1a.

5.4 Subordinate Routines

Subordinate routines ua, inital_1, uxxx7c and simp are considered next. The analytical solution of Eq. (5.2) is programmed in function ua.

```
ua=function(x,t){
#
# Function ua computes the exact solution of the KdV
# equation for comparison with the numerical solution.
#
```

```
# Analytical solution
  expm=exp(-1/2*sqrt(c1)*(x-c1*t));
  expp=exp( 1/2*sqrt(c1)*(x-c1*t));
  ua=(1/2)*c1*(2/(expp+expm))^2;
#
# Return analytical solution
  return(c(ua));
}
```

Listing 5.3 Function ua for the analytical solution of Eq. (5.2).

The coding in Listing 5.3 follows directly from Eq. (5.2). The numerical value of u_a is returned as ua.

inital_1 implements IC (5.3).

```
  inital_1=function(x,t0){
#
# Function inital_1 sets the initial condition for the
# KdV equation
#
  u0=rep(0,n);
#
# Case 1 - single pulse
  if(ncase==1){
    for(i in 1:n){
      u0[i]=ua(x[i],0);
    }
  }
#
# Case 2 - two pulses
  if(ncase==2){
    for(i in 1:n){
      expm=exp(-1/2*sqrt(c1)*(x[i]+15));
      expp=exp( 1/2*sqrt(c1)*(x[i]+15));
      pulse1=(1/2)*c1*(2/(expp+expm))^2;
      expm=exp(-1/2*sqrt(c2)*(x[i]-15));
      expp=exp( 1/2*sqrt(c2)*(x[i]-15));
      pulse2=(1/2)*c2*(2/(expp+expm))^2;
      u0[i]=pulse1+pulse2;
    }
  }
#
```

```
# Return IC
  return(c(u0));
  }
```

Listing 5.4 Function inital_1 for the IC of Eq. (5.3) (ncase=1) or the two pulse IC (ncase=2).

For ncase=1, ua of Listing 5.3 is called (with $t = 0$). For ncase=2, two pulses are computed, then added to give the IC at grid point i, u0[i]=pulse1+pulse2. The first pulse is centered at x=15 and the second is centered at x=-15. Note the two different pulse velocities, c1, c2 (defined numerically in the main program of Listing 5.1).

uxxx7c for the third derivative in Eq. (5.1) follows.

```
  uxxx7c=function(xl,xu,n,u) {
#
# Function uxxx7c computes the derivative uxxx
# in the KdV equation
#
# Coefficient
  r8dx3=1/(8*(dx^3));
#
# uxxx
  uxxx=rep(0,n);
  for(i in 1:n) {
#
#     At the left end, uxxx = 0
     if(i<4){uxxx[i]=0;}
#
#     At the right end, uxxx = 0
     if(i>(n-3)){uxxx[i]=0;}
#
#     Interior points
     if(((i>=4)&(i<=(n-3)))){
     uxxx[i]=r8dx3*
         (   1*u[i-3]-
             8*u[i-2]+
            13*u[i-1]+
             0*u[i  ]-
            13*u[i+1]+
             8*u[i+2]-
             1*u[i+3]);
      }
   }
```

```
#
# Return derivative
   return(c(uxxx));
   }
```

Listing 5.5 Function uxxx7c for $\dfrac{\partial^3 u}{\partial x^3}$ of Eq. (5.1).

The third derivative uxxx[i] at grid point i is a weighted sum of seven values of u at i-3, i-2, i-1, i, i+1, i+2, i+3. The corresponding weighting coefficients [2] are 1, -8, 13, 0, -13, 8, -1. Two points about these weighting coefficients can be noted: (i) they are antisymmetric around the center point at i since the computed derivative is of odd (third) order and (ii) they sum to zero, which is a basic requirement to differentiate a constant to zero. Also, the third derivative is set to zero at i=1, 2, 3, n-2, n-1, n (rather than using noncentered FDs at the boundaries x=x1, xu, which are also available from [2]). These zero boundary values are reflected in the numerical solutions that follow.

Equation (5.1) has three integral invariants.

1) Conservation of mass:

$$u_1(t) = \int_{-\infty}^{\infty} u(x, t)dx \qquad (5.4a)$$

2) Conservation of energy:

$$u_2(t) = \int_{-\infty}^{\infty} \frac{1}{2}u^2(x, t)dx \qquad (5.4b)$$

3) Proposed by Whitham [4]:

$$u_3(t) = \int_{-\infty}^{\infty} 2u^3(x, t) - u_x^2(x, t)dx \qquad (5.4c)$$

The exact values of these invariants for Eq. (5.2) are $u_1(t) = 2, u_2(t) = 1/3, u_3(t) = 0.4$, which are subsequently compared with the values computed by numerical integration from the routine simp. Note that $u_3(t)$ requires the derivative $\dfrac{\partial u}{\partial x} = u_x$, which is computed numerically in simp by the library routine dss004.

The three integrals in Eqs. (5.4) are evaluated numerically with Simpson's rule.

$$\int_{-\infty}^{\infty} u(x, t)dx \approx \frac{h}{3}\left[u_1 + \sum_{i=2}^{n-2}(4u_i + 2u_{i+1}) + u(n)\right] \qquad (5.4d)$$

Function simp that implements Eq. (5.4d) follows.

```
  simp=function(xl,xu,n,u){
#
# Function simp computes three integral invariants
# by Simpson's rule
#
  uint=rep(0,3);
  h=(xu-xl)/(n-1);
#
  for(int in 1:3){
#
#    Conservation of mass
     if(int==1){
        uint[1]=u[1]-u[n];
        iv=seq(from=3,to=n,by=2);
        for(i in iv){
          uint[1]=uint[1]+4*u[i-1]+2*u[i];
        }
        uint[1]=h/3*uint[1];
     }
#
#    Conservation of energy
     if(int==2){
        uint[2]=u[1]^2-u[n]^2;
        iv=seq(from=3,to=n,by=2);
        for(i in iv){
          uint[2]=uint[2]+4*u[i-1]^2+2*u[i]^2;
        }
        uint[2]=(1/2)*h/3*uint[2];
     }
#
#    Whitham conservation
     if(int==3){
        ux=dss004(xl,xu,n,u);
        uint[3]=2*u[1]^3-ux[1]^2-(2*u[n]^3-ux[n]^2);
        iv=seq(from=3,to=n,by=2);
        for(i in iv){
          uint[3]=uint[3]+4*(2*u[i-1]^3-ux[i-1]^2)+
                          2*(2*u[i]^3  -ux[i]^2);
        }
        uint[3]=h/3*uint[3];
     }
  }
}
```

```
#
# Return vector of integrals
  return(c(uint));
  }
```

Listing 5.6 Function simp for the integral constraints of Eqs. (5.4a–5.4c) of Eq. (5.1).

We can note the following details of simp:

- The function is defined.

```
  simp=function(xl,xu,n,u){
#
# Function simp computes three integral invariants
# by Simpson's rule
```

- The integration interval h in Eq. (5.4d) is computed. For the present application, xl=-30, xu=70, n=201, 301 for ncase=1, 2 set in the main program of Listing 5.1.

```
#
  uint=rep(0,3);
  h=(xu-xl)/(n-1);
```

- The invariants of Eqs. (5.4a–5.4c) are applied with a for. Equation (5.4a) is programmed for int=1.

```
#
  for(int in 1:3){
#
#    Conservation of mass
     if(int==1){
        uint[1]=u[1]-u[n];
        iv=seq(from=3,to=n,by=2);
        for(i in iv){
           uint[1]=uint[1]+4*u[i-1]+2*u[i];
        }
        uint[1]=h/3*uint[1];
     }
```

The sum in Eq. (5.4d) includes two terms through the use of seq. The final result is the integral by including the factor h/3.
- Equation (5.4b) is programmed for int=2.

```
#
#    Conservation of energy
     if(int==2){
```

```
           uint[2]=u[1]^2-u[n]^2;
           iv=seq(from=3,to=n,by=2);
           for(i in iv){
              uint[2]=uint[2]+4*u[i-1]^2+2*u[i]^2;
           }
           uint[2]=(1/2)*h/3*uint[2];
       }
```

- Equation (5.4c) is programmed for int=3.

```
   #
   #      Whitham conservation
       if(int==3){
           ux=dss004(xl,xu,n,u);
           uint[3]=2*u[1]^3-ux[1]^2-(2*u[n]^3-ux[n]^2);
           iv=seq(from=3,to=n,by=2);
           for(i in iv){
              uint[3]=uint[3]+4*(2*u[i-1]^3-ux[i-1]^2)+
                              2*(2*u[i]^3  -ux[i]^2);
           }
           uint[3]=h/3*uint[3];
       }
   }
```

Note the use of ux computed by the library routine dss004 (differentiation in space subroutine). The final } concludes simp.

This completes the discussion of the R routines for Eqs. (5.1)–(5.4). The numerical and graphical output follows.

5.5 Model Output

Execution of the preceding main program and subordinate routines produces the following abbreviated numerical output in Table 5.1 for ncase = 1 set in the main program of Listing 5.1.

As expected, the solution array out has the dimensions 4 × 202 for 4 output points and 201 ODEs, with t included as an additional element, that is, 201 + 1 = 202. The computational effort is substantial, ncall = 6666. The agreement of the numerical and analytical solutions is satisfactory (note the last column). This agreement is confirmed by Figure 5.1.

Also, the integral invariants of Eqs. (5.4a–5.4c) are accurate, that is, $u_1(t) = 2, u_2(t) = 1/3, u_3(t) = 0.4$, even at $t = 30$. Note also the accuracy of the peak value u(x=30,t=30) = 0.501964 (the exact value is 0.500000 from Eq. (5.2)). The nearby value of maximum error (0.009146 for u(x=32.5,t=30)) does not appear to be substantial in Figure 5.1 because the solution is changing rapidly (the slope is large).

Abbreviated output for ncase=2 is in Table 5.2.

Table 5.2 Abbreviated output for Eqs. (5.1)–(5.3), ncase=2.

```
[1]  4

[1]  302

ncase =   2    c1 =   2.00    c2 =   0.50
     t          x        u(i,it)
   0.00    -30.000     0.000000
   0.00    -28.333     0.000000
   0.00    -26.667     0.000000
   0.00    -25.000     0.000003
   0.00    -23.333     0.000030
   0.00    -21.667     0.000322
   0.00    -20.000     0.003392
   0.00    -18.333     0.035239
   0.00    -16.667     0.316101
   0.00    -15.000     1.000000
   0.00    -13.333     0.316101
   0.00    -11.667     0.035239
   0.00    -10.000     0.003392
     .          .          .
     .          .          .
     .          .          .
Output for x=-8.333 to 8.333
            removed
     .          .          .
     .          .          .
     .          .          .
   0.00     10.000     0.027516
   0.00     11.667     0.079025
   0.00     13.333     0.179944
   0.00     15.000     0.250000
   0.00     16.667     0.179944
   0.00     18.333     0.079025
   0.00     20.000     0.027516
```

(Continued)

Table 5.2 (Continued)

```
            .                .
            .                .
            .                .
Output for x=21.667 to 63.333
                removed
            .                .
            .                .
            .                .
   0.00       65.000      0.000000
   0.00       66.667      0.000000
   0.00       68.333      0.000000
   0.00       70.000      0.000000

Invariants at t =   0.00
    I1 =       4.2426    Mass conservation
    I2 =       1.0607    Energy conservation
    I3 =       2.3358    Whitham invariant
            .                .
            .                .
            .                .
  Solution output for t = 10, 20
                removed
            .                .
            .                .
            .                .
Invariants at t = 10.00
    I1 =       4.2412    Mass conservation
    I2 =       1.0606    Energy conservation
    I3 =       2.3356    Whitham invariant

Invariants at t = 20.00
    I1 =       4.2409    Mass conservation
    I2 =       1.0607    Energy conservation
    I3 =       2.3354    Whitham invariant

      t          x         u(i,it)
  30.00     -30.000      0.000000
  30.00     -28.333     -0.000126
  30.00     -26.667     -0.000028
  30.00     -25.000      0.000090
```

Table 5.2 (Continued)

```
                    .                    .
                    .                    .
                    .                    .
        Output for x=-23.333 to 23.333
                  removed

                    .                    .
                    .                    .
                    .                    .
        30.00      25.000       0.163470
        30.00      26.667       0.247962
        30.00      28.333       0.196648
        30.00      30.000       0.090326
        30.00      31.667       0.032586
        30.00      33.333       0.010267
        30.00      35.000       0.003322
        30.00      36.667       0.001581
        30.00      38.333       0.000063
        30.00      40.000       0.000210
        30.00      41.667       0.003668
        30.00      43.333       0.037950
        30.00      45.000       0.333641
        30.00      46.667       1.002108
        30.00      48.333       0.296536
        30.00      50.000       0.032781
                    .                    .
                    .                    .
                    .                    .
        Output for x=51.667 to 63.333
                  removed

                    .                    .
                    .                    .
                    .                    .
        30.00      65.000      -0.000165
        30.00      66.667      -0.000268
        30.00      68.333      -0.000189
        30.00      70.000       0.000000
```

(Continued)

Table 5.2 (Continued)

```
Invariants at t = 30.00
    I1 =      4.2436    Mass conservation
    I2 =      1.0607    Energy conservation
    I3 =      2.3356    Whitham invariant

ncall = 26310
```

As expected, the solution array out has the dimensions 4 × 302 for 4 output points and 301 ODEs, with t included as an additional element, that is, $301 + 1 = 302$. The computational effort is very substantial, ncall = 26310.

Also, the integral invariants of Eqs. (5.4a–5.4c) are nearly constant through $t = 0, 10, 20, 30$. Note also the peak values starting at $t = 0$

```
0.00    -15.000    1.000000

0.00     15.000    0.250000

30.00    26.667    0.247962

30.00    46.667    1.002108
```

are nearly retained at $t = 30$ (even after the peaks have merged, then reemerged, as indicated in Figure 5.2). There is a small phase shift (shift in x), but the shapes of the peaks are retained, one of the major discoveries about solitons[3].

5.6 Summary and Conclusions

Listing 5.2 indicates that the coding of KdV equation (5.1) is straightforward, and the numerical solution is acceptable. This is particularly noteworthy since the nonlinearity of Eq. (5.1), $u\dfrac{\partial u}{\partial x}$, and the higher order derivative, $\dfrac{\partial^3 u}{\partial x^3}$, are easily included in the SCMOL solution.

---

3 As Eq. (5.2) indicates, the speed of the soliton (moving left to right) is determined by the traveling wave (Lagrange) variable, (x-ct). The velocity c is a constant set in the main program of Listing 5.1. For ncase=2, the two solitons have velocities if (ncase==2){c1=2; c2=0.5;n=301;}. Thus, the pulse initially to the left has the higher velocity and it therefore overtakes the pulse to the right. After the two pulses merge and reemerge, as indicated in Figure 5.2, they continue on in t with their original shapes. A small phase shift (in x) occurs as a result of the collision.

This suggests an extension of the preceding analysis. The two RHS nonlinear terms and the LHS derivative $\dfrac{\partial u}{\partial t}$ of Eq. (5.1) can be computed and plotted against x and parametrically in t to observe the details of how the numerical solution $u(x, t)$ evolves from the IC. This procedure of examining the individual terms in the PDE can provide insight into the solution beyond just considering the solution, and is easily accomplished. In particular, the contribution of the RHS terms is clarified, which possibly indicates that some terms contribute relatively little to the solution.

References

1 Debnath, L. (1997), *Nonlinear Partial Differential Equations for Scientists and Engineers*, Chapter 9, Birkhauser, Boston, MA.
2 Fornberg, B. (1998), Calculation of weights in finite difference formulas, *SIAM Review*, **40**, no. 3, 685–691.
3 Korteweg, D.J., and F. de Vries (1895), On the change of form of long waves advancing in a rectangular canal, and on a new type of long stationary waves, *Philos. Mag.*, **39**, 422–443.
4 Strang, G. (1986), *Introduction to Applied Mathematics*, Wellesley-Cambridge Press, Wellesley, MA, pp. 599–602.

6

Maxwell Equations

The Maxwell equations (ME) for electromagnetic field theory are first stated in coordinate-free form, and then specialized to 1D in Cartesian coordinates to give a damped wave equation (DWE) for spline collocation method of lines (SCMOL) analysis.

6.1 PDE Model

The ME are

$$\nabla \times \mathbf{H} = \mathbf{J} + \mathbf{J_d} = \mathbf{J} + \frac{\partial \mathbf{D}}{\partial t} \tag{6.1a}$$

$$\nabla \times \mathbf{E} = -\frac{\partial \mathbf{B}}{\partial t} \tag{6.1b}$$

$$\nabla \bullet \mathbf{D} = \rho \tag{6.1c}$$

$$\nabla \bullet \mathbf{B} = 0 \tag{6.1d}$$

$$\frac{\partial \rho}{\partial t} + \nabla \bullet \mathbf{J} = 0 \tag{6.1e}$$

where boldface indicates a vector (except for the vector dot product • which we use for clarity in place of the less apparent dot ·). The variables, parameters, and operators in Eqs. (6.1), with a set of representative units, are as follows:

$\mathbf{H}$	magnetic field intensity ($A\,m^{-1}$)
$\mathbf{E}$	electric field intensity ($V\,m^{-1}$)
$\mathbf{D}$	electric flux density ($C\,m^{-2}$)
$\mathbf{B}$	magnetic flux density ($Wb\,m^{-2}$)
$\mathbf{J}$	electric current density ($A\,m^{-2}$)
$\mathbf{J_d}$	displacement current density ($A\,m^{-2}$)
ρ	charge density ($C\,m^{-3}$)
t	time (s)

Spline Collocation Methods for Partial Differential Equations: With Applications in R, First Edition.
William E. Schiesser.
© 2017 John Wiley & Sons, Inc. Published 2017 by John Wiley & Sons, Inc.
Companion website: www.wiley.com/go/Spline_Collocation

×	curl vector operator
•	vector dot product
∇	del vector operator $(1\ m^{-1})$

PDE system (6.1) clearly has more dependent variables than do equations, so that constitutive equations are used to provide the required additional relationships between the dependent variables.

$$D = \epsilon E; \quad B = \mu H; \quad J = \sigma E \qquad (6.1f,g,h)$$

where

ϵ	capacitivity or permittivity $(F\ m^{-1})$
μ	inductivity or permeability $(H\ m^{-1})$
σ	conductivity $(mhos\ m^{-1})$

Equations (6.1) are a complete set of PDEs (number of dependent variables = number of equations). They can be specialized to 1D in Cartesian coordinates to give a PDE system in $E(x, t)$ [1] (boldface is now dropped).

$$\frac{\partial^2 E}{\partial t^2} + \frac{\sigma}{\epsilon}\frac{\partial E}{\partial t} = \frac{1}{\mu\epsilon}\frac{\partial^2 E}{\partial x^2} \qquad (6.2a)$$

Equation (6.2a) is a linear (constant coefficient), DWE. It is second order in t and x and therefore requires two initial conditions (ICs) and two boundary conditions (BCs)[1]

$$E(x, t = 0) = \cos(\pi x); \quad \frac{\partial E(x, t = 0)}{\partial t} = Re\lambda\cos(\pi x) \qquad (6.2b,c)$$

($Re\lambda$ is defined in Eq. (6.3b)).

$$\frac{\partial E(x = 0, t)}{\partial x} = 0; \quad \frac{\partial E(x = 1, t)}{\partial x} = 0 \qquad (6.2d,e)$$

Equations (6.2d,e) are homogeneous Neumann BCs.

The analytical solution to Eq. (6.2a) is[2]

$$E(x, t) = e^{Re\lambda t}\cos(Im\lambda t)\cos(\pi x) \qquad (6.3a)$$

$$Re\lambda = -\frac{1}{2}\frac{\sigma}{\epsilon} \qquad (6.3b)$$

$$Im\lambda = \frac{1}{2}\left(\sqrt{4\pi^2\frac{1}{\mu\epsilon} - \left(\frac{\sigma}{\epsilon}\right)^2}\right) \qquad (6.3c)$$

---

1 Equation (6.2a) is a form of the telegraph equation, that is, the wave equation with damping (through $\sigma > 0$).
2 The analytical solution is derived in Appendix 6.1.

Equations (6.3) are used to verify the accuracy of the numerical solution of Eqs. (6.2).

Equations (6.2) and (6.3) constitute the PDE problem for the SCMOL analysis. The main program follows (again, the dependent variable is designated $u(x, t)$ in accordance with the convention of the numerical analysis literature).

6.2 Main Program

The following main program is for two cases, ncase=1 with no damping (sigma=0) and ncase=2 with damping (sigma=1).

```
#
# Maxwell
#
# Delete previous workspaces
  rm(list=ls(all=TRUE))
#
# Access ODE integrator, routines
  library(deSolve);
  setwd("f:/collocation/chap6");
  source("pde_1a.R");
#
# Select case
#
# ncase=1 - no damping; ncase=2 - damping
  ncase=1;
#
# Parameters
  if(ncase==1){eps=1;mu=1;sigma=0;}
  if(ncase==2){eps=1;mu=1;sigma=1;}
  Re_lam=-0.5*sigma/eps;
  Im_lam= 0.5*sqrt(4*pi^2/(mu*eps)-(sigma/eps)^2);
#
# Spatial interval
  n=51;xl=0;xu=1;
  x=seq(from=xl,to=xu,by=(xu-xl)/(n-1));
#
# Time interval
  t0=0;tf=1;nout=11;
  tout=seq(from=t0,to=tf,by=(tf-t0)/(nout-1));
  ncall=0;
```

```
#
# Initial conditions (ICs)
  u0=rep(0,2*n);
  u0[1:n]=cos(pi*x);
  u0[(n+1):(2*n)]=Re_lam*u0[1:n];
#
# ODE integration
  out=lsodes(y=u0,times=tout,func=pde_1a,
      sparsetype="sparseint",rtol=1e-6,atol=1e-6,
      maxord=5);
  nrow(out)
  ncol(out)
#
# Store numerical, analytical solutions for plotting
    u1=matrix(0,nrow=n,ncol=nout);
    u2=matrix(0,nrow=n,ncol=nout);
  u1a=matrix(0,nrow=n,ncol=nout);
  t=rep(0,nout);
  for(it in 1:nout){
    t[it]=out[it,1];
  for(i  in 1:n){
    u1[i,it]=out[it,i+1];
    u2[i,it]=out[it,i+1+n];
   u1a[i,it]=exp(Re_lam*tout[it])*
              cos(Im_lam*tout[it])*
              cos(pi*x[i]);
  }
  }
#
# Numerical, analytical solutions
  cat(sprintf("\n      t     x     u(x,t)    ua(x,t)
              diff"));
  for(it in 1:nout){
  iv=seq(from=1,to=n,by=5);
  for(i  in iv){
      cat(sprintf("\n %6.2f%6.2f%10.4f%10.4f,
              %12.6f",tout[it],x[i],u1[i,it],
              u1a[i,it],u1[i,it]-u1a[i,it]));
  }
  }
  cat(sprintf("\n ncall = %4d\n",ncall));
#
# Plot numerical, analytical solutions
```

```
  matplot(x,u1,type="l",lwd=2,col="black",lty=1,
    xlab="x",ylab="u(x,t)",main="Maxwell");
  matpoints(x,u1a,pch="o",col="black");
#
# Plot 3D numerical solution
  persp(x,tout,u1,theta=55,phi=45,xlim=c(xl,xu),
        ylim=c(t0,tout[nout]),zlim=c(-1.1,1.1),
        xlab="x",ylab="t",zlab="u(x,t)");
```

Listing 6.1 Main program for Eqs. (6.2).

We can note the following details about Listing 6.1:

- Previous workspaces are removed. Then the ODE integrator library deS-
  olve is accessed. Note that the setwd (set working directory) uses / rather
  than the usual \. Also, setwd requires editing for the local computer (to
  define the working directory).

```
#
# Maxwell equation
#
# Delete previous workspaces
  rm(list=ls(all=TRUE))
#
# Access ODE integrator, routines
  library(deSolve);
  setwd("f:/collocation/chap6");
  source("pde_1a.R");
```

pde_1a is the routine for the MOL/ODEs (discussed subsequently).
- Two cases are programmed for no damping (sigma=0) and damping
  (sigma=1).

```
#
# Select case
#
# ncase=1 - no damping; ncase=2 - damping
  ncase=1;
#
# Parameters
  if(ncase==1){eps=1;mu=1;sigma=0;}
  if(ncase==2){eps=1;mu=1;sigma=1;}
  Re_lam=-0.5*sigma/eps;
  Im_lam= 0.5*sqrt(4*pi^2/(mu*eps)-(sigma/eps)^2);
```

Equations (6.3b,c) are programmed for two sets of constants (sigma and related constants).

- A uniform grid in x of 51 points is defined with the seq utility for the interval $x_l = 0 \le x \le x_u = 1$. Therefore, the vector x has the values $x = 0, 0.02, \ldots, 1$.

```
#
# Spatial interval
  n=51;xl=0;xu=1;
  x=seq(from=xl,to=xu,by=(xu-xl)/(n-1));
```

- A uniform grid in t of 11 output points is defined with the seq utility for the interval $0 \le t \le t_f = 1$. Therefore, the vector tout has the values $t = 0, 0.1, \ldots, 1$.

```
#
# Time interval
  t0=0;tf=1;nout=11;
  tout=seq(from=t0,to=tf,by=(tf-t0)/(nout-1));
  ncall=0;
```

At this point, the intervals of x and t in Eqs. (6.2) are defined. Also, the counter for the calls to pde_1a is initialized (and passed to pde_1a without a special designation).

- The ICs, Eqs. (6.2b,c), are defined.

```
#
# Initial conditions (ICs)
  u0=rep(0,2*n);
  u0[1:n]=cos(pi*x);
  u0[(n+1):(2*n)]=Re_lam*u0[1:n];
```

These ICs reflect the use of two dependent variable u_1, u_2, defined as

$$E(x, t) = u(x, t) = u_1(x, t); \quad \frac{\partial E(x, t)}{\partial t} = \frac{\partial u(x, t)}{\partial t} = u_2(x, t) \qquad (6.2f,g)$$

This approach to PDEs second order in t is discussed in Section (2.5). The total number of first-order ODEs is therefore $2(51) = 102$.

- The system of 102 MOL/ODEs is integrated by the library integrator lsodes (available in deSolve) with the sparse matrix option specified. As expected, the inputs to lsodes are the ODE function, pde_1a, the IC vector u0, and the vector of output values of t, tout. The length of u0 (102) informs lsodes how many ODEs are to be integrated. func, y, times are reserved names.

```
#
# ODE integration
  out=lsodes(y=u0,times=tout,func=pde_1a,
```

```
  sparsetype="sparseint",rtol=1e-6,atol=1e-6,
  maxord=5);
nrow(out)
ncol(out)
```

The numerical solution to the ODEs is returned in matrix out. In this case, out has the dimensions $nout \times (n + 1) = 11 \times 103$. The offset $n + 1$ is required since the first element of each column has the output t (also in tout), and the $2, \ldots, 103$ column elements have the 102 ODE solutions. This indexing of out is used next.

- The numerical and analytical solutions for u_1, and the numerical solution for u_2 are taken from the solution matrix out returned by lsodes and Eqs. (6.3). The intervals in t and x are covered by two nested fors (with indices it and i for t and x, respectively). The offset i+1 accounts for t placed in the first column position of out (t[it]=out[it,1]).

```
#
# Store numerical, analytical solutions for plotting
  u1=matrix(0,nrow=n,ncol=nout);
  u2=matrix(0,nrow=n,ncol=nout);
 u1a=matrix(0,nrow=n,ncol=nout);
 t=rep(0,nout);
 for(it in 1:nout){
   t[it]=out[it,1];
   for(i in 1:n){
     u1[i,it]=out[it,i+1];
     u2[i,it]=out[it,i+1+n];
   u1a[i,it]=exp(Re_lam*tout[it])*
             cos(Im_lam*tout[it])*
             cos(pi*x[i]);

 }
}
```

Equations (6.3) define the analytical solution. Equations (6.3f,g) define u_1, u_2.

- The numerical and analytical solutions for $u = u_1$, and the difference, are displayed.

```
#
# Numerical, analytical solutions
  cat(sprintf("\n       t     x     u(x,t)    ua(x,t)
             diff"));
  for(it in 1:nout){
  iv=seq(from=1,to=n,by=5);
  for(i  in iv){
    cat(sprintf("\n %6.2f%6.2f%10.4f%10.4f,
```

```
                         %12.6f",tout[it],x[i],u1[i,it],
                         u1a[i,it],u1[i,it]-u1a[i,it]));
    }
  }
  cat(sprintf("\n ncall = %4d\n",ncall));
```

To conserve space, only every fifth value in x of the solutions is displayed numerically (using the subscript iv). Thus, the displayed spacing in x is $5(0.02) = 0.1$. Finally, the number of calls to the ODE/MOL routine pde_1a (considered next) is displayed as a measure of the computational effort required for computing the numerical solution.

- The numerical and analytical solutions for $u = u_1$ are plotted with the matplot, matpoints utilities.

```
#
# Plot numerical, analytical solutions
  matplot(x,u1,type="l",lwd=2,col="black",lty=1,
    xlab="x",ylab="u(x,t)",main="Maxwell equation\n
    line - num; points - anal");
  matpoints(x,u1a,pch="o",col="black");
```

- The numerical solution $u = u_1$ is plotted in 3D perspective with persp.

```
#
# Plot 3D numerical solution
  persp(x,tout,u1,theta=55,phi=45,xlim=c(xl,xu),
        ylim=c(t0,tout[nout]),zlim=c(-1.1,1.1),
        xlab="x",ylab="t",zlab="u(x,t)");
```

This concludes the discussion of the main program in Listing 6.1. The output is displayed in Table 6.1 and Figure 6.1a,b for ncase=1 (no damping) and in Table 6.2 and Figure 6.2a,b for ncase=2 (with damping).

The ODE/MOL routine pde_1a called by lsodes in the main program of Listing 6.1 is considered next.

6.3 ODE Routine

pde_1a for Eq. (6.2a) follows.

```
  pde_1a=function(t,u,parms){
#
# Function pde_1a computes the t derivative vectors
# of u1(x,t),u2(x,t)
#
# One vector to two vectors
```

```
   u1=rep(0,n);u2=rep(0,n);
   for(i in 1:n){
    u1[i]=u[i];
    u2[i]=u[i+n];
    }
#
# u1x
   table1=splinefun(x,u1);
   u1x=table1(x,deriv=1);
#
# Neumann BCs
   u1x[1]=0;u1x[n]=0;
#
# u1xx
   table2=splinefun(x,u1x);
   u1xx=table2(x,deriv=1);
#
# PDE
   mueps=1/(mu*eps);
   sigeps=sigma/eps;
   u1t=rep(0,n);u2t=rep(0,n);
   for(i in 1:n){
     u1t[i]=u2[i];
     u2t[i]=mueps*u1xx[i]-sigeps*u2[i];
   }
#
# Two vectors to one vector
   ut=rep(0,2*n);
   for(i in 1:n){
       ut[i]=u1t[i];
     ut[i+n]=u2t[i];
   }
#
# Increment calls to pde_1a
   ncall <<- ncall+1;
#
# Return derivative vector
   return(list(c(ut)));
#
# End of pde_1a
}
```

Listing 6.2 ODE/MOL routine pde_1a for Eq. (6.2a).

We can note the following details about pde_1a.

- The function is defined.

```
pde_1a=function(t,u,parms){
#
# Function pde_1a computes the t derivative vectors
# of u1(x,t),u2(x,t)
```

t is the current value of t in Eq. (6.2a). u is the $2(51) = 102$-vector of ODE-dependent variables. parm is an argument to pass parameters to pde_1a (unused, but required in the argument list). The arguments must be listed in the order stated to properly interface with lsodes called in the main program. The derivative vector of the LHS of Eq. (6.2a) is calculated next and returned to lsodes.

- u is placed in two vectors, u1, u2 in accordance with Eqs. (6.2f,g).

```
#
# One vector to two vectors
  u1=rep(0,n);u2=rep(0,n);
  for(i in 1:n){
   u1[i]=u[i];
   u2[i]=u[i+n];
   }
```

- $\dfrac{\partial u_1}{\partial x}$ in Eq. (6.2a) is computed with splinefun ($u_1 = u$ from Eq. (6.2f)).

```
#
# u1x
  table1=splinefun(x,u1);
  u1x=table1(x,deriv=1);
```

- BCs (6.2d,e) are defined.

```
#
# Neumann BCs
  u1x[1]=0;u1x[n]=0;
```

- $\dfrac{\partial^2 u_1}{\partial x^2}$ in Eq. (6.2a) is computed with splinefun by differentiating $\dfrac{\partial u_1}{\partial x}$ (stagewise differentiation).

```
#
# u1xx
  table2=splinefun(x,u1x);
  u1xx=table2(x,deriv=1);
```

- Equation (6.2a) is programmed. First the constants in Eq. (6.2a), μ, ϵ, σ set in the main program of Listing 6.1, are placed in two groups, mueps, sigeps, for use in the for that steps through x.

```
#
#  PDE
   mueps=1/(mu*eps);
   sigeps=sigma/eps;
   u1t=rep(0,n);u2t=rep(0,n);
   for(i in 1:n){
     u1t[i]=u2[i];
     u2t[i]=mueps*u1xx[i]-sigeps*u2[i];
   }
```

The derivatives $\dfrac{\partial u_1}{\partial t}, \dfrac{\partial u_2}{\partial t}$ (u1t, u2t) follow from Eqs. (6.2a,f,g).

- u1t, u2t are placed in one 102-vector, ut, to return to lsodes called in the main program of Listing 6.1.

```
#
#  Two vectors to one vector
   ut=rep(0,2*n);
   for(i in 1:n){
       ut[i]=u1t[i];
     ut[i+n]=u2t[i];
   }
```

- The counter for the calls to pde_1a is incremented and its value is returned to the calling program (of Listing 6.1) with the <<- operator.

```
#
#  Increment calls to pde_1a
   ncall <<- ncall+1;
```

- The derivative vector (LHS of Eq. (6.2a)) is returned to lsodes, which requires a list. c is the R vector utility. The combination of return, list, c gives lsodes (the ODE integrator called in the main program of Listing 6.1) the required derivative vector for the next step along the solution.

```
#
#  Return derivative vector
   return(list(c(ut)))
#
#  End of pde_1a
   }
```

The final } concludes pde_1a.

This completes the programming of Eqs. (6.2). The numerical and graphical output is considered next.

6.4 Model Output

Execution of the preceding main program and subordinate routine produces the following abbreviated numerical output in Table 6.1 for ncase = 1 set in the main program of Listing 6.1.

As expected, the solution array out has the dimensions 11 x 103 for 11 output points and 102 ODEs, with t included as an additional element, that is, 102 + 1 = 103. The computational effort is modest, ncall = 405. The agreement of the numerical and analytical solutions is confirmed (note the last column). This agreement is also confirmed in Figure 6.1a.

The solution for $t = 1$ is the same as for $t = 0$ (Table 6.1), but with a sign reversal. This follows from the analytical solution of Eqs. (6.3) with $\sigma = 0$ (no damping so that the solution is periodic in t).

Abbreviated output for ncase=2 is displayed in Table 6.2.

Again, as with Table 6.1, the accuracy of the numerical solution is confirmed. This agreement is also confirmed in Figure 6.2a. The computational effort is modest, ncall = 399. Also, the solution is not periodic as a result of damping (sigma=1), as reflected in Figure 6.2a,b.

ncase=2 is the more physically realistic case since waves will experience some form of internal losses, including electromagnetic waves described by the ME. The loss through damping is reflected in Eq. (6.3) and, in particular, the exponential $e^{Re\lambda t}$ with $Re\lambda = -\frac{1}{2}\frac{\sigma}{\epsilon}$ from Eq. (6.3b). The exponential decay of the solution to Eq. (6.2a) with t with $\sigma > 0$ is clear.

6.5 Summary and Conclusions

Listing 6.2 indicates that the coding of the ME (6.2a) is straightforward, and Tables 6.1 and 6.2 indicate that the numerical solution is acceptable. Also, the coding in Listing 6.2 indicates that the SCMOL analysis of PDEs second order in t can be handled by converting into a system of ODEs first order in t (e.g., Eqs. (6.2f,g)).

As a final point, Eq. (6.2a) with $\sigma = 0$ is the classical linear wave equation

$$\frac{\partial^2 E}{\partial t^2} = c^2 \frac{\partial^2 E}{\partial x^2} \tag{6.2h}$$

Table 6.1 Abbreviated output for Eqs. (6.2), ncase=1.

```
[1]  11

[1]  103
```

t	x	u(x,t)	ua(x,t)	diff
0.00	0.00	1.0000	1.0000	0.000000
0.00	0.10	0.9511	0.9511	0.000000
0.00	0.20	0.8090	0.8090	0.000000
0.00	0.30	0.5878	0.5878	0.000000
0.00	0.40	0.3090	0.3090	0.000000
0.00	0.50	0.0000	0.0000	0.000000
0.00	0.60	-0.3090	-0.3090	0.000000
0.00	0.70	-0.5878	-0.5878	0.000000
0.00	0.80	-0.8090	-0.8090	0.000000
0.00	0.90	-0.9511	-0.9511	0.000000
0.00	1.00	-1.0000	-1.0000	0.000000

```
                        .              .
                        .              .
                        .              .
             Output for t = 0.1 to 0.9 removed
                        .              .
                        .              .
                        .              .
```

t	x	u(x,t)	ua(x,t)	diff
1.00	0.00	-1.0000	-1.0000	-0.000025
1.00	0.10	-0.9511	-0.9511	-0.000021
1.00	0.20	-0.8090	-0.8090	-0.000020
1.00	0.30	-0.5878	-0.5878	-0.000016
1.00	0.40	-0.3090	-0.3090	-0.000002
1.00	0.50	-0.0000	-0.0000	-0.000000
1.00	0.60	0.3090	0.3090	0.000002
1.00	0.70	0.5878	0.5878	0.000016
1.00	0.80	0.8090	0.8090	0.000020
1.00	0.90	0.9511	0.9511	0.000021
1.00	1.00	1.0000	1.0000	0.000025

```
ncall =  405
```

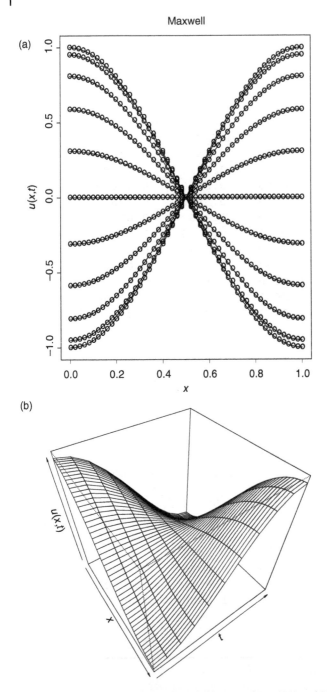

Figure 6.1 (a) Numerical and analytical solutions of Eqs. (6.2) and (6.3) ncase=1, lines – num, points – anal. (b) Numerical solution of Eqs. (6.2), ncase=1.

Table 6.2 Abbreviated output for Eqs. (6.2), ncase=2.

[1] 11

[1] 103

t	x	u(x,t)	ua(x,t)	diff
0.00	0.00	1.0000	1.0000	0.000000
0.00	0.10	0.9511	0.9511	0.000000
0.00	0.20	0.8090	0.8090	0.000000
0.00	0.30	0.5878	0.5878	0.000000
0.00	0.40	0.3090	0.3090	0.000000
0.00	0.50	0.0000	0.0000	0.000000
0.00	0.60	-0.3090	-0.3090	0.000000
0.00	0.70	-0.5878	-0.5878	0.000000
0.00	0.80	-0.8090	-0.8090	0.000000
0.00	0.90	-0.9511	-0.9511	0.000000
0.00	1.00	-1.0000	-1.0000	0.000000

.
.
.

Output for t = 0.1 to 0.9 removed

.
.
.

t	x	u(x,t)	ua(x,t)	diff
1.00	0.00	-0.6061	-0.6060	-0.000013
1.00	0.10	-0.5764	-0.5764	-0.000011
1.00	0.20	-0.4903	-0.4903	-0.000011
1.00	0.30	-0.3562	-0.3562	-0.000009
1.00	0.40	-0.1873	-0.1873	-0.000001
1.00	0.50	0.0000	-0.0000	0.000000
1.00	0.60	0.1873	0.1873	0.000001
1.00	0.70	0.3562	0.3562	0.000009
1.00	0.80	0.4903	0.4903	0.000011
1.00	0.90	0.5764	0.5764	0.000011
1.00	1.00	0.6061	0.6060	0.000013

ncall = 399

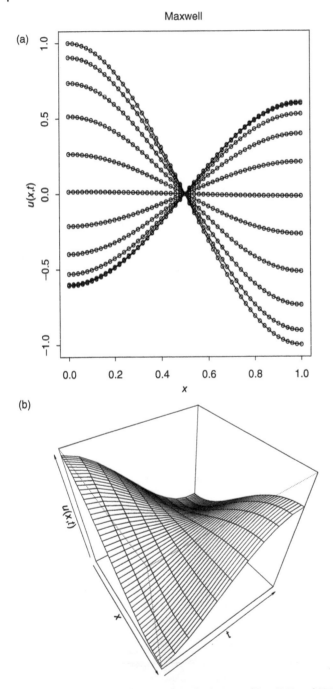

Figure 6.2 (a) Numerical and analytical solutions of Eqs. (6.2) and (6.3) ncase=2, lines – num, points – anal. (b) Numerical solution of Eqs. (6.2), ncase=2.

where c is the velocity of a wave with amplitude $E(x, t)$. Comparing Eqs. (6.2a,h), we have

$$c = \frac{1}{\sqrt{\mu\epsilon}} \tag{6.2i}$$

In other words, electromagnetic waves travel at the velocity c given by Eq. (6.2i).[3]

If we use values for μ and ϵ for free space, $\mu = 4\pi \times 10^{-7}$ H m^{-1}, $\epsilon = 10^{-9}/(36\pi)$ F m^{-1}, we have from Eq. (6.2i)

$$c = \frac{1}{\sqrt{(4\pi \times 10^{-7})(10^{-9}/(36\pi))}} = 3 \times 10^{8}$$

which is the speed of light in free space in meter per second. Thus, we come to the remarkable conclusion that the ME predict the speed of electromagnetic waves is the speed of light. In fact, light waves are electromagnetic waves.

Appendix A6.1. Derivation of the Analytical Solution

The derivation of the analytical solution, Eqs. (6.3), of Eqs. (6.2) follows. If we assume a solution of the form

$$E(x, t) = f(t)\cos(\pi x) \tag{A6.1}$$

then $f(t)$ is a function to be determined.

The choice of Eq. (A6.1) is guided by the following ideas:

1. A product solution in x and t is a commonly used form of a solution for linear PDEs (the separation of variables approach).
2. When substituted in Eq. (6.2a), $\cos(\pi x)$ appears as a factor in all of the terms, that is

$$\cos(\pi x)\frac{d^2 f(t)}{dt^2} + \frac{\sigma}{\epsilon}\cos(\pi x)\frac{df(t)}{dt} = -(\pi^2)\frac{1}{\mu\epsilon}\cos(\pi x)f(t)$$

Thus, we require

$$\left(\frac{d^2 f(t)}{dt^2} + \frac{\sigma}{\epsilon}\frac{df(t)}{dt} + \pi^2\frac{1}{\mu\epsilon}f(t)\right)\cos(\pi x) = 0$$

or

$$\frac{d^2 f(t)}{dt^2} + \frac{\sigma}{\epsilon}\frac{df(t)}{dt} + \pi^2\frac{1}{\mu\epsilon}f(t) = 0 \tag{A6.2}$$

for all x.[4]

---

3 Here the electric field $E(x, t)$ is considered, but Eq. (6.2h) also applies to the magnetic field $H(x, t)$; in other words, Eq. (6.2h) applies to electromagnetic fields.

4 Here we include $x = 1/2$ for which $\cos(\pi x) = 0$. This point is clear in Figures 6.1 and 6.2, and does not cause a problem with respect to continuity or smoothness.

3. The trial solution of Eq. (A6.1) satisfies BCs (6.2d,e).

Thus, by assuming a product solution of the form of Eq. (A6.1), we have reduced the original PDE problem, Eq. (6.2a), to an ODE problem, Eq. (A6.2).

Since Eq. (A6.2) is a linear, constant coefficient ODE, we assume an exponential solution

$$f(t) = ae^{bt} \tag{A6.3}$$

where a, b are constants to be determined.

Substitution of Eq. (A6.3) in Eq. (A6.2) gives

$$ab^2 e^{bt} + \frac{\sigma}{\epsilon} abe^{bt} + \frac{1}{\mu\epsilon} \pi^2 ae^{bt} = 0$$

or after division by ae^{bt}

$$b^2 + \frac{\sigma}{\epsilon} b + \frac{1}{\mu\epsilon} \pi^2 = 0$$

b is computed from the quadratic formula

$$b_\pm = \frac{1}{2}\left(-\frac{\sigma}{\epsilon} \pm \sqrt{\left(\frac{\sigma}{\epsilon}\right)^2 - 4\pi^2 \frac{1}{\mu\epsilon}}\right)$$

$$= \frac{1}{2}\left(-\frac{\sigma}{\epsilon} \pm i\sqrt{4\pi^2 \frac{1}{\mu\epsilon} - \left(\frac{\sigma}{\epsilon}\right)^2}\right) = Re\lambda \pm iIm\lambda$$

where

$$Re\lambda = \frac{1}{2}\left(-\frac{\sigma}{\epsilon}\right); \quad Im\lambda = \frac{1}{2}\left(\sqrt{4\pi^2 \frac{1}{\mu\epsilon} - \left(\frac{\sigma}{\epsilon}\right)^2}\right); \quad i = \sqrt{-1}$$

Thus, $f(t)$ is

$$f(t) = a_1 e^{Re\lambda t} e^{iIm\lambda t} + a_2 e^{Re\lambda t} e^{-iIm\lambda t} \tag{A6.4}$$

(a linear combination of two exponentials). If we take

$$a_1 = 1/2; \quad a_2 = -1/2$$

Equation (A6.4) becomes

$$f(t) = e^{Re\lambda t} \frac{e^{iIm\lambda t} - e^{-iIm\lambda t}}{2} = e^{Re\lambda t} \cos(Im\lambda t)$$

and Eq. (A6.1) is

$$E(x, t) = e^{Re\lambda t} \cos(Im\lambda t) \cos(\pi x) \tag{A6.5}$$

Equation (A6.5) satisfies IC (6.2b)

$$E(x, t = 0) = \cos(\pi x)$$

and IC (6.2c)

$$\frac{\partial E(x,t)}{\partial t} = (-Im\,\lambda e^{Re\lambda t}\sin(Im\,\lambda t) + Re\,\lambda e^{Re\lambda t}\cos(Im\,\lambda t))\cos(\pi x)$$

$$\frac{\partial E(x,t=0)}{\partial t} = Re\,\lambda\cos(\pi x)$$

Thus, Eq. (A6.5) is the solution to Eqs. (6.2) (and is stated as Eqs. (6.3)).

Reference

1 Schiesser, W.E., and G.W. Griffiths (2009), *A Compendium of Partial Differential Equation Models*, Chapter 9, Cambridge University Press, Cambridge, UK.

7

Poisson–Nernst–Planck Equations

The Poisson–Nernst–Planck (PNP) equations describe the diffusion of charged particles (ions) in solution resulting from an electric field, so-called *electrodiffusion*. A PNP/PDE model is developed next, followed by a numerical solution.

7.1 PDE Model

The Nernst–Planck extension of Fick's first law for the flux of charged particles, including the effect of an electric field is ([1], 1311–1312)

$$J = -D\left(\nabla c + \frac{Ze}{k_B T} c \nabla \phi\right) = -D\left(\nabla c + \frac{Ze}{k_B T} c(-E)\right) \tag{7.1a}$$

The variables and parameters (constants) in Eq. (7.1a) are defined in Table 7.1 (MKS units).[1]

In particular, the third RHS of Eq. (7.1a) with E contributes the effect of the electric field to the flux J. We note, in particular, in Eq. (7.1a) that this third RHS includes the nonlinearity cE, that is, the contribution to the electrodiffusion flux is proportional to the product of two dependent variables (the second dependent variable, the electric field E, is subsequently defined by a second PDE).

Equation (7.1a) is for one diffusing species. Multiple species are included as a sum over all of the various species for the third RHS term.

The PNP analog of Fick's second law (with convection) is

$$\frac{\partial c}{\partial t} = \nabla \cdot \left[D\nabla c - uc - \left(\frac{DZe}{k_B T}\right)(cE)\right] \tag{7.1b}$$

---

1 Rather than using u_1, u_2 according to the usual naming convention for PDE-dependent variables, the dependent variables in Eq. (7.1a) are c, E to reflect the physical application. Also, the relation between ϕ and E, $\nabla \phi = -E$, has been used in Eq. (7.1a).

Spline Collocation Methods for Partial Differential Equations: With Applications in R, First Edition.
William E. Schiesser.
© 2017 John Wiley & Sons, Inc. Published 2017 by John Wiley & Sons, Inc.
Companion website: www.wiley.com/go/Spline_Collocation

E in Eq. (7.1b) is a second dependent variable (the first is the ion concentration c). Thus, a second PDE is required, which is usually taken as Poisson's equation.

$$\frac{Ze}{\epsilon}c = \nabla^2\phi \qquad (7.1c)$$

Equation (7.1c) indicates the effect of the concentration c on the electric potential ϕ (and also, the electric field E through $\nabla\phi = -E$).

Table 7.1 Variables and parameters for the Poisson–Nernst–Planck equations.

Variable parameter	Interpretation units, value
t	time (s)
c	ion concentration ($\mathrm{kg\,mol\,m^{-3}}$)
c_0	reference ion concentration ($100\,\mathrm{mM} = 0.1\,\mathrm{M}$ $= 0.1\,\mathrm{g\,mol\,cc^{-1}} = (0.1)(10^3) = 10^2\,\mathrm{kg\,mol\,m^{-3}}$)
J	charged particle flux ($\mathrm{kg\,mol\,s^{-1}\,m^{-2}}$)
D	diffusivity ($\mathrm{m^2\,s^{-1}}$)
u	fluid velocity ($\mathrm{m\,s^{-1}}$)
Z	valence
e	proton charge ($1.6021766208(98) \times 10^{-19}\,\mathrm{C}$ (C = coulomb))
k_B	Boltzmann constant ($1.38064852(79) \times 10^{-23}\,\mathrm{m^2\,kg\,s^{-2}\,K^{-1}}$)
T	temperature (K)
E	electric field ($\mathrm{V\,m^{-1}} = \mathrm{kg\,m\,C^{-1}\,s^{-2}}$)
E_0	reference channel electric field $= -5 \times 10^6\,\mathrm{V\,m^{-1}}$
ϕ	electric potential ($\mathrm{V} = \mathrm{kg\,m^2\,C^{-1}\,s^{-2}}$)
ϕ_0	reference electric potential ($\mathrm{V} = \mathrm{kg\,m^2\,C^{-1}\,s^{-2}}$)
∇	coordinate-free spatial differential operator ($\mathrm{m^{-1}}$)
$\nabla\phi = -E$	electric field ($\mathrm{V\,m^{-1}}$)
ϵ	permittivity (for water, $710 \times 10^{-12}\,\mathrm{m^{-3}\,kg^{-1}\,s^2\,C^2}$)
x_L	reference channel length $= 10^{-8}\,\mathrm{m}$
k_m	transfer coefficient $= 1\,\mathrm{m^{-1}}$

Equations (7.1) constitute the PDE model in coordinate-free format. They are next considered in 1D in Cartesian coordinates with the single spatial (boundary value) independent variable x. This choice of a coordinate system reflects the approximate geometry of an ion channel in a cell membrane ([1], Fig. 1, [2], Figs. 1,9). For constant D, Eqs. (7.1b,c) are[2]

$$\frac{\partial c}{\partial t} = D\frac{\partial^2 c}{\partial x^2} - u\frac{\partial c}{\partial x} - \left(\frac{DZe}{k_B T}\right)\frac{\partial(cE)}{\partial x} \tag{7.2a}$$

$$0 = \frac{\partial^2 \phi}{\partial x^2} - \left(\frac{Ze}{\epsilon}\right)c \tag{7.2b}$$

Because of the wide variation in the magnitudes of the dependent and independent variables in Eqs. (7.2), variables scaled with reference values (dimensionless variables) are convenient for analysis. Dimensionless variables are defined as primed.

$$x' = x/x_L; \quad t' = tD/x_L^2; \quad c' = c/c_0; \quad \phi' = \phi/\phi_0; \quad E' = E/E_0$$

If the dimensionless variables are substituted in Eqs. (7.2), and the primes are dropped,

$$\left(\frac{D}{x_L^2}\right)\frac{\partial c}{\partial t} = \left(\frac{D}{x_L^2}\right)\frac{\partial^2 c}{\partial x^2} - \left(\frac{u}{x_L}\right)\frac{\partial c}{\partial x} - \left(\frac{Ze}{k_B T}\right)\left(\frac{Dc_0 E_0}{x_L}\right)\frac{\partial(cE)}{\partial x}$$

$$0 = \left(\frac{\phi_0}{x_L^2}\right)\frac{\partial^2 \phi}{\partial x^2} - \left(\frac{Ze}{\epsilon}\right)(c_0)c$$

or

$$\frac{\partial c}{\partial t} = \frac{\partial^2 c}{\partial x^2} - \left(\frac{ux_L}{D}\right)\frac{\partial c}{\partial x} - \left(\frac{Ze}{k_B T}\right)(x_L c_0 E_0)\frac{\partial(cE)}{\partial x} \tag{7.3a}$$

$$0 = \frac{\partial^2 \phi}{\partial x^2} - \left(\frac{Ze}{\epsilon}\right)\left(\frac{x_L^2 c_0}{\phi_0}\right)c \tag{7.3b}$$

Also, the BC at $x = x_u$ for Eq. (7.3a) (Eq. (7.4c)) requires the cell interior concentration c_i, which is given by an ordinary differential equation (ODE)

$$\tau\frac{dc_i}{dt} = c(x = x_L, t) - c_i \tag{7.3d}$$

τ is a time constant or mixing time for the cell interior.

---

2 Equation (7.2a) is parabolic–hyperbolic (with a first-order initial value derivative in t, a first-order boundary value derivative in x (hyperbolic) and a second-order boundary value derivative in x (parabolic)). Equation (7.2b) is elliptic (with only a second-order boundary value derivative in x). Thus, Eqs. (7.2) are a 2×2 (two equations in two unknowns, c, E) parabolic–hyperbolic–elliptic PDE system.

Equations (7.3) constitute the PDEs to be analyzed. Equation (7.3a) is first order in t and second order in x, and therefore requires one initial condition (IC) and two boundary conditions (BCs).

$$c(x, t = 0) = f(x) \qquad (7.4a)$$

$$c(x = x_l, t) = c_l(t); \qquad \frac{\partial c(x = x_u, t)}{\partial x} = k_m(c_i - c(x = x_u, t)) \qquad (7.4b,c)$$

Equation (7.4c) is a Robin BC equating the LHS diffusion flux at the end of the channel ($x = x_u = x_L$) to the RHS mass transfer rate into the interior of the cell. k_m is a ratio of a mass transfer coefficient to D, with units m^{-1}.

Equation (7.3b) is second order in x and therefore requires two BCs.

$$\phi(x = x_l, t) = \phi_l(t); \qquad \frac{\partial \phi(x = x_u, t)}{\partial x} = 0 \qquad (7.4d,e)$$

BC (7.4e) with $\nabla \phi = -E$ or in 1D $\dfrac{\partial \phi}{\partial x} = -E$ gives $E(x = x_u, t) = 0$, that is, the electric field is zero at the right boundary $x = x_u = x_L$. From the numerical solution discussed subsequently, $\dfrac{\partial \phi}{\partial x} \approx 0$ near the right boundary indicating a small electric field near the cell (small effect of electrodiffusion).

Equation (7.3d) requires an IC.

$$c_i(t = 0) = c_{i0} \qquad (7.4f)$$

Equations (7.3) and (7.4) constitute the ODE/PDE system that is subsequently analyzed numerically by the method of lines (MOL).

Since MOL is oriented toward the solution of initial value PDEs, and Eq. (7.3b) does not have a derivative in t, it will be augmented with a t derivative to fit within the MOL framework.[3]

$$\mu \frac{\partial c}{\partial c} = \frac{\partial^2 \phi}{\partial x^2} - \left(\frac{Ze}{\epsilon} \right) \left(\frac{x_L^2 c_0}{\phi_0} \right) c \qquad (7.3c)$$

Equation (7.3c) is first order in t and therefore requires on IC.

$$\phi(x, t = 0) = 0 \qquad (7.4g)$$

The solution of Eqs. (7.3) and (7.4) is programmed in the following routines.

---

3 The use of the derivative in t in Eq. (7.3c) is an application of the method of false or pseudo transients. μ is a small parameter, which means the LHS of Eq. (7.3c) is small (close to zero) in accordance with Eq. (7.3b). In other words, the solution to Eq. (7.3c) should be close to the solution to Eq. (7.3b), and Eq. (7.3c) fits within the MOL framework as demonstrated experimentation in the subsequent computer analysis. A value of μ can be determined by some numerical experimentation.

7.2 Main Program

A main program for Eqs. (7.3) and (7.4) follows.

```
#
# Poisson-Nernst-Planck
#
# Delete previous workspaces
  rm(list=ls(all=TRUE))
#
# Access ODE integrator
  library("deSolve");
#
# Access functions for numerical solution
  setwd("f:/collocation/chap7");
  source("pde_1a.R");
#
# Parameters
     u=0.001;    xL=1.0e-08;         D=1.0e-10;
          Z=1; e=1.6022e-19; kB=1.3806e-23;
        T=300;    c0=1.0e+02;      E0=5.0e+06;
      mu=0.1;    phi0=5e-02; eps=710.0e-12;
        km=1;      tau=0.01;
#
# Constants
  c11=u*xL/D;
  c12=((Z*e)/(kB*T))*(xL*c0*E0);
  c21=((Z*e)/eps)*((xL^2*c0)/phi0);
#
# Select case
  ncase=2;
#
# No electrodiffusion
  if(ncase==1){
    n=21;c12=0;
    t0=0;tf=0.1
  }
#
# Electrodiffusion
  if(ncase==2){
    n=21;
    t0=0;tf=0.2
  }
```

```
  cat(sprintf("\n c11 = %10.3e\n c12 = %10.3e\n
              c21 = %10.3e\n",c11,c12,c21));
#
# Grid (in x)
  xl=0;xu=1;
  x=seq(from=xl,to=xu,by=(xu-xl)/(n-1));
#
# Independent variable for ODE integration
  nout=6;
  tout=seq(from=t0,to=tf,by=(tf-t0)/(nout-1));
#
# Initial conditions
  c=rep(0,n);phi=rep(0,n);
  c[1]=1;phi[1]=1;
  u0=rep(0,2*n+1);
  for(i in 1:n){
   u0[i]=c[i];
   u0[i+n]=phi[i];
  }
  u0[2*n+1]=0;
  ncall=0;
#
# ODE integration
  out=lsodes(y=u0,times=tout,func=pde_1a,
      sparsetype="sparseint",rtol=1e-6,atol=1e-6,
      maxord=5);
  nrow(out)
  ncol(out)
#
# Arrays for display
    c=matrix(0,nrow=n,ncol=nout);
  phi=matrix(0,nrow=n,ncol=nout);
  for(it in 1:nout){
  for(i  in 1:n){
      c[i,it]=out[it,i+1];
    phi[i,it]=out[it,i+1+n];
  }
  }
#
# Numerical solutions
  cat(sprintf("\n       t        x     c(x,t)
              phi(x,t)"));
```

```
  for(it in 1:nout){
  for(i   in 1:n){
    cat(sprintf("\n %10.5f%10.4f%10.4f%10.4f",
      tout[it],x[i],c[i,it],phi[i,it]));

  }
    cat(sprintf("\n"));
  }
  matplot(x,c,type="l",lwd=2,col="black",lty=1,
          xlab="x",ylab="c(x,t)",main=
          Poisson-Nernst-Planck");
  matplot(x,phi,type="l",lwd=2,col="black",lty=1,
          xlab="x",ylab="phi(x,t)",main=
          "Poisson-Nernst-Planck");
#
# Calls to ODE routine
  cat(sprintf("\n\n  ncall = %3d\n",ncall));
```

Listing 7.1 Main program for Eqs. (7.3) and (7.4).

We can note the following details about Listing 7.1:

- Previous workspaces are removed. Then the ODE integrator library deS-olve is accessed. Note that the setwd (set working directory) uses / rather than the usual \.

```
#
# Poisson-Nernst-Planck
#
# Delete previous workspaces
  rm(list=ls(all=TRUE))
#
# Access ODE integrator
  library("deSolve");
#
# Access functions for numerical solution
  setwd("f:/collocation/chap7");
  source("pde_1a.R");
```

pde_1a is the routine for the spline collocation method of lines (SCMOL) ODEs (discussed subsequently). The setwd requires editing to define the directory with the R routines.

- The parameters of Eqs. (7.3) and (7.4) are defined numerically. These values are taken from Table 7.1, which also indicates the units.

```
#
# Parameters
     u=0.001;    xL=1.0e-08;      D=1.0e-10;
         Z=1;  e=1.6022e-19;  kB=1.3806e-23;
       T=300;   c0=1.0e+02;     E0=5.0e+06;
      mu=0.1;   phi0=5e-02;  eps=710.0e-12;
       km=1;      tau=0.01;
```

- The constants in Eqs. (7.3) are precomputed and then passed through to the ODE/MOL routine pde_1a (discussed subsequently) without a special designation (a feature of R).

```
#
# Constants
    c11=u*xL/D;
    c12=((Z*e)/(kB*T))*(xL*c0*E0);
    c21=((Z*e)/eps)*((xL^2*c0)/phi0);
```

c11 is a dimensionless group termed the Peclet number. It can be considered as a ratio of the convection to diffusion in the ion channel. c12 defines the contribution of electrodiffusion to the mass conservation balance of Eq. (7.3a). c21 defines the contribution of electrodiffusion to the electric potential ϕ in Eqs. (7.3b,c).

- Two cases are programmed. For the first case, electrodiffusion is not included in Eq. (7.3a) (c12=0). For the second case, electrodiffusion is included in Eq. (7.3a), which substantially changes the features of the solution. Also, the interval in t is extended from $t_f = 0.1$ to $t_f = 0.2$ to give more of the solution transient. These two cases are discussed subsequently.

```
#
# Select case
    ncase=2;
#
# No electrodiffusion
    if(ncase==1){
      n=21;c12=0;
      t0=0;tf=0.1
    }
#
# Electrodiffusion
    if(ncase==2){
      n=21;
```

```
       t0=0;tf=0.2
       }
     cat(sprintf("\n c11 = %10.3e\n c12 = %10.3e\n
                  c21 = %10.3e\n",c11,c12,c21));
```

- A uniform grid in x is defined with the seq utility for the interval $x_l = 0 \leq x \leq x_u = 1$. Therefore, the vector x has the values $x = 0, 0.05, \ldots, 1$ (with n=21).

```
#
# Grid (in x)
  xl=0;xu=1;
  x=seq(from=xl,to=xu,by=(xu-xl)/(n-1));
```

- A uniform grid in t of 6 output points is defined with the seq utility for the interval $0 \leq t \leq t_f = 0.1$. Therefore, the vector tout has the values $t = 0, 0.02, \ldots, 0.1$. This interval in t was determined by trial and error to define the numerical solution variation with t.

```
#
# Independent variable for ODE integration
  nout=6;
  tout=seq(from=t0,to=tf,by=(tf-t0)/(nout-1));
```

At this point, the intervals of x and t in Eqs. (7.3) are defined.
- The ICs of Eqs. (7.4) are defined.

```
#
# Initial conditions
  c=rep(0,n);phi=rep(0,n);
  c[1]=1;phi[1]=1;
  u0=rep(0,2*n+1);
  for(i in 1:n){
    u0[i]=c[i];
    u0[i+n]=phi[i];
  }
  u0[2*n+1]=0;
  ncall=0;
```

These ICs reflect the use of the two dependent variables c, ϕ of Eqs. (7.3a,c) and the IC of Eq. (7.3d) (a total of $2(21) + 1 = 43$ initial values with n=21). The ICs at $x = 0$, c[1]=1, phi[1]=1, are assigned nonzero values to be consistent with the BCs used in pde_1a. Also, the counter for the calls to pde_1a is initialized.

- The system of 43 ODEs is integrated by the library integrator `lsodes` (available in `deSolve`) with the sparse matrix option specified. As expected, the inputs to `lsodes` are the ODE function, `pde_1a`, the IC vector `u0`, and the vector of output values of t, `tout`. The length of `u0` (43) informs `lsodes` how many ODEs are to be integrated. `func,y,times` are reserved names.

```
#
# ODE integration
  out=lsodes(y=u0,times=tout,func=pde_1a,
      sparsetype="sparseint",rtol=1e-6,atol=1e-6,
      maxord=5);
  nrow(out)
  ncol(out)
```

The numerical solution to the ODEs is returned in matrix `out`. In this case, `out` has the dimensions 6×44. The offset $43 + 1 = 44$ is required since the first element of each column has the output t (also in `tout`), and the 2, ... , 44 column elements have the 43 ODE solutions. This indexing of `out` is used next.

- The numerical solutions for c, ϕ are taken from the solution matrix `out` returned by `lsodes`. The intervals in t and x are covered by two nested `for`s (with indices `it` and `i` for t and x, respectively). The offset `i+1` accounts for t placed in the first column position of `out` (`out[it,1]=tout[it]`).

```
#
# Arrays for display
    c=matrix(0,nrow=n,ncol=nout);
  phi=matrix(0,nrow=n,ncol=nout);
  for(it in 1:nout){
  for(i   in 1:n){
      c[i,it]=out[it,i+1];
    phi[i,it]=out[it,i+1+n];
  }
  }
  }
```

- The solutions for c, ϕ are displayed numerically and graphically. A pair of nested `for`s step through t and x. `matplot` is used to plot $c(x,t), \phi(x,t)$ against x with t as a parameter.

```
#
# Numerical solutions
  cat(sprintf("\n              t         x      c(x,t)
                    phi(x,t)"));
```

```
      for(it in 1:nout){
      for(i   in 1:n){
        cat(sprintf("\n %10.5f%10.4f%10.4f%10.4f",
          tout[it],x[i],c[i,it],phi[i,it]));
      }
        cat(sprintf("\n"));
      }
      matplot(x,c,type="l",lwd=2,col="black",lty=1,
              xlab="x",ylab="c(x,t)",main=
              Poisson-Nernst-Planck");
      matplot(x,phi,type="l",lwd=2,col="black",lty=1,
              xlab="x",ylab="phi(x,t)",main=
              "Poisson-Nernst-Planck");
  #
  # Calls to ODE routine
    cat(sprintf("\n\n  ncall = %3d\n",ncall));
```

The number of calls to pde_1a is displayed at the end of the solution.

The ODE routine pde_1a called by lsodes in the main program of Listing 7.1 is considered next.

7.3 ODE Routine

pde_1a for Eqs. (7.3) follows.

```
  pde_1a=function(t,u,parms){
#
# Function pde_1a computes the t derivative
# vector of c(x,t), phi(x,t), ci(t)
#
# One vector to two vectors and a scalar
  c=rep(0,n);phi=rep(0,n);
  for(i in 1:n){
    c[i]=u[i];
  phi[i]=u[i+n];
    }
  ci=u[2*n+1];
#
# BCs
  c[1]=1;phi[1]=1;
#
# cx
  table=splinefun(x,c);
```

```
   cx=table(x,deriv=1);
#
# BC
   cx[n]=km*(ci-c[n]);
#
# cxx
   table=splinefun(x,cx);
   cxx=table(x,deriv=1);
#
# phix
   table=splinefun(x,phi);
   phix=table(x,deriv=1);
#
# BC
   phix[n]=0;
#
# phixx
   table=splinefun(x,phix);
   phixx=table(x,deriv=1);
#
# c*phix
   cphix=rep(0,n);
   for(i in 1:n){
     cphix[i]=c[i]*phix[i];
   }
#
# (c*phix)x
   table=splinefun(x,cphix);
   cphixx=table(x,deriv=1);
#
# PDEs
   ct=rep(0,n);phit=rep(0,n);
   for(i in 1:n){
       ct[i]=cxx[i]-c11*cx[i]+c12*cphixx[i];
     phit[i]=(1/mu)*(phixx[i]-c21*c[i]);
   }
   ct[1]=0;phit[1]=0;
   cit=(1/tau)*(c[n]-ci);
#
# Increment calls to pde_1a
   ncall <- ncall+1;
#
# Two vectors and a scalar to one vector
```

```
    ut=rep(0,2*n+1);
    for(i in 1:n){
        ut[i]=ct[i];
      ut[i+n]=phit[i];
    }
    ut[2*n+1]=cit;
#
# Return derivative vector
    return(list(c(ut)));
    }
```

Listing 7.2 ODE/PDE routine pde_1a for Eqs. (7.3).

We can note the following details about pde_1a:

- The function is defined.

```
    pde_1a=function(t,u,parms){
#
# Function pde_1a computes the t derivative
# vector of c(x,t), phi(x,t), ci(t)
```

t is the current value of t in Eqs. (7.3). u is the $2(21) + 1 = 43$-vector. parm is an argument to pass parameters to pde_1a (unused, but required in the argument list). The arguments must be listed in the order stated to properly interface with lsodes called in the main program. The derivative vector of the LHS of Eqs. (7.3) is calculated next and returned to lsodes.
- u is placed in two vectors, c, phi and one scalar ci, in accordance with Eqs. (7.3).

```
#
# One vector to two vectors and a scalar
    c=rep(0,n);phi=rep(0,n);
    for(i in 1:n){
       c[i]=u[i];
      phi[i]=u[i+n];
    }
    ci=u[2*n+1];
```

Note that 2*n+1 = 43.
- BCs (7.4b,d) are defined numerically.

```
#
# BCs
    c[1]=1;phi[1]=1;
```

- $\dfrac{\partial c}{\partial x}$ is computed with `splinefun`.

```
#
# cx
  table=splinefun(x,c);
  cx=table(x,deriv=1);
```

- BC (7.4c) redefines $\dfrac{\partial c(x,t)}{\partial x}$ at $x = x_u$.

```
#
# BC
  cx[n]=km*(ci-c[n]);
```

The mass transfer coefficient k_m in Eq. (7.4c) is defined numerically in the main program of Listing 7.1.

- The derivative $\dfrac{\partial^2 c}{\partial x^2}$ in Eq. (7.3a) is computed by stagewise differentiation.

```
#
# cxx
  table=splinefun(x,cx);
  cxx=table(x,deriv=1);
```

- $\dfrac{\partial \phi}{\partial x}$ is computed.

```
#
# phix
  table=splinefun(x,phi);
  phix=table(x,deriv=1);
```

- BC (7.4e) redefines $\dfrac{\partial \phi(x,t)}{\partial x}$ at $x = x_u$.

```
#
# BC
  phix[n]=0;
```

- The derivative $\dfrac{\partial^2 \phi}{\partial x^2}$, in Eqs. (7.3b,c) is computed by stagewise differentiation.

```
#
# phixx
  table=splinefun(x,phix);
  phixx=table(x,deriv=1);
```

- The product (cE) in Eq. (7.3a) is computed (from Table 7.1, in 1D, $E = -\dfrac{\partial \phi}{\partial x}$).

Then the derivative $\dfrac{\partial(cE)}{\partial x} = -\dfrac{\partial\left(c\dfrac{\partial \phi}{\partial x}\right)}{\partial x}$ (cphixx) is computed.

```
#
# c*phix
  cphix=rep(0,n);
  for(i in 1:n){
    cphix[i]=c[i]*phix[i];
  }
#
# (c*phix)x
  table=splinefun(x,cphix);
  cphixx=table(x,deriv=1);
```

- The LHSs of Eqs. (7.3a,c), $\dfrac{\partial c}{\partial t}$ and $\dfrac{\partial \phi}{\partial t}$, are computed in a for. $E = -\dfrac{\partial \phi}{\partial x}$ introduces a sign change in the (cE) $(c12)$ term of Eq. (7.3a)).

```
#
# PDEs
  ct=rep(0,n);phit=rep(0,n);
  for(i in 1:n){
      ct[i]=cxx[i]-c11*cx[i]+c12*cphixx[i];
    phit[i]=(1/mu)*(phixx[i]-c21*c[i]);
  }
  ct[1]=0;phit[1]=0;
  cit=(1/tau)*(c[n]-ci);
```

The convergence parameter in Eq. (7.3c) is $\mu = 0.1$ set in the main program of Listing 7.1. This is an arbitrary parameter selected to achieve approximate convergence of the solution of Eq. (7.3c) to the solution of Eq. (7.3b). The derivatives $\dfrac{\partial c(x = x_l = 0, t)}{\partial t} = 0$, $\dfrac{\partial \phi(x = x_l = 0, t)}{\partial t} = 0$ from BCs (7.4b,d) since the functions $c_l(t), \phi_l(t)$ are constants $(c_l(t) = 1, \ \phi_l(t) = 1$ from the BCs set above). Finally, $\dfrac{dc_i}{dt}$ of Eq. (7.3d) is programmed (cit). The mixing (residence) time $\tau = 0.01$ is one-tenth of the timescale for the PDEs $(0 \le t \le 0.1)$, and can be adjusted in the main program of Listing 7.1 (based on an estimate for a cell).

- The counter for the calls to pde_1a is incremented and its value is returned to the calling program (of Listing 7.1) with the <<- operator.

```
#
# Increment calls to pde_1a
  ncall <- ncall+1;
```

- The derivatives $\dfrac{\partial c}{\partial t}, \dfrac{\partial \phi}{\partial t}$, and $\dfrac{dc_i}{dt}$ are placed in a single derivative vector, ut, for return to lsodes (the ODE integrator called in the main program of Listing 7.1).

```
#
# Two vectors and a scalar to one vector
   ut=rep(0,2*n+1);
   for(i in 1:n){
        ut[i]=ct[i];
     ut[i+n]=phit[i];
   }
   ut[2*n+1]=cit;
```

- The derivative vector is returned to lsodes, which requires a list. c is the R vector utility. The combination of return, list, c gives lsodes the required derivative vector for the next step along the solution.

```
#
# Return derivative vector
   return(list(c(ut)));
   }
```

The final } concludes pde_1a.

This completes the programming of Eqs. (7.3) and (7.4). The numerical and graphical output is considered next.

7.4   Model Output

Execution of the preceding main program and subordinate routine pde_1a produces the following abbreviated numerical output in Table 7.2 for no electrodiffusion (c12=0) set in the main program of Listing 7.1.

```
#
# No electrodiffusion
   if(ncase==1){
      n=21;c12=0;
      t0=0;tf=0.1
   }
```

with ncase=1.

We can note the following details about this output:

- c11 = 1.000e-01 (the Peclet number in Eq. (7.3a)) indicates a small ratio of convection to diffusion in the channel.
- c12 = 0.000e+00 indicates the effect of electrodiffusion on $c(x, t)$ (from Eq. (7.3a)) is not included for this particular case.
- c21 = 4.513e-23 indicates the contribution of $c(x, t)$ to the electric potential $\phi(x, t)$ (from Eq. (7.3c)) is small.

Table 7.2 Abbreviated output for Eqs. (7.3), c12=0.

```
c11  =    1.000e-01
c12  =    0.000e+00
c21  =    4.513e-23
```

[1] 6

[1] 44

t	x	c(x,t)	phi(x,t)
0.00000	0.0000	1.0000	1.0000
0.00000	0.0500	0.0000	0.0000
0.00000	0.1000	0.0000	0.0000
0.00000	0.1500	0.0000	0.0000
0.00000	0.2000	0.0000	0.0000
0.00000	0.2500	0.0000	0.0000
0.00000	0.3000	0.0000	0.0000
0.00000	0.3500	0.0000	0.0000
0.00000	0.4000	0.0000	0.0000
0.00000	0.4500	0.0000	0.0000
0.00000	0.5000	0.0000	0.0000
0.00000	0.5500	0.0000	0.0000
0.00000	0.6000	0.0000	0.0000
0.00000	0.6500	0.0000	0.0000
0.00000	0.7000	0.0000	0.0000
0.00000	0.7500	0.0000	0.0000
0.00000	0.8000	0.0000	0.0000
0.00000	0.8500	0.0000	0.0000
0.00000	0.9000	0.0000	0.0000
0.00000	0.9500	0.0000	0.0000
0.00000	1.0000	0.0000	0.0000

```
                  .            .
                  .            .
                  .            .
       Output for t = 0.02 to 0.08 removed
                  .            .
                  .            .
                  .            .
```

| 0.10000 | 0.0000 | 1.0000 | 1.0000 |
| 0.10000 | 0.0500 | 0.9133 | 0.9915 |

(Continued)

Table 7.2 (Continued)

0.10000	0.1000	0.8272	0.9831
0.10000	0.1500	0.7429	0.9748
0.10000	0.2000	0.6613	0.9666
0.10000	0.2500	0.5834	0.9587
0.10000	0.3000	0.5100	0.9510
0.10000	0.3500	0.4417	0.9436
0.10000	0.4000	0.3789	0.9365
0.10000	0.4500	0.3219	0.9299
0.10000	0.5000	0.2709	0.9236
0.10000	0.5500	0.2259	0.9179
0.10000	0.6000	0.1868	0.9126
0.10000	0.6500	0.1533	0.9079
0.10000	0.7000	0.1251	0.9038
0.10000	0.7500	0.1019	0.9002
0.10000	0.8000	0.0834	0.8973
0.10000	0.8500	0.0693	0.8950
0.10000	0.9000	0.0592	0.8934
0.10000	0.9500	0.0530	0.8924
0.10000	1.0000	0.0504	0.8920

```
ncall = 279
```

- As expected, the solution array out has the dimensions 6 x 44 for 6 output points and 43 ODEs, with t included as an additional element, that is, 43 + 1 = 44.
- The interval in x is $((1 - 0)/(21 - 1) = 0.05$ as expected.
- The solution in Table 7.2 confirms ICs (7.4a,f,g) as coded in Listing 7.1 (for $t = 0$). Also, BCs (7.4b,d) from Listing 7.1 are confirmed.
- The variation in $c(x, t = 0.1)$ is a monotonic increase in x. This contrasts with a significantly different variation when electrodiffusion is included ($c_{12} > 0$) as discussed next.
- The computational effort is modest, ncall = 279.

The solution is displayed graphically in Figure 7.1a,b. Note in particular the discontinuity in the ICs of $c(x = 0, t = 0)$ and $\phi(x = 0, t = 0)$.[4]

---

4 The discontinuities in Figure 7.1a,b (at $t = 0$) can be removed by using c[,2:nout], phi[,2:nout] in the calls to matplot. The first columns corresponding to $c(x, t = 0)$ and $\phi(x, t = 0)$ are thereby circumvented.

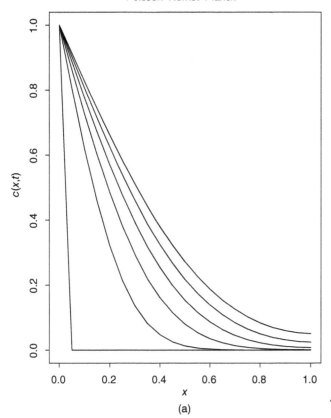

Poisson–Nernst–Planck

Figure 7.1 (a) Numerical $c(x, t)$ from Eqs. (7.3) and (7.4), c12=0.

The solution when electrodiffusion is included ($c_{12} > 0$ in Eq. (7.3a)) with

```
#
# Electrodiffusion
  if(ncase==2){
    n=21;
    t0=0;tf=0.2
  }
```

Abbreviated numerical output is in Table 7.3.

We can note the following details about this output:

- c12 = 1.934e+02 so that a large effect of electrodiffusion on $c(x, t)$ (from Eq. (7.3a)) can be expected (and is confirmed by the solution for $t > 0$).
- The ICs and BCs (for $t = 0$) are again confirmed as with Table 7.2.

Poisson–Nernst–Planck

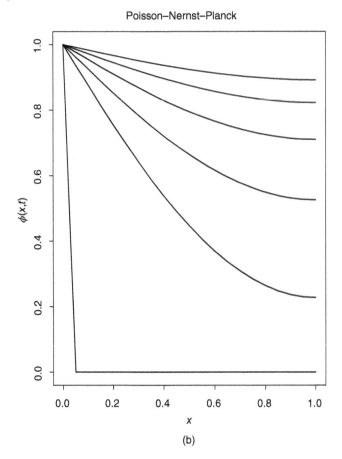

Figure 7.1 (b) Numerical $\phi(x, t)$ from Eqs. (7.3) and (7.4), c12=0.

- $c(x, t)$ near $x = 0$ has a pronounced large transient that goes through a maximum, for example, 0.04000 1.0000 125.6358 0.5255, then decreases, 0.20000 1.0000 42.9708 0.9908.
- The total calls to pde_1a from lsodes has increased substantially, ncall = 7107.

Clearly the effect of electrodiffusion is substantial as reflected in the change of the solutions with c12, for example, comparison of Figure 7.1a,b with Figure 7.2a,b. The reasons for the variations in the solutions can be further elucidated by computing and displaying the RHS terms of Eqs. (7.3) and then the LHS derivatives in t. This is readily accomplished by using the numerical solutions $c(x, t)$ and $\phi(x, t)$ ([3], 27–41).

Table 7.3 Abbreviated output for Eqs. (7.3), `c12>0`.

```
c11 =   1.000e-01
c12 =   1.934e+02
c21 =   4.513e-23

[1]  6

[1]  44
```

t	x	c(x,t)	phi(x,t)
0.00000	0.0000	1.0000	1.0000
0.00000	0.0500	0.0000	0.0000
0.00000	0.1000	0.0000	0.0000
0.00000	0.1500	0.0000	0.0000
0.00000	0.2000	0.0000	0.0000
0.00000	0.2500	0.0000	0.0000
0.00000	0.3000	0.0000	0.0000
0.00000	0.3500	0.0000	0.0000
0.00000	0.4000	0.0000	0.0000
0.00000	0.4500	0.0000	0.0000
0.00000	0.5000	0.0000	0.0000
0.00000	0.5500	0.0000	0.0000
0.00000	0.6000	0.0000	0.0000
0.00000	0.6500	0.0000	0.0000
0.00000	0.7000	0.0000	0.0000
0.00000	0.7500	0.0000	0.0000
0.00000	0.8000	0.0000	0.0000
0.00000	0.8500	0.0000	0.0000
0.00000	0.9000	0.0000	0.0000
0.00000	0.9500	0.0000	0.0000
0.00000	1.0000	0.0000	0.0000
.	.		
.	.		
.	.		

(Continued)

Table 7.3 (Continued)

```
Output for t = 0.04, x = 0 to 0.85 removed
                     .             .
                     .             .
                     .             .
      0.04000      0.9000      40.5972       0.5313
      0.04000      0.9500      90.2836       0.5270
      0.04000      1.0000     125.6358       0.5255
                     .             .
                     .             .
                     .             .
Output for t = 0.08, x = 0 to 0.85 removed
                     .             .
                     .             .
                     .             .
      0.08000      0.9000      66.2572       0.8253
      0.08000      0.9500      93.1980       0.8237
      0.08000      1.0000     102.9498       0.8231
                     .             .
                     .             .
                     .             .
Output for t = 0.12, x = 0 to 0.85 removed
                     .             .
                     .             .
                     .             .
      0.12000      0.9000      63.0178       0.9349
      0.12000      0.9500      73.3511       0.9343
      0.12000      1.0000      76.7880       0.9341
                     .             .
                     .             .
                     .             .
Output for t = 0.16, x = 0 to 0.85 removed
                     .             .
                     .             .
                     .             .
```

Table 7.3 (Continued)

0.16000	0.9000	51.0088	0.9757
0.16000	0.9500	55.0389	0.9755
0.16000	1.0000	56.4271	0.9754
0.20000	0.0000	1.0000	1.0000
0.20000	0.0500	1.2377	0.9993
0.20000	0.1000	1.5311	0.9986
0.20000	0.1500	1.9241	0.9979
0.20000	0.2000	2.4516	0.9972
0.20000	0.2500	3.1780	0.9965
0.20000	0.3000	4.1477	0.9958
0.20000	0.3500	5.4443	0.9952
0.20000	0.4000	7.1108	0.9946
0.20000	0.4500	9.2282	0.9941
0.20000	0.5000	11.8074	0.9935
0.20000	0.5500	14.8848	0.9930
0.20000	0.6000	18.3938	0.9926
0.20000	0.6500	22.2804	0.9922
0.20000	0.7000	26.3663	0.9918
0.20000	0.7500	30.4924	0.9915
0.20000	0.8000	34.3839	0.9913
0.20000	0.8500	37.8249	0.9911
0.20000	0.9000	40.5267	0.9910
0.20000	0.9500	42.3221	0.9909
0.20000	1.0000	42.9708	0.9908

```
ncall = 7107
```

Poisson–Nernst–Planck

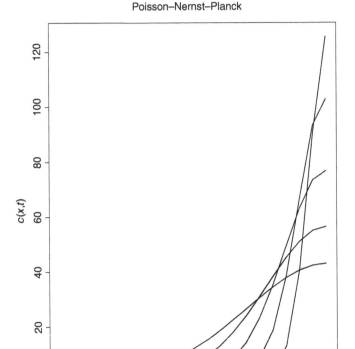

(a)

Figure 7.2 (a) Numerical $c(x, t)$ from Eqs. (7.3) and (7.4), c12=1.934e+02.

7.5 Summary and Conclusions

The PNP/ODE/PDE system of Eqs. (7.3) and (7.4) is accommodated within the SCMOL framework provided the elliptic PDE, Eq. (7.3b), is changed to a parabolic PDE, Eq. (7.3c). Also, Eq. (7.3d) demonstrates how an ODE can be used in a PDE BC. The nonlinearity $\dfrac{\partial(cE)}{\partial x} = -\dfrac{\partial(c\frac{\partial\phi}{\partial x})}{\partial x}$ in Eq. (7.3a), and programmed in pde_1a of Listing 7.2, is particularly noteworthy.

With the routines in Listings 7.1 and 7.2 available, numerical experiments can be carried out to gain further insights into the characteristics of the PNP/ODE/PDE model. This use of the routines might include the following:

Poisson–Nernst–Planck

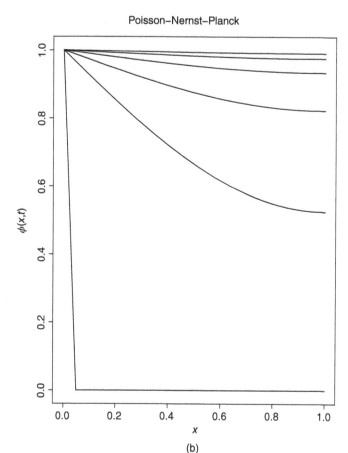

(b)

Figure 7.2 (b) Numerical $\phi(x, t)$ from Eqs. (7.3) and (7.4), c12=1.934e+02.

- Variation in the parameters, for example, c12 (Eq. (7.3a)), k_m (Eq. (7.4c)), D (Eq. (7.3a)), μ (Eq. (7.3c)), τ (Eq. (7.3d)). As a word of caution, the routines cannot be guaranteed in advance to give a stable, accurate numerical solution with an acceptable level of computation. Some experimentation may be required to achieve this outcome, for example, variation in the number of spatial grid points, adjustment of the timescale, variation of the options and parameters of the ODE integrator such as the error tolerances, and selection of another ODE integrator.
- Additional output, for example, the RHS and LHS terms of Eqs. (7.3), other combinations of variables such as the chemical flux at $x = x_u = x_L$ (from Eq. (7.1a)) and the solution to Eq. (7.3d).
- Addition of chemical species to reflect a multicomponent system.

Ideally, a model and R implementation will result that produces solutions in agreement with experimental data. A process of trial and error may be required to achieve convergence to an acceptable and useful degree of agreement between the model and data.

References

1 Barcilon, V., D.-P. Chen, and R.S. Eisenberg (1992), Ion flow through narrow membrane channels - Part II, *SIAM Journal on Applied Mathematics*, **52**, no. 5, 1405–1425.

2 Coalson, R.D., and M.G. Kurnikova (2005), Poisson-Nernst-Planck theory approach to the calculation of current through biological ion channels, *IEEE Transactions Nanobioscience*, **4**, no. 1, 81–93.

3 Schiesser, W.E. (2014), *Differential Equation Analysis in Biomedical Science and Engineering: Partial Differential Equation Applications in R*, John Wiley and Sons, Hoboken, NJ.

8

Fokker–Planck Equation

The Fokker–Planck (FP) equation has several interpretations and applications, particularly for stochastic dynamic systems. For example, it can describe the time evolution of the probability density function of particle Brownian motion [1]. A 1D version of the FP equation is considered next. The spline collocation method of lines (SCMOL) solution is implemented in a series of R routines.

8.1 PDE Model

The FP in 1D is

$$\frac{\partial}{\partial t}u(x,t) = -\frac{\partial}{\partial x}[v(x,t)u(x,t)] + \frac{\partial^2}{\partial x^2}[D(x,t)u(x,t)] \qquad (8.1a)$$

where $v(x,t)$ and $D(x,t)$ are the Ito drift and diffusion, respectively (see Table 8.1 for the definition of the model variables and parameters). Equation (8.1a) is a *convection–diffusion* equation. If a source term is included in the RHS, it is termed a *convection–diffusion–reaction* equation.

Since Eq. (8.1a) is first order in t and second order in x, it requires one initial condition (IC) and two boundary conditions (BCs), for example,

$$u(x, t = 0) = f(x) \qquad (8.1b)$$

$$u(x \to \pm\infty, t) = 0 \qquad (8.1c)$$

$$\frac{\partial u(x \to \pm\infty, t)}{\partial x} = 0 \qquad (8.1d)$$

BCs (8.1c) are homogeneous Dirichlet and BCs (8.1d) are homogeneous Neumann (two cases considered subsequently). $f(x)$ is an IC function to be specified. For the example that follows, $f(x) = e^{-cx^2}$ (a Gaussian function centered at $x = 0$).

If the drift is zero and the diffusion coefficient is constant at D, Eq. (8.1a) becomes

$$\frac{\partial}{\partial t}u(x,t) = D\frac{\partial^2}{\partial x^2}u(x,t) \qquad (8.2)$$

which is the 1D diffusion equation (Fick's or Fourier's second law).

Spline Collocation Methods for Partial Differential Equations: With Applications in R, First Edition.
William E. Schiesser.
© 2017 John Wiley & Sons, Inc. Published 2017 by John Wiley & Sons, Inc.
Companion website: www.wiley.com/go/Spline_Collocation

In the subsequent example, v, D are taken as constants, but in principle, the variable coefficient cases $v(x, t), D(x, t)$ can be accommodated within the SCMOL framework. Also, the problem can be extended to the nonlinear, variable coefficient case $v(x, t, u(x, t)), D(x, t, u(x, t))$.

Equations (8.1) constitute the PDE model to be considered.[1] The solution of Eqs. (8.1) is programmed in the following routines.

Table 8.1 Variables and parameters for the FP Eqs. (8.1).

Variable parameter	Interpretation
u	dependent variable (e.g., probability density function)
t	time
x	distance from the mean value
$v(x, t)$	drift (or velocity) of $u(x, t)$
$D(x, t)$	diffusivity for the spread of $u(x, t)$
c	constant in the IC Gaussian IC function

8.2 Main Program

A main program for Eqs. (8.1) follows.

```
#
# Fokker-Planck
#
# Delete previous workspaces
  rm(list=ls(all=TRUE))
#
# Access ODE integrator
  library("deSolve");
#
# Access functions for numerical solution
  setwd("f:/collocation/chap8");
  source("pde_1a.R");
#
# Grid (in x)
```

---

1 Equation (8.1a) is parabolic–hyperbolic (with a first-order initial value derivative in t, a first-order boundary value derivative in x (hyperbolic) and a second-order boundary value derivative in x (parabolic)).

```
  n=151;xl=-5;xu=10;
  x=seq(from=xl,to=xu,by=(xu-xl)/(n-1));
#
# Parameters
  D=0.1;v=1;c=2;
#
# Independent variable for ODE integration
  nout=6;t0=0;tf=5;
  tout=seq(from=t0,to=tf,by=(tf-t0)/(nout-1));
#
# Initial condition
  u0=rep(0,n);
  for(i in 1:n){
    u0[i]=exp(-c*x[i]^2);
  }
  ncall=0;
#
# ODE integration
  out=lsodes(y=u0,times=tout,func=pde_1a,
      sparsetype="sparseint",rtol=1e-6,atol=1e-6,
      maxord=5);
  nrow(out)
  ncol(out)
#
# Arrays for display
   u=matrix(0,nrow=n,ncol=nout);
  for(it in 1:nout){
  for(i   in 1:n){
    u[i,it]=out[it,i+1];
  }
  }
#
# Numerical solution
  cat(sprintf("\n        t      x      u(x,t)"));
  for(it in 1:nout){
  iv=seq(from=1,to=n,by=10);
  for(i   in iv){
    cat(sprintf("\n %6.2f%6.2f%10.4f",
        tout[it],x[i],u[i,it]));
    }
    cat(sprintf("\n"));
    }
#
```

```
# Graphical output
  matplot(x,u,type="l",lwd=2,col="black",lty=1,
          xlab="x",ylab="u(x,t)",main=
          "Fokker-Planck");
#
# Calls to ODE routine
  cat(sprintf("\n\n  ncall = %3d\n",ncall));
```

Listing 8.1 Main program for Eqs. (8.1).

We can note the following details about Listing 8.1.

- Previous workspaces are removed. Then the ODE integrator library deSolve is accessed. Note that the setwd (set working directory) uses / rather than the usual /. The setwd requires editing for the local computer (to specify the directory with the R files).

```
#
# Fokker-Planck
#
# Delete previous workspaces
  rm(list=ls(all=TRUE))
#
# Access ODE integrator
  library("deSolve");
#
# Access functions for numerical solution
  setwd("f:/collocation/chap8");
  source("pde_1a.R");
```

pde_1a is the routine for the MOL/ODEs (discussed subsequently).
- A uniform grid in x with 151 points is defined with the seq utility for the interval $x_l = -5 \le x \le x_u = 10$. Therefore, the vector x has the values $x = -5, -4.9, \ldots, 10$.

```
#
# Grid (in x)
  n=151;xl=-5;xu=10;
  x=seq(from=xl,to=xu,by=(xu-xl)/(n-1));
```

An approximation to BCs (8.1c,d) is used.

$$u(x \pm \infty, t) \approx u(x = -5, 10, t) = 0$$
$$\frac{\partial u(x \pm \infty, t)}{\partial x} \approx \frac{\partial u(x = -5, 10, t)}{\partial x} = 0$$

that is, $x = -5, 10$ is used in place of $x \pm \infty$. The accuracy of this approximation is discussed when considering the numerical solution. The interval in x is not symmetric with respect to $x = 0$ since the Gaussian pulse IC travels left to right in x.

- Parameters (constants) used in Eqs. (8.1) are defined numerically.

```
#
# Parameters
  D=0.1;v=1;c=2;
```

This choice of values is explained when the numerical solution to Eqs. (8.1) is discussed.

- A uniform grid in t of 6 output points is defined with the seq utility for the interval $0 \le t \le t_f = 5$. Therefore, the vector tout has the values $t = 0, 1, \ldots, 5$. This interval in t was determined by trial and error to define the numerical solution variation with t.

```
#
# Independent variable for ODE integration
  nout=6;t0=0;tf=5;
  tout=seq(from=t0,to=tf,by=(tf-t0)/(nout-1));
```

At this point, the intervals of x and t in Eqs. (8.1) are defined.

- The IC of Eq. (8.1b) is defined (a Gaussian function centered at $x = 0$).

```
#
# Initial condition
  u0=rep(0,n);
  for(i in 1:n){
    u0[i]=exp(-c*x[i]^2);
  }
  ncall=0;
```

The center of the Gaussian pulse is at $x = 0$. The IC pulse will move left to right in x as a result of the drift (convection) term in Eq. (8.1a). Also, the counter for the calls to pde_1a is initialized.

- The system of 151 ODEs is integrated by the library integrator lsodes (available in deSolve) with the sparse matrix option specified. As expected, the inputs to lsodes are the ODE function, pde_1a, the IC vector u0, and the vector of output values of t, tout. The length of u0 (151) informs lsodes how many ODEs are to be integrated. func,y,times are reserved names.

```
#
# ODE integration
  out=lsodes(y=u0,times=tout,func=pde_1a,
      sparsetype="sparseint",rtol=1e-6,atol=1e-6,
      maxord=5);
```

```
nrow(out)
ncol(out)
```

The numerical solution to the ODEs is returned in matrix out. In this case, out has the dimensions 6×152. The offset $151 + 1 = 152$ is required since the first element of each column has the output t (also in tout), and the $2, \ldots, 152$ column elements have the 151 ODE solutions. This indexing of out is used next.

- The numerical solution for $u(x, t)$ is taken from the solution matrix out returned by lsodes. The intervals in t and x are covered by two nested fors (with indices it and i for t and x, respectively). The offset i+1 accounts for t placed in the first column position of out.

```
#
# Arrays for display
   u=matrix(0,nrow=n,ncol=nout);
   for(it in 1:nout){
   for(i   in 1:n){
     u[i,it]=out[it,i+1];
   }
   }
```

- The solution for u is displayed numerically. A pair of nested fors step through t and x.

```
#
# Numerical solution
   cat(sprintf("\n        t      x      u(x,t)"));
   for(it in 1:nout){
   iv=seq(from=1,to=n,by=10);
   for(i   in iv){
     cat(sprintf("\n %6.2f%6.2f%10.4f",
         tout[it],x[i],u[i,it]));
   }
   cat(sprintf("\n"));
   }
```

The subscript in the inner for steps by 10 (rather than 1) to conserve space.

- The numerical solution is displayed with matplot (for $u(x, t)$ against x with t as a parameter).

```
#
# Graphical output
   matplot(x,u,type="l",lwd=2,col="black",lty=1,
           xlab="x",ylab="u(x,t)",main=
           "Fokker-Planck");
```

```
#
# Calls to ODE routine
  cat(sprintf("\n\n  ncall = %3d\n",ncall));
```

The number of calls to pde_1a is displayed at the end of the solution.

The ODE routine pde_1a called by lsodes in the main program of Listing 8.1 is considered next.

8.3 ODE Routine

pde_1a for Eqs. (8.1) follows.

```
  pde_1a=function(t,u,parms){
#
# Function pde_1a computes the t derivative
# vector of u(x,t)
#
# BCs
  u[1]=0;u[n]=0;
#
# ux
  tablex=splinefun(x,u);
  ux=tablex(x,deriv=1);
#
# uxx
  tablexx=splinefun(x,u);
  uxx=tablexx(x,deriv=2);
#
# PDE
  ut=rep(0,n);
  for(i in 2:(n-1)){
    ut[i]=-v*ux[i]+D*uxx[i];
  }
  ut[1]=0;ut[n]=0;
#
# Increment calls to pde_1a
  ncall <<- ncall+1;
#
# Return derivative vector
  return(list(c(ut)));
  }
```

Listing 8.2 ODE/PDE routine pde_1a for Eqs. (8.1), Dirichlet BCs.

We can note the following details about pde_1a.

• The function is defined.

```
pde_1a=function(t,u,parms){
#
# Function pde_1a computes the t derivative
# vector of u(x,t)
```

t is the current value of t in Eqs. (8.1). u is the 151-vector. parm is an argument to pass parameters to pde_1a (unused but required in the argument list). The arguments must be listed in the order stated to properly interface with lsodes called in the main program. The derivative vector of the LHS of Eqs. (8.1) is calculated next and returned to lsodes.

• Homogeneous Dirichlet BCs (8.1c) are implemented.

```
#
# BCs
  u[1]=0;u[n]=0;
```

• $\dfrac{\partial u}{\partial x}$ of Eq. (8.1a) is computed with splinefun.

```
#
# ux
  tablex=splinefun(x,u);
  ux=tablex(x,deriv=1);
```

• $\dfrac{\partial^2 u}{\partial x^2}$ of Eq. (8.1a) is computed with splinefun.

```
#
# uxx
  tablexx=splinefun(x,u);
  uxx=tablexx(x,deriv=2);
```

Note the specification of a second derivative with deriv=2.

• The LHS derivative of Eq. (8.1a), $\dfrac{\partial}{\partial t} u(x,t)$, is computed.

```
#
# PDE
  ut=rep(0,n);
  for(i in 2:(n-1)){
    ut[i]=-v*ux[i]+D*uxx[i];
  }
  ut[1]=0;ut[n]=0;
```

Homogeneous BCs (8.1c) have zero derivatives in t (note the use of subscripts 1,n corresponding to $x = -5, 10$).

- The counter for the calls to pde_1a is returned to the main program of Listing 8.1 by the <<- operator.

```
#
# Increment calls to pde_1a
  ncall <<- ncall+1;
```

- The derivative in ut is returned to lsodes (the ODE integrator called in the main program of Listing 8.1).

```
#
# Return derivative vector
  return(list(c(ut)));
  }
```

The final } concludes pde_1a.

This completes the programming of Eqs. (8.1). The numerical and graphical output is considered next.

8.4 Model Output

Execution of the preceding main program and subordinate routine pde_1a produces the following abbreviated numerical output in Table 8.2.

Table 8.2 Abbreviated output for Eqs. (8.1), Dirichlet BCs.

```
[1] 6

[1] 152

        t     x      u(x,t)
     0.00 -5.00     0.0000
     0.00 -4.00     0.0000
     0.00 -3.00     0.0000
     0.00 -2.00     0.0003
     0.00 -1.00     0.1353
     0.00  0.00     1.0000
     0.00  1.00     0.1353
```

(Continued)

Table 8.2 (Continued)

```
0.00   2.00      0.0003
0.00   3.00      0.0000
0.00   4.00      0.0000
0.00   5.00      0.0000
0.00   6.00      0.0000
0.00   7.00      0.0000
0.00   8.00      0.0000
0.00   9.00      0.0000
0.00  10.00      0.0000

         .          .
         .          .
         .          .

Output for t=1,2,3,4,
        x=1,2,3,4
1.00   1.00      0.7444
2.00   2.00      0.6194
3.00   3.00      0.5418
4.00   4.00      0.4875

         .          .
         .          .
         .          .

5.00  -5.00      0.0000
5.00  -4.00      0.0000
5.00  -3.00      0.0000
5.00  -2.00      0.0000
5.00  -1.00      0.0000
5.00   0.00      0.0000
5.00   1.00      0.0007
5.00   2.00      0.0122
5.00   3.00      0.0904
5.00   4.00      0.2999
5.00   5.00      0.4468
5.00   6.00      0.2999
5.00   7.00      0.0904
5.00   8.00      0.0122
5.00   9.00      0.0007
5.00  10.00      0.0000

ncall = 402
```

We can note the following details about this output:

- As expected, the solution array out has the dimensions 6 x 152 for 6 output points and 151 ODEs, with t included as an additional element, that is, $151 + 1 = 152$.
- The interval in x is $((10 - (-5))/(151 - 1) = 0.1$ as expected (with every tenth value in Table 8.2).
- The approximation to BCs (8.1c)

$$u(x = -5, 10, t) = 0$$

is confirmed (for $t = 5$) and the neighboring values remain approximately at zero also, so the interval $-5 \leq x \leq 10$ is effectively $-\infty \leq x \leq \infty$ (using a finite interval or spatial domain in place of an infinite interval is a requirement of a numerical approach since $\pm\infty$ is not possible on a computer).
- The IC of Eq. (8.1b) appears to be confirmed with the maximum value $u(x = 0, t = 0) = 1$. Also, the symmetry of the Gaussian pulse around $x = 0$ is clear.
- The peak value of $u(x = 5, t = 5) = 0.4468$ demonstrates the drift (convection, movement left to right) in x. The other peak values are indicated in Table 8.2 corresponding to $v = 1$ (so that the peaks occur at $x = t$).
- The peak values decrease with t due to the effect of the diffusion (dispersion) from $D = 1$.
- The computational effort is modest, ncall = 402.

The solution is displayed graphically in Figure 8.1. Note in particular the movement of the initial Gaussian peak centered at $x = 0$ for $v = 1$, with diffusion from $D = 1$, that is, the peaks for $t = 0, 1, 2, 3, 4, 5$ corresponding to $x = 0, 1, 2, 3, 4, 5$ for $v = 1$.[2] which will accommodate the pulse as it arrives at $x = x_u = 10$. BC (8.3) is discussed in detail in [2].

If BCs (8.1d) (homogeneous Neumann) are used, pde_1a of Listing 8.2 is modified to

```
    pde_1b=function(t,u,parms){
#
# Function pde_1b computes the t derivative
```

---

2 As a word of caution, if the pulse reaches the right boundary $x = x_u = 10$ (as a result of convection), BCs (8.1c,d) cannot be used. That is, the pulse cannot move through the boundary at $x = 10$ with zero value (BC (8.1c)) or zero slope (BC (8.1d)). Rather, an *outflow* BC is required. One possibility is

$$\frac{\partial u(x = x_u, t)}{\partial t} = -v\frac{\partial u(x = x_u, t)}{\partial x} \qquad (8.3)$$

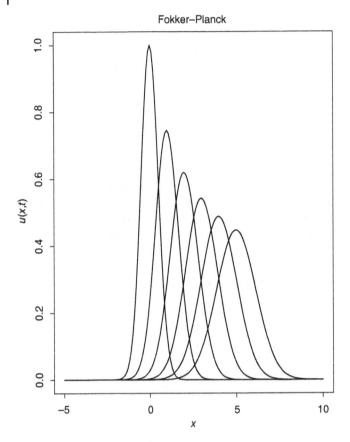

Figure 8.1 Numerical $u(x, t)$ from Eqs. (8.1).

```
# vector of u(x,t)
#
# ux
  tablex=splinefun(x,u);
  ux=tablex(x,deriv=1);
#
# BCs
  ux[1]=0;ux[n]=0;
#
# uxx
  tablexx=splinefun(x,ux);
  uxx=tablexx(x,deriv=1);
```

```
#
# PDE
  ut=rep(0,n);
  for(i in 1:n){
    ut[i]=-v*ux[i]+D*uxx[i];
  }
#
# Increment calls to pde_1b
  ncall <<- ncall+1;
#
# Return derivative vector
  return(list(c(ut)));
  }
```

Listing 8.3 ODE/PDE routine pde_1b for Eqs. (8.1), Neumann BCs.

pde_1b is similar to pde_1a of Listing 8.2 and does not require a detailed explanation. Note the change from BCs (8.1c) to BCs (8.1d).

The main program of Listing 8.1 remains unchanged (except for the use of pde_1b rather than pde_1a). Abbreviated output from pde_1b is given in Table 8.3.

Table 8.3 Abbreviated output for Eqs. (8.1), Neumann BCs.

```
[1]  6

[1]  152

        t       x      u(x,t)
     0.00  -5.00     0.0000
     0.00  -4.00     0.0000
     0.00  -3.00     0.0000
     0.00  -2.00     0.0003
     0.00  -1.00     0.1353
     0.00   0.00     1.0000
     0.00   1.00     0.1353
     0.00   2.00     0.0003
     0.00   3.00     0.0000
     0.00   4.00     0.0000
```

(Continued)

Table 8.3 (Continued)

```
0.00   5.00      0.0000
0.00   6.00      0.0000
0.00   7.00      0.0000
0.00   8.00      0.0000
0.00   9.00      0.0000
0.00  10.00      0.0000
          .          .
          .          .
          .          .
Output for t=1,2,3,4,
       x=1,2,3,4
1.00   1.00      0.7454
2.00   2.00      0.6202
3.00   3.00      0.5423
4.00   4.00      0.4879
          .          .
          .          .
          .          .
5.00  -5.00     -0.0000
5.00  -4.00      0.0000
5.00  -3.00      0.0000
5.00  -2.00      0.0000
5.00  -1.00      0.0000
5.00   0.00      0.0000
5.00   1.00      0.0007
5.00   2.00      0.0122
5.00   3.00      0.0903
5.00   4.00      0.2998
5.00   5.00      0.4472
5.00   6.00      0.2998
5.00   7.00      0.0903
5.00   8.00      0.0122
5.00   9.00      0.0007
5.00  10.00      0.0000

ncall = 413
```

Table 8.3 is similar to Table 8.2, in particular, for the peak values.

```
Table 8.2, Dirichlet BCs
   0.00   0.00     1.0000
   1.00   1.00     0.7444
   2.00   2.00     0.6194
   3.00   3.00     0.5418
   4.00   4.00     0.4875
   5.00   5.00     0.4468

Table 8.3, Neumann BCs
   0.00   0.00     1.0000
   1.00   1.00     0.7454
   2.00   2.00     0.6202
   3.00   3.00     0.5423
   4.00   4.00     0.4879
   5.00   5.00     0.4472
```

The graphical output is indistinguishable from Figure 8.1 and is not presented here.

8.5 Summary and Conclusions

The FP/PDE system of Eqs. (8.1) is readily accommodated within the SCMOL framework. With the routines in Listings 8.1–8.3 available, numerical experiments can be carried out to gain further insight into the characteristics of the FP/PDE model, or extensions of this model.

For example, an FP system with $D = D(t)$ (the diffusivity varies with t) could be programmed as (from pde_1a or pde_1b)

```
#
# PDE
  ut=rep(0,n);
  for(i in 1:n){
    ut[i]=-v*ux[i]+D*(1+t)*uxx[i];
  }
```

where D increases linearly with t (and t is available as an input argument of pde_1a or pde_b). We would expect this time variable D to give greater diffusion (dispersion) in Figure 8.1.

Similar considerations apply for the cases $v = v(x, t, u), D = D(x, u)$ so that the numerical approach provides a flexible format for variants of the FP pde.

As another example, a multicomponent model for a convection–diffusion–reaction system could be programmed as (an extension of pde_1b of Listing 8.3)

```
pde_1c=function(t,u,parms){
#
# Function pde_1c computes the t derivative vector of
# u1(x,t),u2(x,t)
#
# One vector to two vectors
  u1=rep(0,n);u2=rep(0,n);
  for(i in 1:n){
    u1[i]=u[i];
    u2[i]=u[i+n];
  }
#
# u1x
  tablex=splinefun(x,u1);
  u1x=tablex(x,deriv=1);
#
# u2x
  tablex=splinefun(x,u2);
  u2x=tablex(x,deriv=1);
#
# BCs
  u1x[1]=0;u1x[n]=0;
  u2x[1]=0;u2x[n]=0;
#
# u1xx
  tablexx=splinefun(x,u1x);
  u1xx=tablexx(x,deriv=1);
#
# u2xx
  tablexx=splinefun(x,u2x);
  u2xx=tablexx(x,deriv=1);
#
# PDEs
  u1t=rep(0,n);u2t=rep(0,n);
  for(i in 1:n){
    u1t[i]=-v1*u1x[i]+D1*u1xx[i]-k1*u1[i]+k2*u2[i];
    u2t[i]=-v2*u2x[i]+D2*u1xx[i]+k1*u1[i]-k2*u2[i];
  }
#
```

```
# Increment calls to pde_1c
  ncall <<- ncall+1;
#
# Two vectors to one vector
  ut=rep(0,2*n);
  for(i in 1:n){
    ut[i]   =u1t[i];
    ut[i+n]=u2t[i];
  }
#
# Return derivative vector
  return(list(c(ut)));
  }
```

Listing 8.4 ODE/PDE routine pde_1c for two components, Neumann BCs.

k1, k2 are kinetic rate constants. u1, u2 will differ primarily through the ICs. This example demonstrates how multicomponent systems can be accommodated within the SCMOL framework, possibly using BC (8.3) for outflow at $x = x_u$ (with BCs (8.1c) or (8.1d) for BCs at $x = x_l$).

This example demonstrates the straightforward application of SCMOL to simultaneous PDE systems, with variations in the IC and BCs.

References

1 http://en.wikipedia.org/wiki/Fokker-Planck_equation (accessed 21 December 2016).
2 Schiesser, W.E. (1996), PDE boundary conditions from minimum reduction of the PDE, *Applied Numerical Mathematics*, **20**, 171–179.

9

Fisher–Kolmogorov Equation

The Fisher–Kolmogorov (FK) equation is the diffusion equation (Fick's second law) with a logistic source term. It has a variety of applications [1] and is an ideal test problem for numerical methods because it is a nonlinear equation with an exact (analytical) solution.

A 1D version of the FK equation is considered next. The spline collocation method of lines (SCMOL) solution is implemented in a series of R routines, with comparison to the analytical solution.

9.1 PDE Model

The FK in 1D is

$$\frac{\partial u}{\partial t} = \frac{\partial^2 u}{\partial x^2} + u(1 - u^q), \quad q > 0 \tag{9.1}$$

with the analytical solution [1]

$$u(x, t) = \frac{1}{(1 + ae^{b\xi})^s}, \quad \xi = x - ct \tag{9.2a}$$

where

$$s = \frac{2}{q}; \quad b = \frac{q}{[2(q + 2)]^{1/2}}; \quad c = \frac{q + 4}{[2(q + 2)]^{1/2}} \tag{9.2b,c,d}$$

We take

$$a = \sqrt{2} - 1 \tag{9.2e}$$

as suggested in [1] although a is arbitrary.

Equations (9.1) and (9.2) constitute the PDE model and are programmed in the following routines.

Spline Collocation Methods for Partial Differential Equations: With Applications in R, First Edition.
William E. Schiesser.
© 2017 John Wiley & Sons, Inc. Published 2017 by John Wiley & Sons, Inc.
Companion website: www.wiley.com/go/Spline_Collocation

9.2 Main Program

A main program for Eqs. (9.1) and (9.2) follows.

```
#
# Fisher-Kolmogorov
#
# Delete previous workspaces
  rm(list=ls(all=TRUE))
#
# Access ODE integrator
  library("deSolve");
#
# Access functions for numerical solution
  setwd("f:/collocation/chap9");
  source("pde_1a.R");source("ua_1.R");
#
# Grid in x
  n=31;xl=-5;xu=10;
  x=seq(from=xl,to=xu,by=(xu-xl)/(n-1));
#
# Parameters
  q=1;
  a=2^(1/2)-1;
  b=q/(2*(q+2))^(1/2);
  c=(q+4)/(2*(q+2))^(1/2);
  s=2/q;
#
# Independent variable for ODE integration
  nout=6;t0=0;tf=5;
  tout=seq(from=t0,to=tf,by=(tf-t0)/(nout-1));
#
# Initial condition
  u0=rep(0,n);
  for(i in 1:n){
    u0[i]=ua_1(x[i],t0);
  }
  ncall=0;
#
# ODE integration
  out=lsodes(y=u0,times=tout,func=pde_1a,
      sparsetype="sparseint",rtol=1e-6,atol=1e-6,
      maxord=5);
```

```
  nrow(out)
  ncol(out)
#
# Arrays for numerical solution
  u=matrix(0,nrow=n,ncol=nout);
  t=rep(0,nout);
  for(it in 1:nout){
  for(i  in 1:n){
    u[i,it]=out[it,i+1];
      t[it]=out[it,1];
  }
  }
#
# Arrays for analytical solution,
# errors in numerical solution
   ua=matrix(0,nrow=n,ncol=nout);
  err=matrix(0,nrow=n,ncol=nout);
  for(it in 1:nout){
    u[1,it]=ua_1(x[1],t[it]);
    u[n,it]=ua_1(x[n],t[it]);
    for(i in 1:n){
       ua[i,it]=ua_1(x[i],t[it]);
       err[i,it]=u[i,it]-ua[i,it];
    }
  }
#
# Display selected output
  cat(sprintf("\n q = %4.2f,  a = %4.2f,  b = %4.2f,
            c = %4.2f  s = %4.2f\n",q,a,b,c,s));
  for(it in 1:nout){
    cat(sprintf("\n     t       x       u(x,t)
                ua(x,t)     err(x,t)\n"));
    iv=seq(from=1,to=n,by=5);
    for(i in iv){
      cat(sprintf("%6.2f%8.3f%12.6f%12.6f
                  %12.6f\n",t[it],x[i],u[i,it],
                  ua[i,it],err[i,it]));
    }
    cat(sprintf("\n"));
  }
  cat(sprintf(" ncall = %4d\n",ncall));
#
# Graphical output
```

```
matplot(x,u,type="l",lwd=2,col="black",
        lty=1,xlab="x",ylab="u(x,t)",
        main="Fisher-Kolmogorov");
matpoints(x,ua,pch="o",col="black");
```

Listing 9.1 Main program for Eqs. (9.1) and (9.2).

We can note the following details about Listing 9.1:

- Previous workspaces are removed. Then the ODE integrator library deSolve is accessed. Note that the setwd (set working directory) uses / rather than the usual \. Also, the setwd requires editing for the local computer (to define the working directory).

```
#
# Fisher-Kolmogorov
#
# Delete previous workspaces
  rm(list=ls(all=TRUE))
#
# Access ODE integrator
  library("deSolve");
#
# Access functions for numerical solution
  setwd("f:/collocation/chap9");
  source("pde_1a.R");source("ua_1.R");
```

pde_1a is the routine for the MOL/ODEs (discussed subsequently). ua_1 implements Eqs. (9.2).

- A uniform grid in x with 31 points is defined with the seq utility for the interval $x_l = -5 \leq x \leq x_u = 10$. Therefore, the vector x has the values $x = -5, -4.5, \ldots, 10$.

```
#
# Grid in x
  n=31;xl=-5;xu=10;
  x=seq(from=xl,to=xu,by=(xu-xl)/(n-1));
```

- The model parameters (constants) are defined numerically.

```
#
# Parameters
  q=1;
  a=2^(1/2)-1;
  b=q/(2*(q+2))^(1/2);
```

```
c=(q+4)/(2*(q+2))^(1/2);
s=2/q;
```

- A uniform grid in t of 6 output points is defined with the `seq` utility for the interval $0 \le t \le t_f = 5$. Therefore, the vector `tout` has the values $t = 0, 1, \ldots, 5$.

```
#
# Independent variable for ODE integration
  nout=6;t0=0;tf=5;
  tout=seq(from=t0,to=tf,by=(tf-t0)/(nout-1));
```

At this point, the intervals of x and t in Eq. (9.1) are defined.
- The IC for Eq. (9.1) is taken from Eqs. (9.2) with $t = 0$.

```
#
# Initial condition
  u0=rep(0,n);
  for(i in 1:n){
    u0[i]=ua_1(x[i],t0);
  }
  ncall=0;
```

ua_1 for Eqs. (9.2) is discussed subsequently. The counter for the calls to the ODE routine pde_1a (discussed subsequently) is initialized.
- The system of 31 ODEs is integrated by the library integrator `lsodes` (available in `deSolve`) with the sparse matrix option specified. As expected, the inputs to `lsodes` are the ODE function, pde_1a, the IC vector u0, and the vector of output values of t, tout. The length of u0 (31) informs `lsodes` how many ODEs are to be integrated. func, y, times are reserved names.

```
#
# ODE integration
  out=lsodes(y=u0,times=tout,func=pde_1a,
      sparsetype="sparseint",rtol=1e-6,atol=1e-6,
      maxord=5);
  nrow(out)
  ncol(out)
```

The numerical solution to the ODEs is returned in matrix out. In this case, out has the dimensions 6×32 (displayed in the numerical output). The offset $31 + 1 = 32$ is required since the first element of each column has the output t (also in tout), and the 2, ..., 32 column elements have the 31 ODE solutions. This indexing of out is used next.
- The numerical solution for $u(x, t)$ is taken from the solution matrix out returned by `lsodes`. The intervals in t and x are covered by two nested `for`s

(with indices it and i for t and x, respectively). The offset i+1 accounts for t returned in the first column position of out and placed in the vector t [it].

```
#
# Arrays for numerical solution
  u=matrix(0,nrow=n,ncol=nout);
  t=rep(0,nout);
  for(it in 1:nout){
  for(i   in 1:n){
    u[i,it]=out[it,i+1];
      t[it]=out[it,1];
  }
  }
```

- The numerical and analytical solutions from Eqs. (9.1) to (9.2), and the difference in these solutions (err), are placed in arrays.

```
#
# Arrays for analytical solution,
# errors in numerical solution
   ua=matrix(0,nrow=n,ncol=nout);
  err=matrix(0,nrow=n,ncol=nout);
  for(it in 1:nout){
    u[1,it]=ua_1(x[1],t[it]);
    u[n,it]=ua_1(x[n],t[it]);
    for(i in 1:n){
        ua[i,it]=ua_1(x[i],t[it]);
        err[i,it]=u[i,it]-ua[i,it];
    }
  }
```

The numerical solution of Eq. (9.1) at $x = -5, 10$, which is also the analytical solution of Eqs. (9.2) from the BCs for Eq. (9.1), is placed in u [1,it], u [n,it] since these values are not returned from the ODE routine pde_1a in array out.

- The parameters of the numerical and analytical solutions are displayed, followed by every fifth value of the solutions and their difference.

```
#
# Display selected output
  cat(sprintf("\n q = %4.2f,   a = %4.2f,   b = %4.2f,
            c = %4.2f   s = %4.2f\n",q,a,b,c,s));
    for(it in 1:nout){
      cat(sprintf("\n     t        x        u(x,t)
```

```
                  ua(x,t)       err(x,t)\n"));
        iv=seq(from=1,to=n,by=5);
        for(i in iv){
          cat(sprintf("%6.2f%8.3f%12.6f%12.6f
                      %12.6f\n",t[it],x[i],u[i,it],
                      ua[i,it],err[i,it]));
        }
        cat(sprintf("\n"));
      }
    cat(sprintf("  ncall = %4d\n",ncall));
```

The number of calls to pde_1a is displayed at the end of the solution.

- The numerical and analytical solutions are plotted with matplot and matpoints, respectively.

```
#
# Graphical output
    matplot(x,u,type="l",lwd=2,col="black",
            lty=1,xlab="x",ylab="u(x,t)",
            main="Fisher-Kolmogorov");
    matpoints(x,ua,pch="o",col="black");
```

The ODE routine pde_1a called by lsodes in the main program of Listing 9.1 is considered next.

9.3 ODE Routine

pde_1a for Eq. (9.1) follows.

```
  pde_1a=function(t,u,parms){
#
# Function pde_1a computes the t derivative
# vector of u(x,t)
#
# BCs
  u[1]=ua_1(x[1],t);
  u[n]=ua_1(x[n],t);
#
# uxx
  tablexx=splinefun(x,u);
  uxx=tablexx(x,deriv=2);
#
# PDE
```

```
    ut=rep(0,n);
    for(i in 1:n){
      ut[i]=uxx[i]+u[i]*(1-u[i]^q);
    }
    ut[1]=0;ut[n]=0;
#
# Increment calls to pde_1a
  ncall <<- ncall+1;
#
# Return derivative vector
  return(list(c(ut)));
  }
```

Listing 9.2 ODE/MOL routine pde_1a for Eq. (9.1).

We can note the following details about pde_1a:

- The function is defined.

```
    pde_1a=function(t,u,parms){
#
# Function pde_1a computes the t derivative
# vector of u(x,t)
```

  t is the current value of t in Eq. (9.1) that is used in the BCs (below). u is the 31-vector of ODE-dependent variables. parm is an argument to pass parameters to pde_1a (unused, but required in the argument list). The arguments must be listed in the order stated to properly interface with lsodes called in the main program. The derivative vector of the LHS of Eq. (9.1) is calculated below and returned to lsodes.

- The analytical solution of Eqs. (9.2) is used for Dirichlet BCs for Eq. (9.1).

```
#
# BCs
  u[1]=ua_1(x[1],t);
  u[n]=ua_1(x[n],t);
```

  Note the subscripts 1, n corresponding to $x = -5, 10$.

- The second derivative $\dfrac{\partial^2 u}{\partial x^2}$ in Eq. (9.1) is computed with splinefun.

```
#
# uxx
  tablexx=splinefun(x,u);
  uxx=tablexx(x,deriv=2);
```

- The LHS derivative of Eq. (9.1) $\frac{\partial u}{\partial t}$ is computed.

```
#
# PDE
  ut=rep(0,n);
  for(i in 1:n){
    ut[i]=uxx[i]+u[i]*(1-u[i]^q);
  }
  ut[1]=0;ut[n]=0;
```

Since the values $u(x = -5, t), u(x = 10, t)$ are set as BCs, the boundary derivatives in t are set to zero so that the ODE integration by lsodes does not move the boundary values away from their prescribed values. Also, these boundary values are not returned to the calling program so they are reset in the main program of Listing 9.1 (discussed previously) for the purpose of displaying the complete numerical solution (only ODE-dependent variables computed by ODEs are returned by pde_1a). The straightforward programming of the nonlinear logistic rate $u(1 - u^q)$ is particularly noteworthy.

- The counter for the calls to pde_1a is returned to the main program of Listing 9.1 by the <<- operator.

```
#
# Increment calls to pde_1a
  ncall <<- ncall+1;
```

- The derivative in ut is returned to lsodes (the ODE integrator called in the main program of Listing 9.1).

```
#
# Return derivative vector
  return(list(c(ut)));
  }
```

Note that a list is used as required by lsodes. The final } concludes pde_1a.

The subordinate routine ua_1 for the analytical solution of Eqs. (9.2) (called by the main program and ODE routine of Listings 9.1 and 9.2) follows.

9.4 Subordinate Routine

```
#
# Function ua_1 computes the exact solution of the
# Fisher-Kolmogorov equation for comparison with the
```

```
# numerical solution
#
# Analytical solution
  z=x-c*t;
  ua=1/(1+a*exp(b*z))^s;
#
# Return solution
  return(c(ua));
  }
```

Listing 9.3 Function ua_1 for the analytical solution of Eqs. (9.2).

ua_1 is a straightforward implementation of Eqs. (9.2) with the constants c, a, b, s transferred from the main program of Listing 9.1. Note the use of the Lagrangian variable z=x-c*t so that this is a traveling wave solution (a solution in terms of z=x-c*t is usually associated with a hyperbolic PDE, but in this case, Eq. (9.1) is parabolic). A numerical value ua is returned (not a list as in Listing 9.2).

This completes the programming of Eqs. (9.1) and (9.2). The numerical and graphical output is considered next.

9.5 Model Output

Execution of the preceding main program and subordinate routines pde_1a ua_1 produces the following abbreviated numerical output in Table 9.1. We can note the following details about this output:

- As expected, the solution array out has the dimensions 6 x 32 for 6 output points and 31 ODEs, with t included as an additional element, that is, 31 + 1 = 32.
- The interval in x is $((10 - (-5))/(31 - 1) = 0.5$ (with every fifth value in Table 9.1).
- The numerical and analytical ICs at $t = 0$ agree exactly since the analytical solution is used in the main program of Listing 9.1 to define the numerical IC.
- The numerical and analytical solutions at $x = -5, 10$ agree exactly from the BCs set in pde_1a.
- The agreement between the numerical and analytical solutions is quite acceptable given that only 31 points were used in x.
- The computational effort is modest, ncall = 114.

Table 9.1 Abbreviated output for Eqs. (9.1) and (9.2).

```
[1]  6

[1]  32

q = 1.00,   a = 0.41,   b = 0.41,   c = 2.04   s = 2.00

    t        x       u(x,t)      ua(x,t)      err(x,t)
  0.00   -5.000    0.900512    0.900512     0.000000
  0.00   -2.500    0.757104    0.757104     0.000000
  0.00    0.000    0.500000    0.500000     0.000000
  0.00    2.500    0.216452    0.216452     0.000000
  0.00    5.000    0.056974    0.056974     0.000000
  0.00    7.500    0.010306    0.010306     0.000000
  0.00   10.000    0.001531    0.001531     0.000000
                       .           .
                       .           .
                       .           .
            Output for t = 1,2,3,4 removed
                       .           .
                       .           .
                       .           .

    t        x       u(x,t)      ua(x,t)      err(x,t)
  5.00   -5.000    0.998334    0.998334     0.000000
  5.00   -2.500    0.995391    0.995387     0.000003
  5.00    0.000    0.987285    0.987279     0.000006
  5.00    2.500    0.965299    0.965290     0.000009
  5.00    5.000    0.907994    0.907981     0.000013
  5.00    7.500    0.773248    0.773236     0.000012
  5.00   10.000    0.524514    0.524514     0.000000

ncall =   114
```

The solution is displayed graphically in Figure 9.1, which again demonstrates the agreement between the numerical and analytical solutions. The solution at $x = -5, 10$ varies with t as defined by the BCs in pde_1a.

The traveling wave characteristic (dependence on only the Lagrangian variable z=x−ct) is clear (the use of "characteristic" has a mathematical interpretation since z=x−ct is the characteristic of Eq. (9.1)).

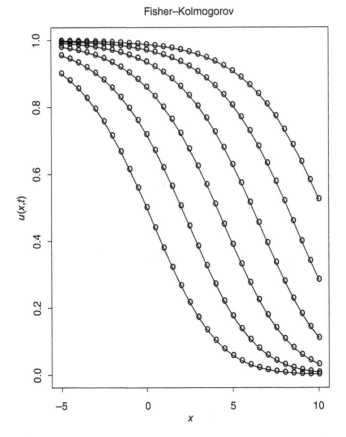

Figure 9.1 Numerical and analytical $u(x, t)$ from Eqs. (9.1) and (9.2) lines – num, points – anal.

9.6 Summary and Conclusions

The programming of the FK equation (9.1) and the analytical solution Eqs. (9.2) within the SCMOL framework is straightforward, and the two solutions agree with acceptable accuracy and computational effort.

Reference

1 Murray, J.D. (2002), *Mathematical Biology, I: An Introduction*, 3rd ed., Springer, New York, pp. 450–451.

10

Klein–Gordon Equation

The Klein–Gordon (KG) equation is an extension of the linear wave equation with additional linear and/or nonlinear terms [1]. It has application, for example, to relativistic fields (waves).

We consider here a linear and a nonlinear form of the KG equation, with a comparison of the numerical spline collocation method of lines (SCMOL) solution with an analytical solution.

10.1 PDE Model, Linear Case

The linear KG equation is

$$\frac{\partial^2 u}{\partial t^2} + a\frac{\partial^2 u}{\partial x^2} + bu = 0 \tag{10.1}$$

where a, b are arbitrary constants.

Equation (10.1) is second order in t and x and, therefore, requires two initial conditions (ICs) and two boundary conditions (BCs), which are taken as

$$u(x, t = 0) = \cos(2\pi x); \quad \frac{\partial u(x, t = 0)}{\partial t} = 0 \tag{10.2a,b}$$

$$\frac{\partial u(x = 0, t)}{\partial x} = \frac{\partial u(x = 1, t)}{\partial x} = 0 \tag{10.2c,d}$$

The analytical solution to Eqs. (10.1) and (10.2) is

$$u_a(x, t) = \cos(ct)\cos(2\pi x); \quad c = \sqrt{b - a(2\pi)^2} \tag{10.3}$$

Equations (10.1)–(10.3) constitute the first (linear) PDE model and are programmed in the following routines.

Spline Collocation Methods for Partial Differential Equations: With Applications in R, First Edition.
William E. Schiesser.
© 2017 John Wiley & Sons, Inc. Published 2017 by John Wiley & Sons, Inc.
Companion website: www.wiley.com/go/Spline_Collocation

10.2 Main Program

A main program for Eqs. (10.1)–(10.3) follows.

```
#
# Klein-Gordon
#
# Delete previous workspaces
  rm(list=ls(all=TRUE))
#
# Access ODE integrator
  library("deSolve");
#
# Access functions for numerical solution
  setwd("f:/collocation/chap10");
  source("pde_1.R");
#
# Grid in x
  n=21;xl=0;xu=1;
  x=seq(from=xl,to=xu,by=(xu-xl)/(n-1));
#
# Parameters
  a=-2.5;b=1;c=sqrt(-a*(2*pi)^2+b);
#
# Independent variable for ODE integration
  nout=11;t0=0;tf=1;
  tout=seq(from=t0,to=tf,by=(tf-t0)/(nout-1));
#
# ICs from analytical solution
  u0=rep(0,2*n);
  for(i in 1:n){
    u0[i]   =cos(2*pi*x[i]);
    u0[i+n]=0;
  }
  ncall=0;
#
# ODE integration
  out=lsodes(y=u0,times=tout,func=pde_1,
      sparsetype="sparseint",rtol=1e-6,atol=1e-6,
      maxord=5);
  nrow(out)
  ncol(out)
#
```

```
# Arrays for numerical solution
  u1=matrix(0,nrow=n,ncol=nout);
  u2=matrix(0,nrow=n,ncol=nout);
  t=rep(0,nout);
  for(it in 1:nout){
  for(i in 1:n){
    u1[i,it]=out[it,i+1];
    u2[i,it]=out[it,i+1+n];
      t[it]=out[it,1];
  }
  }
#
# Arrays for analytical solution,
# errors in numerical solution
  u1a=matrix(0,nrow=n,ncol=nout);
  err=matrix(0,nrow=n,ncol=nout);
  for(it in 1:nout){
    for(i in 1:n){
      u1a[i,it]=cos(c*t[it])*cos(2*pi*x[i]);
      err[i,it]=u1[i,it]-u1a[i,it];
    }
  }
#
# Display selected output
  cat(sprintf("\n a = %4.2f,   b = %4.2f,
             c = %4.2f\n",a,b,c));
  for(it in 1:nout){
    cat(sprintf("\n       t         x        u1(x,t)
               u1a(x,t)      err(x,t)\n"));
    iv=seq(from=1,to=n,by=5);
    for(i in iv){
      cat(sprintf("%6.2f%8.3f%12.6f%12.6f
                 %12.6f\n",t[it],x[i],u1[i,it],
                 u1a[i,it],err[i,it])));
    }
    cat(sprintf("\n"));
  }
  cat(sprintf(" ncall = %4d\n",ncall));
#
# Plot 2D numerical solution
    matplot(x,u1,type="l",lwd=2,col="black",lty=1,
      xlab="x",ylab="u(x,t)",main="Klein-Gordon");
    matpoints(x,u1a,pch="o",col="black");
```

```
#
# Plot 3D numerical solution
    persp(x,t,u1,theta=55,phi=45,xlim=c(xl,xu),
          ylim=c(t0,t[nout]),xlab="x",ylab="t",
          zlab="u(x,t)");
```

Listing 10.1 Main program for Eqs. (10.1)–(10.3).

We can note the following details about Listing 10.1:

- Previous workspaces are removed. Then the ODE integrator library deSolve is accessed. Note that the setwd (set working directory) uses / rather than the usual \. The setwd requires editing for the local computer to access the working directory.

```
#
# Klein-Gordon
#
# Delete previous workspaces
    rm(list=ls(all=TRUE))
#
# Access ODE integrator
    library("deSolve");
#
# Access functions for numerical solution
    setwd("f:/collocation/chap10");
    source("pde_1.R");
```

  pde_1 is the routine for the MOL/ODEs (discussed subsequently).
- A uniform grid in x with 21 points is defined with the seq utility for the interval $x_l = 0 \le x \le x_u = 1$. Therefore, the vector x has the values $x = 0, 0.05, \ldots, 1$.

```
#
# Grid in x
    n=21;xl=0;xu=1;
    x=seq(from=xl,to=xu,by=(xu-xl)/(n-1));
```

- The model parameters (constants) in Eq. (10.1) are defined numerically.

```
#
# Parameters
    a=-2.5;b=1;c=sqrt(-a*(2*pi)^2+b);
```

- A uniform grid in t of 11 output points is defined with the seq utility for the interval $0 \le t \le t_f = 1$. Therefore, the vector tout has the values $t = 0, 0.1, \ldots, 1$.

```
#
# Independent variable for ODE integration
  nout=11;t0=0;tf=1;
  tout=seq(from=t0,to=tf,by=(tf-t0)/(nout-1));
```

At this point, the intervals of x and t in Eq. (10.1) are defined.
- The IC for Eq. (10.1), Eqs. (10.2a,b), is placed in a vector u0 of length 2(21) = 42.[1]

```
#
# ICs from analytical solution
  u0=rep(0,2*n);
  for(i in 1:n){
    u0[i]   =cos(2*pi*x[i]);
    u0[i+n]=0;
  }
  ncall=0;
```

The counter for the number of calls to pde_1 is initialized.
- The system of 42 ODEs is integrated by the library integrator lsodes (available in deSolve) with the sparse matrix option specified. As expected, the inputs to lsodes are the ODE function, pde_1, the IC vector u0, and the vector of output values of t, tout. The length of u0 (42) informs lsodes how many ODEs are to be integrated. func,y,times are reserved names.

```
#
# ODE integration
  out=lsodes(y=u0,times=tout,func=pde_1,
      sparsetype="sparseint",rtol=1e-6,atol=1e-6,
      maxord=5);
  nrow(out)
  ncol(out)
```

The numerical solution to the ODEs is returned in matrix out. In this case, out has the dimensions 11 × 43 (displayed in the numerical output). The offset $42 + 1 = 43$ is required since the first element of each column has the output t (also in tout), and the 2, ... , 43 column elements have the 42 ODE solutions. This indexing of out is used next.

---

1 This IC is based on the use of two variables, u_1, u_2, defined as

$$u_1(x, t) = u(x, t); \quad u_2(x, t) = \frac{\partial u}{\partial t} = \frac{\partial u_1}{\partial t}$$

- The numerical solution for $u(x, t)$, consisting of u_1, u_2, is taken from the solution matrix out returned by lsodes. The intervals in t and x are covered by two nested fors (with indices it and i for t and x, respectively). The offset i+1 accounts for t returned in the first column position of out and placed in the vector t[it].

```
#
# Arrays for numerical solution
  u1=matrix(0,nrow=n,ncol=nout);
  u2=matrix(0,nrow=n,ncol=nout);
  t=rep(0,nout);
  for(it in 1:nout){
  for(i in 1:n){
    u1[i,it]=out[it,i+1];
    u2[i,it]=out[it,i+1+n];
      t[it]=out[it,1];
  }
  }
```

- The analytical solution from Eq. (10.3) and the difference between the numerical and analytical solutions (err) are placed in arrays.

```
#
# Arrays for analytical solution,
# errors in numerical solution
  u1a=matrix(0,nrow=n,ncol=nout);
  err=matrix(0,nrow=n,ncol=nout);
  for(it in 1:nout){
    for(i in 1:n){
      u1a[i,it]=cos(c*t[it])*cos(2*pi*x[i]);
      err[i,it]=u1[i,it]-u1a[i,it];
    }
  }
```

- The parameters of the numerical and analytical solutions are displayed, followed by every fifth value of the solutions and their difference.

```
#
# Display selected output
  cat(sprintf("\n a = %4.2f,   b = %4.2f,
              c = %4.2f\n",a,b,c));
  for(it in 1:nout){
    cat(sprintf("\n       t          x        u1(x,t)
                u1a(x,t)      err(x,t)\n"));
    iv=seq(from=1,to=n,by=5);
```

```
  for(i in iv){
    cat(sprintf("%6.2f%8.3f%12.6f%12.6f
                 %12.6f\n",t[it],x[i],u1[i,it],
                 u1a[i,it],err[i,it])));
  }
  cat(sprintf("\n"));
}
cat(sprintf(" ncall = %4d\n",ncall));
```

The number of calls to pde_1 is displayed at the end of the solution.

- The numerical and analytical solutions are plotted with matplot and matpoints, respectively.

```
#
# Plot 2D numerical solution
    matplot(x,u1,type="l",lwd=2,col="black",lty=1,
      xlab="x",ylab="u(x,t)",main="Klein-Gordon");
    matpoints(x,u1a,pch="o",col="black");
#
# Plot 3D numerical solution
    persp(x,t,u1,theta=55,phi=45,xlim=c(xl,xu),
          ylim=c(t0,t[nout]),xlab="x",ylab="t",
          zlab="u(x,t)");
```

The numerical solution is also plotted in 3D with persp.

The ODE routine pde_1 called by lsodes in the main program of Listing 10.1 is considered next.

10.3 ODE Routine

pde_1 for Eq. (10.1) follows.

```
  pde_1=function(t,u,parm){
#
# Function pde_1 computes the t derivative vectors
# for u1(x,t), u2(x,t)
#
# One vector to two vectors
  u1=rep(0,n);u2=rep(0,n);
  for(i in 1:n){
    u1[i]=u[i];
    u2[i]=u[i+n];
  }
```

```
#
# ux
  tablex=splinefun(x,u1);
  u1x=tablex(x,deriv=1);
#
# BCs
  u1x[1]=0;u1x[n]=0;
#
# uxx
  tablexx=splinefun(x,u1x);
  u1xx=tablexx(x,deriv=1);
#
# PDEs
  u1t=rep(0,n);u2t=rep(0,n);
  for(i in 1:n){
    u1t[i]=u2[i];
    u2t[i]=-a*u1xx[i]-b*u1[i];
  }
#
# Two vectors to one vector
  ut=rep(0,2*n);
  for(i in 1:n){
    ut[i]  =u1t[i];
    ut[i+n]=u2t[i];
  }
#
# Increment calls to pde_1
  ncall <<- ncall+1;
#
# Return derivative vector
  return(list(c(ut)));
  }
```

Listing 10.2 ODE/MOL routine pde_1 for Eq. (10.1).

We can note the following details about pde_1:

- The function is defined.

```
  pde_1=function(t,u,parm){
#
# Function pde_1 computes the t derivative vectors
# for u1(x,t), u2(x,t)
```

t is the current value of *t* in Eq. (10.1). u is the dependent variable 42-vector. parm is an argument to pass parameters to pde_1 (unused, but required in the argument list). The arguments must be listed in the order stated to properly interface with lsodes called in the main program. The derivative vector of the LHS of Eq. (10.1) is calculated below and returned to lsodes.

- u is placed in two 21-vectors, u1, u2, to facilitate the programming of Eq. (10.1).

```
#
# One vector to two vectors
  u1=rep(0,n);u2=rep(0,n);
  for(i in 1:n){
    u1[i]=u[i];
    u2[i]=u[i+n];
  }
```

- $\dfrac{\partial u}{\partial x} = \dfrac{\partial u_1}{\partial x}$ is computed with splinefun.

```
#
# ux
  tablex=splinefun(x,u1);
  u1x=tablex(x,deriv=1);
```

- Homogeneous Neumann BCs (10.2c,d) are set. The subscripts 1, n correspond to $x = 0, 1$.

```
#
# BCs
  u1x[1]=0;u1x[n]=0;
```

- The second derivative $\dfrac{\partial^2 u}{\partial x^2} = \dfrac{\partial^2 u_1}{\partial x^2}$ in Eq. (10.1) is computed with splinefun by differentiating the first derivative (i.e., stagewise differentiation).

```
#
# uxx
  tablexx=splinefun(x,u1x);
  u1xx=tablexx(x,deriv=1);
```

- The LHS derivative of Eq. (10.1) $\dfrac{\partial^2 u(x,t)}{\partial t^2} = \dfrac{\partial u_2(x,t)}{\partial t}$ is computed using u_1, u_2. The two 21-derivative vectors are placed in u1t, u2t.

```
#
# PDE
```

```
    u1t=rep(0,n);u2t=rep(0,n);
    for(i in 1:n){
      u1t[i]=u2[i];
      u2t[i]=-a*u1xx[i]-b*u1[i];
    }
```

- u1t,u2t are placed in a 42-vector ut.

```
  #
  # Two vectors to one vector
    ut=rep(0,2*n);
    for(i in 1:n){
      ut[i]   =u1t[i];
      ut[i+n]=u2t[i];
    }
```

- The counter for the calls to pde_1 is returned to the main program of Listing 10.1 by the <<- operator.

```
  #
  # Increment calls to pde_1
    ncall <<- ncall+1;
```

- The derivative in ut is returned to lsodes (the ODE integrator called in the main program of Listing 10.1).

```
  #
  # Return derivative vector
    return(list(c(ut)));
    }
```

Note that a list is used as required by lsodes. The final } concludes pde_1.

10.4 Model Output

Execution of the preceding main program and ODE routine pde_1 produces the following abbreviated numerical output in Table 10.1.
We can note the following details about this output:

- As expected, the solution array out has the dimensions 11 x 43 for 11 output points and 42 ODEs, with t included as an additional element, that is, $42 + 1 = 43$.
- The interval in x is $((1 - 0)/(21 - 1) = 0.05$ (with every fifth value in Table 10.1).

Table 10.1 Abbreviated output for Eqs. (10.1)–(10.3).

```
[1]  11

[1]  43

a = -2.50,   b = 1.00,   c = 9.98
```

t	x	ul(x,t)	ula(x,t)	err(x,t)
0.00	0.000	1.000000	1.000000	0.000000
0.00	0.250	0.000000	0.000000	0.000000
0.00	0.500	-1.000000	-1.000000	0.000000
0.00	0.750	-0.000000	-0.000000	0.000000
0.00	1.000	1.000000	1.000000	0.000000

t	x	ul(x,t)	ula(x,t)	err(x,t)
0.10	0.000	0.539211	0.541582	-0.002371
0.10	0.250	-0.000002	0.000000	-0.000002
0.10	0.500	-0.541575	-0.541582	0.000006
0.10	0.750	-0.000002	-0.000000	-0.000002
0.10	1.000	0.539211	0.541582	-0.002371

```
            .                    .
            .                    .
            .                    .
      Output for t = 0.2 to 0.9 removed
            .                    .
            .                    .
            .                    .
```

t	x	ul(x,t)	ula(x,t)	err(x,t)
1.00	0.000	-0.846597	-0.847248	0.000652
1.00	0.250	-0.000197	-0.000000	-0.000197
1.00	0.500	0.847001	0.847248	-0.000247
1.00	0.750	-0.000197	0.000000	-0.000197
1.00	1.000	-0.846597	-0.847248	0.000652

```
ncall =   292
```

- The numerical and analytical ICs at $t = 0$ agree exactly since the analytical solution is used in the main program of Listing 10.1 to define the numerical IC.
- The maximum error occurs at the boundaries $x = 0, 1$, which is a characteristic of the splines in `splinefun`.
- The agreement between the numerical and analytical solutions is quite acceptable given that only 21 points were used in x.
- The computational effort is modest, `ncall = 292`.

The solution is displayed graphically in Figure 10.1a, which again demonstrates the agreement between the numerical and analytical solutions.

The 3D plot of the solution in Figure 10.1b reflects the oscillatory characteristic of the solution in x and t (see Eq. (10.3)). Equation (10.1) is linear. We next consider a nonlinear extension, the cubic KG equation.

10.5 PDE Model, Nonlinear Case

The nonlinear KG equation is (with $m \neq 0, 1$)

$$\frac{\partial^2 u}{\partial t^2} + a\frac{\partial^2 u}{\partial x^2} + bu + gu^m = 0 \tag{10.4}$$

Since Eq. (10.4) is Eq. (10.1) with the addition of the term gu^m, a comparison of the solutions of Eqs. (10.1) and (10.4) gives a direct indication of the effect of this nonlinear term. For $m = 3$ (the cubic KG), an analytical solution is available [1] that can be used to test the numerical solution.

Equation (10.4) requires two ICs and two BCs.

$$u(x, t = 0) = u_a(x, t = 0); \quad \frac{\partial u(x, t = 0)}{\partial t} = \frac{\partial u_a(x, t = 0)}{\partial t} \tag{10.5a,b}$$

$$u(x = 0, t) = u_a(x = 0, t); \quad u(x = 1, t) = u_a(x = 1, t) \tag{10.5c,d}$$

where $u_a(x, t)$ is the analytical solution

$$u_a(x, t) = B \tan(K(x + ct)) \tag{10.6a}$$

with the derivative in t

$$\frac{\partial u_a(x, t)}{\partial t} = \frac{BKc}{\cos\left(K(x + ct)\right)^2} \tag{10.6b}$$

with

$$B = \sqrt{\frac{b}{g}}; \quad K = \sqrt{\frac{-b}{2(a + c^2)}} \tag{10.6c,d}$$

where a, b, c, g are arbitrary constants.

Equations (10.4)–(10.6) constitute the second (nonlinear) PDE model and are programmed in the following routines.

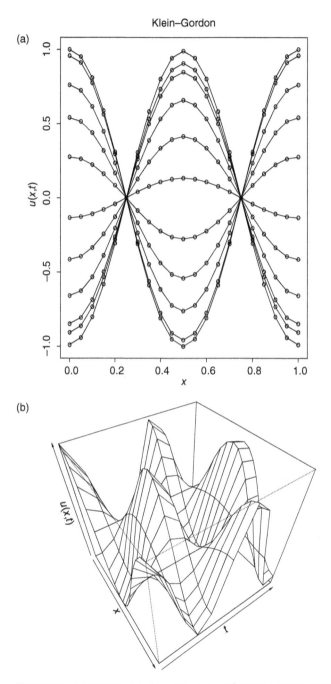

Figure 10.1 (a) Numerical and analytical $u(x, t)$ from Eqs. (10.1)–(10.3) lines – num, points – anal. (b) Numerical $u(x, t)$ from Eqs. (10.1)–(10.3).

10.6   Main Program

A main program for Eqs. (10.4)–(10.6) follows (it is similar to the main program in Listing 10.1 so the differences are emphasized).

```
#
# Klein-Gordon
#
# Delete previous workspaces
  rm(list=ls(all=TRUE))
#
# Access ODE integrator
  library("deSolve");
#
# Access functions for numerical solution
  setwd("f:/collocation/chap10");
  source("pde_1a.R");source("ua_1.R");
  source("uat_1.R");
#
# Grid in x
  n=21;xl=0;xu=1;
  x=seq(from=xl,to=xu,by=(xu-xl)/(n-1));
#
# Parameters
  m=3;a=-2.5;b=1;g=1.5;c=0.5;
  B=(b/g)^0.5;K=(-b/(2*(a+c^2)))^0.5;
#
# Independent variable for ODE integration
  nout=9;t0=0;tf=4;
  tout=seq(from=t0,to=tf,by=(tf-t0)/(nout-1));
#
# ICs from analytical solution
  u0=rep(0,2*n);
  for(i in 1:n){
    u0[i]   = ua_1(x[i],t0);
    u0[i+n] =uat_1(x[i],t0);
  }
  ncall=0;
#
# ODE integration
  out=lsodes(y=u0,times=tout,func=pde_1a,
      sparsetype="sparseint",rtol=1e-6,atol=1e-6,
      maxord=5);
```

```
  nrow(out)
  ncol(out)
#
# Arrays for numerical solution
  u1=matrix(0,nrow=n,ncol=nout);
  u2=matrix(0,nrow=n,ncol=nout);
  t=rep(0,nout);
  for(it in 1:nout){
  for(i in 1:n){
    u1[i,it]=out[it,i+1];
    u2[i,it]=out[it,i+1+n];
      t[it]=out[it,1];
  }
  }
#
# Arrays for analytical solution,
# errors in numerical solution
  u1a=matrix(0,nrow=n,ncol=nout);
  err=matrix(0,nrow=n,ncol=nout);
  for(it in 1:nout){
    u1[1,it]=ua_1(x[1],t[it]);
    u1[n,it]=ua_1(x[n],t[it]);
    for(i in 1:n){
        u1a[i,it]=ua_1(x[i],t[it]);
        err[i,it]=u1[i,it]-u1a[i,it];
    }
  }
#
# Display selected output
  cat(sprintf("\n a = %4.2f,   b = %4.2f,
              g = %4.2f,   c = %4.2f\n",
              a,b,g,c));
  for(it in 1:nout){
    cat(sprintf("\n      t         x   K*(x+ct)
                u1(x,t)       u1a(x,t)
                err(x,t)\n"));
    iv=seq(from=1,to=n,by=5);
    for(i in iv){
      cat(sprintf("%6.2f%7.3f%10.6f%10.6f
                  %10.6f%10.6f\n",t[it],x[i],
                  K*(x[i]+c*t[it]),u1[i,it],
                  u1a[i,it],err[i,it]));
    }
```

```
   cat(sprintf("\n"));
  }
  cat(sprintf("  ncall = %4d\n",ncall));
#
# Plot 2D numerical solution
   matplot(x,u1,type="l",lwd=2,col="black",lty=1,
     xlab="x",ylab="u(x,t)",main="Klein-Gordon");
   matpoints(x,u1a,pch="o",col="black");
#
# Plot 3D numerical solution
   persp(x,t,u1,theta=55,phi=45,xlim=c(xl,xu),
          ylim=c(t0,t[nout]),xlab="x",ylab="t",
          zlab="u(x,t)");
```

Listing 10.3 Main program for Eqs. (10.4)–(10.6).

We can note the following details about Listing 10.3:

- Previous workspaces are removed, and the ODE integrator lsodes is accessed via the library deSolve. Three functions are then accessed. pde_1a is the routine for the MOL/ODEs (discussed subsequently). ua_1, uat_1 are the routines for the analytical solution of Eqs. (10.6a–d).

```
#
# Klein-Gordon
#
# Delete previous workspaces
  rm(list=ls(all=TRUE))
#
# Access ODE integrator
  library("deSolve");
#
# Access functions for numerical solution
  setwd("f:/collocation/chap10");
  source("pde_1a.R");source("ua_1.R");
  source("uat_1.R");
```

- The grid in x is the same as in Listing 10.1.
- The model parameters (constants) in Eq. (10.4) are defined numerically.

```
#
# Parameters
  m=3;a=-2.5;b=1;g=1.5;c=0.5;
  B=(b/g)^0.5;K=(-b/(2*(a+c^2)))^0.5;
```

`a`, `b` are the same as in Listing 10.1. `g` is the coefficient of the term gu^m in Eq. (10.4). `c`, `B`, `K` are used in Eqs. (10.6a–d).

- A uniform grid in t of 9 output points is defined with the `seq` utility for the interval $0 \le t \le t_f = 4$. Therefore, the vector `tout` has the values $t = 0, 0.5, \ldots, 4$.

```
#
# Independent variable for ODE integration
  nout=9;t0=0;tf=4;
  tout=seq(from=t0,to=tf,by=(tf-t0)/(nout-1));
```

At this point, the intervals of x and t in Eq. (10.4) are defined.

- The ICs for Eq. (10.4), Eqs. (10.5a,b), are placed in a vector `u0` of length $2(21) = 42$. These ICs are based on the use of functions `ua_1`, `uat_1` (discussed subsequently) with `t0=0`.

```
#
# ICs from analytical solution
  u0=rep(0,2*n);
  for(i in 1:n){
    u0[i]   = ua_1(x[i],t0);
    u0[i+n] =uat_1(x[i],t0);
    }
  ncall=0;
```

The counter for the number of calls to `pde_1a` is initialized.

- The system of 42 ODEs is integrated by the library integrator `lsodes` (available in `deSolve`) with the sparse matrix option specified. The inputs to `lsodes` are the ODE function, `pde_1a`, the IC vector `u0`, and the vector of output values of t, `tout`. The length of `u0` (42) informs `lsodes` how many ODEs are to be integrated. `func`, `y`, `times` are reserved names.

```
#
# ODE integration
  out=lsodes(y=u0,times=tout,func=pde_1a,
      sparsetype="sparseint",rtol=1e-6,atol=1e-6,
      maxord=5);
  nrow(out)
  ncol(out)
```

The numerical solution to the ODEs is returned in matrix `out`. In this case, `out` has the dimensions 9×43 (displayed in the numerical output). The offset $42 + 1 = 43$ is required since the first element of each column has the output t (also in `tout`), and the 2, $\ldots$, 43 column elements have the 42 ODE solutions.

- The numerical solution $u(x, t)$ of Eq. (10.4), consisting of u_1, u_2, is taken from the solution matrix out returned by lsodes and placed in arrays u1, u2, as in Listing 10.1.
- The analytical solution from Eqs. (10.6), and the difference between the numerical and analytical solutions are placed in arrays u1a, err.

```
#
# Arrays for analytical solution,
# errors in numerical solution
  u1a=matrix(0,nrow=n,ncol=nout);
  err=matrix(0,nrow=n,ncol=nout);
  for(it in 1:nout){
    u1[1,it]=ua_1(x[1],t[it]);
    u1[n,it]=ua_1(x[n],t[it]);
    for(i in 1:n){
      u1a[i,it]=ua_1(x[i],t[it]);
      err[i,it]=u1[i,it]-u1a[i,it];
    }
  }
```

Also, $u_1(x = 0, t), u_1(x = 1, t)$ are set by the analytical solutions from ua_1 since these values are not returned from the ODE routine pde_1a (only dependent variables computed by integrating ODEs are returned from pde_1a). These boundary values are required for the numerical and graphical output discussed as follows.

- The parameters of the numerical and analytical solutions are displayed, followed by every fifth value of the solutions and their difference.

```
#
# Display selected output
  cat(sprintf("\n a = %4.2f,   b = %4.2f,
              g = %4.2f,   c = %4.2f\n",
              a,b,g,c));
  for(it in 1:nout){
    cat(sprintf("\n     t        x   K*(x+ct)
                u1(x,t)      u1a(x,t)
                err(x,t)\n"));
    iv=seq(from=1,to=n,by=5);
    for(i in iv){
      cat(sprintf("%6.2f%7.3f%10.6f%10.6f
                  %10.6f%10.6f\n",t[it],x[i],
                  K*(x[i]+c*t[it]),u1[i,it],
                  u1a[i,it],err[i,it]));
    }
  }
```

```
    cat(sprintf("\n"));
    }
    cat(sprintf(" ncall = %4d\n",ncall));
```

The Lagrangian variable $K(x + ct)$ in Eqs. (10.6a,b) is also computed and included in the output. The significance of this variable is discussed subsequently. The number of calls to pde_1a is displayed at the end of the solution.

- The numerical and analytical solutions are plotted in 2D with matplot and matpoints, respectively, and in 3D with persp, as in Listing 10.1.

The ODE routine pde_1a called by lsodes in the main program of Listing 10.3 is considered next.

10.7   ODE Routine

pde_1a for Eq. (10.4) follows.

```
  pde_1a=function(t,u,parm){
#
# Function pde_1a computes the t derivative vectors
# for u1(x,t),  u2(x,t)
#
# One vector to two vectors
  u1=rep(0,n);u2=rep(0,n);
  for(i in 1:n){
    u1[i]=u[i];
    u2[i]=u[i+n];
  }
#
# BCs
  u1[1]=ua_1(x[1],t);
  u1[n]=ua_1(x[n],t);
#
# uxx
  tablexx=splinefun(x,u1);
  u1xx=tablexx(x,deriv=2);
#
# PDE
  u1t=rep(0,n);u2t=rep(0,n);
  for(i in 1:n){
    u1t[i]=u2[i];
    u2t[i]=-a*u1xx[i]-b*u1[i]-g*u1[i]^m;
  }
```

```
  u1t[1]=0;u1t[n]=0;
#
# Two vectors to one vector
  ut=rep(0,2*n);
  for(i in 1:n){
    ut[i]   =u1t[i];
    ut[i+n] =u2t[i];
  }
#
# Increment calls to pde_1a
  ncall <<- ncall+1;
#
# Return derivative vector
  return(list(c(ut)));
  }
```

Listing 10.4 ODE/MOL routine pde_1a for Eq. (10.4).

We can note the following details about pde_1a:

- The function is defined.

  ```
  pde_1a=function(t,u,parm){
  #
  # Function pde_1a computes the t derivative vectors
  # for u1(x,t), u2(x,t)
  ```

  The input arguments, t, u, parm, are discussed under Listing 10.2.
- u is placed in two 21-vectors, u1, u2, to facilitate the programming of Eq. (10.4).

  ```
  #
  # One vector to two vectors
    u1=rep(0,n);u2=rep(0,n);
    for(i in 1:n){
      u1[i]=u[i];
      u2[i]=u[i+n];
    }
  ```

- BCs (10.5c,d) are implemented via function ua_1.

  ```
  #
  # BCs
    u1[1]=ua_1(x[1],t);
    u1[n]=ua_1(x[n],t);
  ```

  Subscripts 1, n correspond to $x = 0, 1$.

- The second derivative $\dfrac{\partial^2 u}{\partial x^2} = \dfrac{\partial^2 u_1}{\partial x^2}$ in Eq. (10.4) is computed with spline-fun.

```
#
# uxx
   tablexx=splinefun(x,u1);
   u1xx=tablexx(x,deriv=2);
```

- The LHS derivative of Eq. (10.4) $\dfrac{\partial^2 u(x,t)}{\partial t^2} = \dfrac{\partial^2 u_1(x,t)}{\partial t^2} = \dfrac{\partial u_2(x,t)}{\partial t}$ is computed using u_1, u_2. The two 21-derivative vectors are placed in u1t, u2t.

```
#
# PDE
   u1t=rep(0,n);u2t=rep(0,n);
   for(i in 1:n){
     u1t[i]=u2[i];
     u2t[i]=-a*u1xx[i]-b*u1[i]-g*u1[i]^m;
   }
   u1t[1]=0;u1t[n]=0;
```

Since the values $u_1(x=0,t), u_1(x=1,t)$ are set through BCs (10.5c,d), their derivatives in t are set to zero so that lsodes does not move these boundary values away from their prescribed values. The inclusion of

```
   -g*u1[i]^m
```

distinguishes Eq. (10.4) from Eq. (10.1).
- u1t, u2t are placed in a 42-vector ut.

```
#
# Two vectors to one vector
   ut=rep(0,2*n);
   for(i in 1:n){
     ut[i]   =u1t[i];
     ut[i+n] =u2t[i];
   }
```

- The counter for the calls to pde_1a is returned to the main program of Listing 10.3 by the <<- operator.

```
#
# Increment calls to pde_1a
   ncall <<- ncall+1;
```

- The derivative in ut is returned to lsodes.

```
#
```

```
# Return derivative vector
  return(list(c(ut)));
  }
```

A list is used as required by lsodes. The final } concludes pde_1a.

10.8 Subordinate Routines

Function ua_1 for Eq. (10.6a) is listed next.

```
ua_1=function(x,t){
#
# Function ua_1 computes the exact solution
# of the cubic Klein-Gordon equation for
# comparison with the numerical solution
#
# Analytical solution
  ua=B*tan(K*(x+c*t));
  return(c(ua));
  }
```

Listing 10.5 Function ua_1 for Eq. (10.6a).

We can note the following details about ua_1:

- Since $B = \sqrt{\dfrac{a}{g}}$ (from Eq. (10.6c)), Eq. (10.6a) cannot be applied to Eq. (10.1) for which $g = 0$.
- Equation (10.6a) is a traveling wave solution since it is a function of only the Lagrangian variable K*(x+c*t).
- Since $\tan(K(x + ct)) = \dfrac{\sin(K(x + ct))}{\cos(K(x + ct))}$, the solution of ua_1 is undefined for $K(x + ct) = \pi/2$. The numerical solution to follow with $K(x + ct)$ included in the output (from Listing 10.3) implies this limiting condition.

Function uat_1 for Eq. (10.6b) follows.

```
uat_1=function(x,t){
#
# Function uat_1 computes the time derivative of the
# exact solution of the cubic Klein-Gordon equation
#
# Analytical solution derivative
```

```
uat=B*(1/cos(K*(x+c*t))^2*(K*c));
return(c(uat));
}
```

Listing 10.6 Function uat_1 for Eq. (10.6b).

uat_1 is a straightforward implementation of Eq. (10.6b), again with the limitation $K(x + ct) \neq \pi/2$.

This completes the coding of Eqs. (10.4)–(10.6). The numerical and graphical output is considered next.

10.9 Model Output

We can note the following details about this output:

- As expected, the solution array out has the dimensions 9 x 43 for 9 output points and 42 ODEs, with t included as an additional element, that is, $42 + 1 = 43$.
- The interval in x is $((1-0)/(21-1) = 0.05$ (with every fifth value in Table 10.2).

Table 10.2 Abbreviated output for Eqs. (10.4)–(10.6).

```
[1] 9

[1] 43

  a = -2.50,  b = 1.00,  g = 1.50,  c = 0.50

    t       x    K*(x+ct)    u1(x,t)    u1a(x,t)    err(x,t)
  0.00   0.000   0.000000   0.000000   0.000000   0.000000
  0.00   0.250   0.117851   0.096673   0.096673   0.000000
  0.00   0.500   0.235702   0.196095   0.196095   0.000000
  0.00   0.750   0.353553   0.301337   0.301337   0.000000
  0.00   1.000   0.471405   0.416196   0.416196   0.000000

    t       x    K*(x+ct)    u1(x,t)    u1a(x,t)    err(x,t)
  0.50   0.000   0.117851   0.096673   0.096673   0.000000
```

(Continued)

Table 10.2 (Continued)

```
0.50   0.250   0.235702   0.196090   0.196095 -0.000005
0.50   0.500   0.353553   0.301329   0.301337 -0.000008
0.50   0.750   0.471405   0.416190   0.416196 -0.000006
0.50   1.000   0.589256   0.545810   0.545810  0.000000
                   .                      .
                   .                      .
                   .                      .
         Output for t = 1 to 3.5 removed
                   .                      .
                   .                      .
                   .                      .
   t       x    K*(x+ct)     ul(x,t)    ula(x,t)    err(x,t)
 4.00   0.000   0.942809   1.124594   1.124594   0.000000
 4.00   0.250   1.060660   1.458229   1.459235 -0.001006
 4.00   0.500   1.178511   1.971394   1.973508 -0.002114
 4.00   0.750   1.296362   2.896968   2.900134 -0.003166
 4.00   1.000   1.414214   5.171787   5.171787   0.000000

ncall =   532
```

- The numerical and analytical ICs at $t = 0$ agree exactly since the analytical solution (Eqs. (10.6)) is used in the main program of Listing 10.3 to define the numerical IC.
- The agreement between the numerical and analytical solutions is quite acceptable given that only 21 points were used in x.
- The solution increases as K*(x+ct) approaches $\pi/2 = 1.570796$. This increase is evident in the graphical output of Figure 10.2a,b.
- The computational effort is modest, ncall = 532.

The solution is displayed graphically in Figure 10.1a, which again demonstrates the agreement between the numerical and analytical solutions, particularly where the solution is changing more rapidly (near $K(x + ct) = \pi/2$).

The 3D plot of the solution in Figure 10.2b reflects the increase of the solution in the neighborhood of $\pi/2$. This completes the discussion of the linear and nonlinear KG equation.

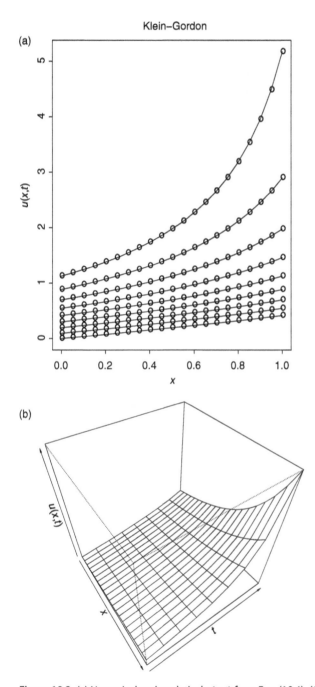

Figure 10.2 (a) Numerical and analytical $u(x, t)$ from Eqs. (10.4)–(10.6) lines –num, points –anal. (b) Numerical $u(x, t)$ from Eqs. (10.4)–(10.6).

10.10 Summary and Conclusions

The programming of the KG equation of Eqs. (10.1) and (10.4) and the analytical solutions Eqs. (10.3) and (10.6) within the SCMOL framework is straightforward, and the two solutions agree with acceptable accuracy and computational effort. Inclusion of the nonlinearity $-gu^m, m = 3$ has a marked effect on the solution.[2]

Reference

1 Dehghan, M., and A. Shokri (2009), Numerical solution of the nonlinear Klein–Gordon equation using radial basis functions, *Journal of Computational and Applied Mathematics*, **230**, no. 2, 400–410.

---

2 For a given interval in x, the solution of Eq. (10.4) is bounded for $(n - 1/2)\pi < K(x + ct) < (n + 1/2)\pi, n = 0, \pm1, \pm2 \ldots$ As $K(x + ct)$ approaches either boundary (at which $\cos(K(x + ct)) = 0$), the solution exhibits so-called blow up.

11

Boussinesq Equation

The Boussinesq equation (BE) is fourth order in space PDE with application, for example, to traveling surface waves [1]. We consider here a form of the BE that has solitary wave solutions. An analytical solution is used for comparison with the spline collocation method of lines (SCMOL) numerical solution.

11.1 PDE Model

The BE to be considered is

$$\frac{\partial^2 u}{\partial t^2} = \frac{\partial^2 u}{\partial x^2} - \frac{\partial^4 u}{\partial x^4} - \frac{\partial^2}{\partial x^2}(u^2) \tag{11.1}$$

with the analytical solution [2]

$$u_a(x,t) = (3/2)(1-c^2)\text{sech}^2\left[\frac{\sqrt{1-c^2}}{2}(x-ct)\right]$$

$$= (3/2)(1-c^2)\cosh^{-2}\left[\frac{\sqrt{1-c^2}}{2}(x-ct)\right] \tag{11.2}$$

where c is an arbitrary constant. We note that Eq. (11.2) is a traveling wave solution with the argument $x - ct$.

Since Eq. (11.1) is second order in t, we require two ICs that are taken from Eq. (11.2).

$$u(x, t = 0) = (3/2)(1-c^2)\cosh^{-2}\left[\frac{\sqrt{1-c^2}}{2}(x)\right] \tag{11.3a}$$

$$\frac{\partial u(x, t = 0)}{\partial t} = (3/2)c(1-c^2)^{3/2}\sinh\left[\frac{\sqrt{1-c^2}}{2}(x)\right]\cosh^{-3}\left[\frac{\sqrt{1-c^2}}{2}(x)\right] \tag{11.3b}$$

Spline Collocation Methods for Partial Differential Equations: With Applications in R, First Edition.
William E. Schiesser.
© 2017 John Wiley & Sons, Inc. Published 2017 by John Wiley & Sons, Inc.
Companion website: www.wiley.com/go/Spline_Collocation

Equation (11.1) is fourth order in x, and the four required boundary conditions (BCs) are taken as

$$u(x = x_l, t) = u(x = x_u, t) = 0 \qquad (11.3c,d)$$

$$\frac{\partial^2 u(x = x_l, t)}{\partial x^2} = \frac{\partial^2 u(x = x_u, t)}{\partial x^2} = 0 \qquad (11.3e,f)$$

Equations (11.1)–(11.3) constitute the problem of interest, and an SCMOL numerical solution is produced with the following routines, starting with a main program.

11.2 Main Program

A main program for Eqs. (11.1)–(11.3) follows.

```
#
# Boussinesq
#
# Delete previous workspaces
  rm(list=ls(all=TRUE))
#
# Access ODE integrator
  library("deSolve");
#
# Access functions for numerical solution
  setwd("f:/collocation/chap11");
  source("pde_1a.R");source("ua_1.R");
  source("uat_1.R");
#
# Grid in x
  n=101;xl=-15;xu=35;
  x=seq(from=xl,to=xu,by=(xu-xl)/(n-1));
#
# Parameters
  c=0.9;
#
# Independent variable for ODE integration
  nout=4;t0=0;tf=15;
  tout=seq(from=t0,to=tf,by=(tf-t0)/(nout-1));
#
# ICs from analytical solution
  u0=rep(0,2*n);
  for(i in 1:n){
    u0[i]   = ua_1(x[i],t0);
```

```
    u0[i+n]=uat_1(x[i],t0);
  }
  ncall=0;
#
# ODE integration
  out=lsodes(y=u0,times=tout,func=pde_1a,
      sparsetype="sparseint",rtol=1e-6,atol=1e-6,
      maxord=5);
  nrow(out)
  ncol(out)
#
# Arrays for numerical solution
  u1=matrix(0,nrow=n,ncol=nout);
  u2=matrix(0,nrow=n,ncol=nout);
  t=rep(0,nout);
  for(it in 1:nout){
  for(i  in 1:n){
    u1[i,it]=out[it,i+1];
    u2[i,it]=out[it,i+1+n];
      t[it]=out[it,1];
  }
  }
#
# Arrays for analytical solution,
# errors in numerical solution
  u1a=matrix(0,nrow=n,ncol=nout);
  err=matrix(0,nrow=n,ncol=nout);
  for(it in 1:nout){
    u1[1,it]=0;
    u1[n,it]=0;
    for(i in 1:n){
      u1a[i,it]=ua_1(x[i],t[it]);
      err[i,it]=u1[i,it]-u1a[i,it];
    }
  }
#
# Display selected output
  cat(sprintf(" xl = %4.1f  xu= %4.1f  c =
              %4.1f\n",xl,xu,c));
  for(it in 1:nout){
    cat(sprintf("\n      t      x    u1(x,t)
                u1a(x,t)   err(x,t)\n"));
    iv=seq(from=1,to=n,by=10);
```

```
    for(i in iv){
      cat(sprintf("%6.2f%7.1f%10.6f%10.6f
                  %10.6f\n",t[it],x[i],u1[i,it],
                  u1a[i,it],err[i,it]));
    }
    cat(sprintf("\n"));
  }
  cat(sprintf("  ncall = %4d\n",ncall));
#
# Plot 2D numerical solution
    matplot(x,u1,type="l",lwd=2,col="black",lty=1,
      xlab="x",ylab="u(x,t)",main="Boussinesq");
    matpoints(x,u1a,pch="o",col="black");
#
# Plot 3D numerical solution
    persp(x,t,u1,theta=55,phi=45,xlim=c(xl,xu),
          ylim=c(t0,t[nout]),xlab="x",ylab="t",
          zlab="u(x,t)");
```

Listing 11.1 Main program for Eqs. (11.1)–(11.3).

We can note the following details about Listing 11.1:

- Previous workspaces are removed. Then the ODE integrator library deSolve is accessed. Note that the setwd (set working directory) uses / rather than the usual \, and will require editing for the local computer (to define the working directory).

```
#
# Boussinesq
#
# Delete previous workspaces
  rm(list=ls(all=TRUE))
#
# Access ODE integrator
  library("deSolve");
#
# Access functions for numerical solution
  setwd("f:/collocation/chap11");
  source("pde_1a.R");source("ua_1.R");
  source("uat_1.R");
```

pde_1a is the routine for the MOL/ODEs (discussed subsequently). ua_1, uat_1 are the routines for the analytical solution of Eq. (11.2).

- A uniform grid in x with 101 points is defined with the seq utility for the interval $x_l = -15 \leq x \leq x_u = 35$. Therefore, the grid spacing is $(35 - (-15)/(101 - 1) = 0.5$ and the vector x has the values $x = -15, -14.5, \ldots, 35$.

```
#
# Grid in x
  n=101;xl=-15;xu=35;
  x=seq(from=xl,to=xu,by=(xu-xl)/(n-1));
```

The interval for $x > 0$ is larger $(0 \leq x \leq 35)$ than for $x < 0$ $(-15 \leq x \leq 0)$ to accommodate the IC pulse moving left to right in x (as demonstrated in the numerical solution that follows). If the pulse reached the right boundary $(x = 35)$, BCs (11.3c–f) would not apply, that is, the solution and its second derivative in x would depart from zero.

- The model parameter c (the wave velocity in Eq. (11.2)) is defined numerically.

```
#
# Parameters
  c=0.9;
```

- A uniform grid in t of 4 output points is defined with the seq utility for the interval $0 \leq t \leq t_f = 15$. Therefore, the vector tout has the values $t = 0, 5, 10, 15$.

```
#
# Independent variable for ODE integration
  nout=4;t0=0;tf=15;
  tout=seq(from=t0,to=tf,by=(tf-t0)/(nout-1));
```

At this point, the intervals of x and t in Eq. (11.1) are defined.
- The ICs for Eq. (11.1), Eqs. (11.3a,b), are placed in a vector u0 of length $2(101) = 202$.[1]

```
#
# ICs from analytical solution
  u0=rep(0,2*n);
  for(i in 1:n){
    u0[i]   = ua_1(x[i],t0);
    u0[i+n] =uat_1(x[i],t0);
```

---

1 This IC is based on the use of two variables, u_1, u_2, defined as

$$u_1(x, t) = u(x, t); \quad u_2(x, t) = \frac{\partial u}{\partial t}$$

```
    }
    ncall=0;
```

`ua_1`, `uat_1` (discussed subsequently) are called for `t0=0`. The counter for the number of calls to `pde_1a` is initialized.

- The system of 202 ODEs is integrated by the library integrator `lsodes` (available in `deSolve`) with the sparse matrix option specified. As expected, the inputs to `lsodes` are the ODE function, `pde_1a`, the IC vector `u0`, and the vector of output values of `t`, `tout`. The length of `u0` (202) informs `lsodes` how many ODEs are to be integrated. `func,y,times` are reserved names.

```
#
# ODE integration
  out=lsodes(y=u0,times=tout,func=pde_1a,
      sparsetype="sparseint",rtol=1e-6,atol=1e-6,
      maxord=5);
  nrow(out)
  ncol(out)
```

The numerical solution to the ODEs is returned in matrix `out`. In this case, `out` has the dimensions 4×203 (displayed in the numerical output). The offset $202 + 1 = 203$ is required since the first element of each column has the output t (also in `tout`), and the 2, ..., 203 column elements have the 202 ODE solutions. This indexing of `out` is used next.

- The numerical solution for $u(x, t)$, consisting of u_1, u_2, is taken from the solution matrix `out` returned by `lsodes`. The intervals in t and x are covered by two nested `for`s (with indices `it` and `i` for t and x, respectively). The offset `i+1` accounts for t returned in the first column position of `out` and placed in the vector `t[it]`.

```
#
# Arrays for numerical solution
  u1=matrix(0,nrow=n,ncol=nout);
  u2=matrix(0,nrow=n,ncol=nout);
  t=rep(0,nout);
  for(it in 1:nout){
  for(i   in 1:n){
    u1[i,it]=out[it,i+1];
    u2[i,it]=out[it,i+1+n];
      t[it]=out[it,1];
  }
  }
```

- The solution at the boundaries, $u_1(x = -15, t) = 0$, $u_1(x = 35, t) = 0$ (u1[1,it],u1[n,it]), is set in accordance with BCs (11.3c,d) since these values are not returned in out (only dependent variables computed as the solutions of ODEs are returned by lsodes, but this is not the case in pde_1a of Listing 11.2 at the boundaries $x = -15, 35$). The analytical solution from Eq. (11.2) (u1a) and the difference between the numerical and analytical solutions (err) are placed in arrays.

```
#
# Arrays for analytical solution,
# errors in numerical solution
  u1a=matrix(0,nrow=n,ncol=nout);
  err=matrix(0,nrow=n,ncol=nout);
  for(it in 1:nout){
    u1[1,it]=0;
    u1[n,it]=0;
    for(i in 1:n){
      u1a[i,it]=ua_1(x[i],t[it]);
      err[i,it]=u1[i,it]-u1a[i,it];
    }
  }
```

- The numerical solution (u1), the analytical solution from Eq. (11.2) (ua), and the difference (err) are displayed.

```
#
# Display selected output
  cat(sprintf("  xl = %4.1f  xu= %4.1f  c =
              %4.1f\n",xl,xu,c));
  for(it in 1:nout){
    cat(sprintf("\n      t        x     u1(x,t)
                u1a(x,t)    err(x,t)\n"));
    iv=seq(from=1,to=n,by=10);
    for(i in iv){
      cat(sprintf("%6.2f%7.1f%10.6f%10.6f
                  %10.6f\n",t[it],x[i],u1[i,it],
                  u1a[i,it],err[i,it]));
    }
    cat(sprintf("\n"));
  }
  cat(sprintf("  ncall = %4d\n",ncall));
```

Every tenth value is displayed with sprintf to conserve space. The number of calls to pde_1a is displayed at the end of the solution.

- The numerical and analytical solutions are plotted with `matplot` and `matpoints`, respectively.

```
#
# Plot 2D numerical solution
    matplot(x,u1,type="l",lwd=2,col="black",lty=1,
      xlab="x",ylab="u(x,t)",main="Boussinesq");
    matpoints(x,u1a,pch="o",col="black");
#
# Plot 3D numerical solution
    persp(x,t,u1,theta=55,phi=45,xlim=c(xl,xu),
          ylim=c(t0,t[nout]),xlab="x",ylab="t",
          zlab="u(x,t)");
```

The numerical solution is also plotted in 3D with `persp`.

The ODE routine `pde_1a` called by `lsodes` in the main program of Listing 11.1 is considered next.

11.3 ODE Routine

`pde_1a` for Eq. (11.1) follows.

```
  pde_1a=function(t,u,parm) {
#
# Function pde_1a computes the t derivative vector
# for u1(x,t), u2(x,t)
#
# One vector to two vectors
  u1=rep(0,n);u2=rep(0,n);
  for(i  in 1:n) {
    u1[i]=u[i];
    u2[i]=u[i+n];
  }
#
# BCs
  u1[1]=0;
  u1[n]=0;
#
# uxx
  tablexx=splinefun(x,u1);
  u1xx=tablexx(x,deriv=2);
```

```
#
# BCs
  u1xx[1]=0;
  u1xx[n]=0;
#
# uxxxx
  tablexxxx=splinefun(x,u1xx);
  u1xxxx=tablexxxx(x,deriv=2);
#
# u1^2
  u1s=rep(0,n);
  for(i in 1:n){
    u1s[i]=u1[i]^2;
  }
#
# u1sxx
  tablexx=splinefun(x,u1s);
  u1sxx=tablexx(x,deriv=2);
#
# PDE
  u1t=rep(0,n);u2t=rep(0,n);
  for(i in 1:n){
    u1t[i]=u2[i];
    u2t[i]=u1xx[i]-u1xxxx[i]-u1sxx[i];
  }
  u1t[1]=0;u1t[n]=0;
#
# Two vectors to one vector
  ut=rep(0,2*n);
  for(i in 1:n){
    ut[i]  =u1t[i];
    ut[i+n]=u2t[i];
  }
#
# Increment calls to pde_1a
  ncall <<- ncall+1;
#
# Return derivative vector
  return(list(c(ut)));
  }
```

Listing 11.2 ODE/MOL routine pde_1a for Eq. (11.1).

We can note the following details about pde_1a:

- The function is defined.

```
pde_1a=function(t,u,parm){
#
# Function pde_1a computes the t derivative vector
# for u1(x,t), u2(x,t)
```

t is the current value of t in Eq. (11.1). u is the 202-vector. parm is an argument to pass parameters to pde_1 (unused, but required in the argument list). The arguments must be listed in the order stated to properly interface with lsodes called in the main program. The derivative vector of the LHS of Eq. (11.1) is calculated below and returned to lsodes.
- Homogeneous Dirichlet BCs (11.3c,d) are implemented.

```
#
# BCs
   u1[1]=0;
   u1[n]=0;
```

- The second derivative $\dfrac{\partial^2 u}{\partial x^2} = \dfrac{\partial^2 u_1}{\partial x^2}$ in Eq. (11.1) is computed with splinefun.

```
#
# uxx
   tablexx=splinefun(x,u1);
   u1xx=tablexx(x,deriv=2);
```

- The second-derivative BCs (11.3e,f) are implemented (at $x = -15, 35$ with subscripts $1, n$).

```
#
# BCs
   u1xx[1]=0;
   u1xx[n]=0;
```

- The fourth derivative $\dfrac{\partial^4 u}{\partial x^4} = \dfrac{\partial^4 u_1}{\partial x^4}$ in Eq. (11.1) is computed with splinefun by differentiating the second derivative (i.e., stagewise differentiation).

```
#
# uxxxx
   tablexxxx=splinefun(x,u1xx);
   u1xxxx=tablexxxx(x,deriv=2);
```

- The square function in Eq. (11.1) is computed.

```
#
# u1^2
   u1s=rep(0,n);
   for(i in 1:n){
     u1s[i]=u1[i]^2;
   }
```

- The square function is then differentiated to give the term $\dfrac{\partial^2}{\partial x^2}(u^2)$.

```
#
# u1sxx
   tablexx=splinefun(x,u1s);
   u1sxx=tablexx(x,deriv=2);
```

- The LHS derivative of Eq. (11.1) $\dfrac{\partial^2 u(x,t)}{\partial t^2} = \dfrac{\partial u_2(x,t)}{\partial t}$ is computed using u_1, u_2. The two 101-derivative vectors are placed in u1t, u2t.

```
#
# PDE
   u1t=rep(0,n);u2t=rep(0,n);
   for(i in 1:n){
     u1t[i]=u2[i];
     u2t[i]=u1xx[i]-u1xxxx[i]-u1sxx[i];
   }
   u1t[1]=0;u1t[n]=0;
```

The derivatives $\dfrac{\partial u_1(x=-15,t)}{\partial t}, \dfrac{\partial u_1(x=35,t)}{\partial t}$ are set to zero in accordance with BCs (11.3c) (so that lsodes does not move the boundary values $u_1(x = -15, t), u_1(x = 35, t)$ from their prescribed zero values).

- u1t, u2t are placed in a 202-vector ut.

```
#
# Two vectors to one vector
   ut=rep(0,2*n);
   for(i in 1:n){
     ut[i]   =u1t[i];
     ut[i+n]=u2t[i];
   }
```

- The counter for the calls to pde_1a is returned to the main program of Listing 11.1 by the «- operator.

```
#
```

```
# Increment calls to pde_1a
  ncall <<- ncall+1;
```

- The derivative in ut is returned to lsodes.

```
#
# Return derivative vector
  return(list(c(ut)));
  }
```

Note that a list is used as required by lsodes. The final } concludes pde_1a.

11.4 Subordinate Routines

Function ua_1 for Eq. (11.2) is listed next.

```
  ua_1=function(x,t){
#
# Function ua_1 computes the analytical solution of
# the Boussinesq equation for comparison with the
# numerical solution
#
# Analytical solution
  xi=(1-c^2)^(0.5)/2*(x-c*t);
  ua=(3/2)*(1-c^2)*(cosh(xi))^(-2);
  return(c(ua));
  }
```

Listing 11.3 Function ua_1 for Eq. (11.2).

$u_a(x, t)$ is a function only of the Lagrangian variable $x - ct$ and therefore is a traveling wave as will be observed in the numerical solution considered next. Also, $u_a(x, t) = 0$ for $c = 1$ (Eq. (11.2)) so that $c = 0.9$ is used for the numerical and analytical solutions of Eq. (11.1).

Function uat_1 for $\dfrac{\partial u_a(x, t)}{\partial t}$ from Eq. (11.2) follows.

```
  uat_1=function(x,t){
#
# Function uat_1 computes the time derivative of the
# analytical solution of the Boussinesq equation
#
# Analytical solution derivative
  xi=(1-c^2)^(0.5)/2*(x-c*t);
```

```
uat= (3/2) *c* (1-c^2)^(3/2) *sinh(xi) * (cosh(xi))^(-3);
return(c(uat));
}
```

Listing 11.4 Function uat_1 from Eq. (11.2).

uat_1 is straightforward and does not require any discussion.

This completes the coding of Eqs. (11.1)–(11.3). The numerical and graphical output is considered next.

11.5 Model Output

Execution of the preceding main program and ODE routine pde_1 and subordinate routines ua_1, uat_1 produces the following abbreviated numerical output in Table 11.1.

Table 11.1 Abbreviated output for Eqs. (11.1)–(11.3).

```
[1] 4

[1] 203

xl = -15.0   xu= 35.0   c =  0.9

   t       x     u1(x,t)   u1a(x,t)    err(x,t)
 0.00   -15.0   0.000000   0.001645  -0.001645
 0.00   -10.0   0.014217   0.014217   0.000000
 0.00    -5.0   0.104066   0.104066   0.000000
 0.00     0.0   0.285000   0.285000   0.000000
 0.00     5.0   0.104066   0.104066   0.000000
 0.00    10.0   0.014217   0.014217   0.000000
 0.00    15.0   0.001645   0.001645   0.000000
 0.00    20.0   0.000186   0.000186   0.000000
 0.00    25.0   0.000021   0.000021   0.000000
 0.00    30.0   0.000002   0.000002   0.000000
 0.00    35.0   0.000000   0.000000  -0.000000
                   .          .
                   .          .
                   .          .
```

(Continued)

Table 11.1 (Continued)

```
     Output for t = 5, 10 removed
                 .              .
                 .              .
                 .              .
      t       x      ul(x,t)   ula(x,t)   err(x,t)
    15.00   -15.0   0.000000   0.000005  -0.000005
    15.00   -10.0   0.000050   0.000041   0.000009
    15.00    -5.0   0.000283   0.000359  -0.000075
    15.00     0.0   0.002590   0.003154  -0.000564
    15.00     5.0   0.026771   0.026712   0.000059
    15.00    10.0   0.166421   0.167267  -0.000845
    15.00    15.0   0.257649   0.256586   0.001063
    15.00    20.0   0.060030   0.059811   0.000219
    15.00    25.0   0.007427   0.007484  -0.000057
    15.00    30.0   0.000891   0.000856   0.000035
    15.00    35.0   0.000000   0.000097  -0.000097

    ncall = 3974
```

We can note the following details about this output:

- As expected, the solution array out has the dimensions 4 x 203 for 4 output points and 202 ODEs, with t included as an additional element, that is, $202 + 1 = 203$.
- The interval in x is $((35 - (-15))/(101 - 1) = 0.5$ (with every tenth value in Table 11.1).
- At $t = 0$, the peak value at $x = 0$, $u(x = 0, t = 0) = 0.285000$, which travels left to right with a velocity $c = 0.9$ according to Eq. (11.2) (not in the abbreviated numerical solution for $t > 0$ but demonstrated in Figure 11.1a discussed next).
- The error at the boundaries $x = -15, 35$ is not zero since the boundary values of the analytical solution in Listing 11.1 are not zero.
- The agreement between the numerical and analytical solutions is quite acceptable.
- The computational effort is appreciable, ncall = 3974.

The solution is displayed graphically in Figure 11.1a, which again demonstrates the agreement between the numerical and analytical solutions.

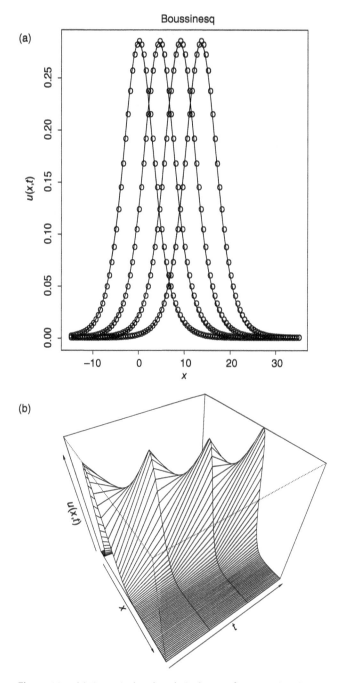

Figure 11.1 (a) Numerical and analytical $u(x, t)$ from Eqs. (11.1)–(11.3) lines – num, points – anal. (b) Numerical $u(x, t)$ from Eqs. (11.1)–(11.3).

The traveling wave solution as a function of only $x - ct$ is clear. Since the peak values occur at $x - ct = 0$, with $c = 0.9$, $t = 15$, the peak value is at $x = (0.9)(15) = 13.5$, which does not appear in Table 11.1.[2]

The 3D plot of the solution in Figure 11.1b reflects the traveling wave characteristic[3] of the solution in x and t (from Eq. (11.2)).

This completes the programming and discussion of the solutions to Eqs. (11.1)–(11.3).

11.6 Summary and Conclusions

The programming of the BE numerical solution of Eqs. (11.1) and (11.3), and the analytical solution Eq. (11.2) within the SCMOL framework is straightforward, and the two solutions agree with acceptable accuracy and computational effort. Inclusion of the nonlinearity $-\dfrac{\partial^2}{\partial x^2}(u^2)$ of Eq. (11.1) (in pde_1a of Listing 11.1) is particularly noteworthy.

References

1 Biswas, A., D. Milovic, and A. Ranasinghe (2009), Solitary waves of Boussinesq equation in a power law media, *Communications in Nonlinear Science and Numerical Simulation*, **14**, 3738–3742.

2 vande Wouwer, A. (2002), University of Mons, Belgium, private communication.

---

2 From $t = 0$, this peak value is 0.285000. The accurate resolution of the peak values is confirmed in Figure 11.1a (within the limits of observed accuracy in a graph or plot). This is a stringent test since the solution changes most rapidly at the peak values of x $(x = (0.9)(5) = 4.5, (0.9)(10) = 9, (0.9)(15) = 13.5)$.

3 The term "characteristic" has a mathematical interpretation. $x - ct$ is termed a characteristic for Eq. (11.1), and the numerical and analytical solutions are defined along this characteristic.

12

Cahn–Hilliard Equation

The Cahn–Hilliard (CH) equation is fourth order in space PDE with application, for example, to simultaneous phase separation in response to a chemical potential [1]. In the following discussion, a steady-state (equilibrium) analytical solution is used for comparison with the spline collocation method of lines (SCMOL) numerical solution.

12.1 PDE Model

The CH equation to be considered is [1]

$$\frac{\partial u}{\partial t} = D\nabla^2(u^3 - u - \gamma\nabla^2 u) \tag{12.1}$$

where

Variable	Interpretation
u	dependent variable
∇	coordinate-free spatial differential operator
t	time
D	diffusivity
γ	characteristic length.

A 1D special case of Eq. (12.1) in Cartesian coordinates is considered for which $\nabla^2 = \frac{\partial^2}{\partial x^2}$.

$$\frac{\partial u}{\partial t} = D\frac{\partial^2\left(u^3 - u - \gamma\frac{\partial^2 u}{\partial x^2}\right)}{\partial x^2} \tag{12.2}$$

Since Eq. (12.1) is first order in t, one IC is required

$$u(x, t = 0) = f(x) \tag{12.3a}$$

where $f(x)$ is a function to be specified.

Spline Collocation Methods for Partial Differential Equations: With Applications in R, First Edition.
William E. Schiesser.
© 2017 John Wiley & Sons, Inc. Published 2017 by John Wiley & Sons, Inc.
Companion website: www.wiley.com/go/Spline_Collocation

Equation (12.1) is fourth order in x and four boundary conditions (BCs) are required.

$$u(x = x_l, t) = -1; \quad u(x = x_u, t) = 1 \qquad (12.3\text{b,c})$$

$$\frac{\partial^2 u(x = x_l, t)}{\partial x^2} = \frac{\partial^2 u(x = x_u, t)}{\partial x^2} = 0 \qquad (12.3\text{d,e})$$

A steady-state (equilibrium) solution is available for Eqs. (12.2) and (12.3) [2].

$$u_e(x) = \tanh\left(\frac{x}{\sqrt{2\gamma}}\right) \qquad (12.4)$$

Equations (12.2) and (12.3) constitute the problem of interest, and an SCMOL numerical solution is produced with the following routines, starting with a main program. The solution of Eq. (12.4) is used to test the numerical solution $u(x, t \rightarrow \infty)$.

12.2 Main Program

A main program for Eqs. (12.2)–(12.4) follows.

```
#
# Cahn-Hilliard
#
# Delete previous workspaces
  rm(list=ls(all=TRUE))
#
# Access ODE integrator
  library("deSolve");
#
# Access functions for numerical solution
  setwd("f:/collocation/chap12");
  source("pde_1a.R");
#
# Select case (for IC)
  ncase=1;
#
# Grid in x
  n=51;xl=-5;xu=5;
  x=seq(from=xl,to=xu,by=(xu-xl)/(n-1));
#
# Parameters
  D=1;gam=0.5;
#
```

```
# Independent variable for ODE integration
  nout=6;t0=0;tf=5;
  tout=seq(from=t0,to=tf,by=(tf-t0)/(nout-1));
#
# IC
  u0=rep(0,n);
  for(i in 1:n){
    if(ncase==1){u0[i]=0};
    if(ncase==2){u0[i]=tanh(x[i]/sqrt(2*gam))};
    if(ncase==3){
      ul=-1;uu=1;
      if(i<11){u0[i]=ul};
      if(i>41){u0[i]=uu};
      if((i>=11)&(i<=41)){
        u0[i]=ul+(uu-ul)*(i-11)/(41-11);
      }
    }
  }
  ncall=0;
#
# ODE integration
  out=lsodes(y=u0,times=tout,func=pde_1a,
      sparsetype="sparseint",rtol=1e-6,atol=1e-6,
      maxord=5);
  nrow(out)
  ncol(out)
#
# Arrays for numerical solution
  u=matrix(0,nrow=n,ncol=nout);
  t=rep(0,nout);
  for(it in 1:nout){
  for(i  in 1:n){
    u[i,it]=out[it,i+1];
    t[it]=out[it,1];
  }
  }
#
# Display selected output
  cat(sprintf(" xl = %4.1f     xu = %4.1f\n D = %4.1f
              gam = %4.1f\n",xl,xu,D,gam));
  for(it in 1:nout){
    cat(sprintf("\n        t       x     u(x,t)\n"));
    iv=seq(from=1,to=n,by=5);
```

```
    for(i in iv){
      cat(sprintf("%7.3f%7.3f%10.6f\n",t[it],x[i],
                     u[i,it]));
    }
    cat(sprintf("\n"));
  }
  cat(sprintf(" ncall = %4d\n",ncall));
#
# Equilibrium solution
  if(ncase>1){
    ue=rep(0,n);
    cat(sprintf("\n        Equilibrium solution\n"));
    for(i in 1:n){
      ue[i]=tanh(x[i]/sqrt(2*gam));
    }
    iv=seq(from=1,to=n,by=5);
    for(i in iv){
      cat(sprintf("\n        %7.3f%10.6f",x[i],ue[i]));
    }
  }
#
# Plot 2D numerical solution
    matplot(x,u,type="l",lwd=2,col="black",lty=1,
      xlab="x",ylab="u(x,t)",main="Cahn-Hilliard");
    if(ncase>1){
    matpoints(x,ue,pch="o",col="black");
    }
#
# Plot 3D numerical solution
    persp(x,t,u,theta=0,phi=55,xlim=c(xl,xu),
          ylim=c(t0,t[nout]),xlab="x",ylab="t",
```

Listing 12.1 Main program for Eqs. (12.2)–(12.4).

We can note the following details about Listing 12.1:

- Previous workspaces are removed. Then the ODE integrator library de-Solve is accessed. Note that the setwd (set working directory) uses / rather than the usual \. The setwd requires editing for the local computer (to specify the directory for the routines).

```
#
# Cahn-Hilliard
#
```

```
# Delete previous workspaces
  rm(list=ls(all=TRUE))
#
# Access ODE integrator
  library("deSolve");
#
# Access functions for numerical solution
  setwd("f:/collocation/chap12");
  source("pde_1a.R");
```

pde_1a is the routine for the MOL/ODEs (discussed subsequently).
- A case is selected, which specifies an IC (considered below).

```
#
# Select case (for IC)
  ncase=3;
```

- A uniform grid in *x* with 51 points is defined with the seq utility for the interval $x_l = -5 \leq x \leq x_u = 5$. Therefore, the grid spacing is $(5 - (-5)/ (51 - 1) = 0.2$ and the vector x has the values $x = -5, -4.8, \ldots, 5$.

```
#
# Grid in x
  n=51;xl=-5;xu=5;
  x=seq(from=xl,to=xu,by=(xu-xl)/(n-1));
```

- The model parameters D, γ (in Eq. (12.2)) are defined numerically.

```
#
# Parameters
  D=1;gam=0.5;
```

- A uniform grid in *t* of 6 output points is defined with the seq utility for the interval $0 \leq t \leq t_f = 5$. Therefore, the vector tout has the values $t = 0, 1, \ldots, 5$.

```
#
# Independent variable for ODE integration
  nout=6;t0=0;tf=5;
  tout=seq(from=t0,to=tf,by=(tf-t0)/(nout-1));
```

At this point, the intervals of *x* and *t* in Eq. (12.2) are defined.
- The IC for Eq. (12.2), Eq. (12.3a), for three different $f(x)$ (ncase=1, 2, 3), is placed in a vector u0 of length 51.

```
#
# IC
  u0=rep(0,n);
```

```
for (i in 1:n) {
  if (ncase==1) {u0 [i]=0};
  if (ncase==2) {u0 [i]=tanh (x [i]/sqrt (2*gam))};
  if (ncase==3) {
    ul=-1;uu=1;
    if (i<11) {u0 [i]=ul};
    if (i>41) {u0 [i]=uu};
    if ((i>=11) & (i<=41)) {
      u0 [i]=ul+ (uu-ul) * (i-11) / (41-11) ;
    }
  }
}
ncall=0;
```

The three cases are discussed subsequently. The counter for the number of calls to pde_1a is initialized.

- The system of 51 ODEs is integrated by the library integrator lsodes (available in deSolve) with the sparse matrix option specified. As expected, the inputs to lsodes are the ODE function, pde_1a, the IC vector u0, and the vector of output values of t, tout. The length of u0 (51) informs lsodes how many ODEs are to be integrated. func,y,times are reserved names.

```
#
# ODE integration
  out=lsodes (y=u0, times=tout, func=pde_1a,
      sparsetype="sparseint", rtol=1e-6, atol=1e-6,
      maxord=5) ;
  nrow (out)
  ncol (out)
```

The numerical solution to the ODEs is returned in matrix out. In this case, out has the dimensions 6×52 (displayed in the numerical output). The offset $51 + 1 = 52$ is required since the first element of each column has the output t (also in tout), and the 2, ..., 52 column elements have the 51 ODE solutions. This indexing of out is used next.

- The numerical solution for $u(x, t)$ is taken from the solution matrix out returned by lsodes. The intervals in t and x are covered by two nested fors (with indices it and i for t and x, respectively). The offset i+1 accounts for t returned in the first column position of out and placed in the vector t [it].

```
#
# Arrays for numerical solution
  u=matrix (0, nrow=n, ncol=nout) ;
```

```
t=rep(0,nout);
for(it in 1:nout){
for(i   in 1:n){
  u[i,it]=out[it,i+1];
    t[it]=out[it,1];
}
}
```

- The parameters D, γ are displayed, followed by the numerical solution to Eq. (12.2), $u(x, t)$.

```
#
# Display selected output
  cat(sprintf(" xl = %4.1f    xu = %4.1f\n  D = %4.1f
               gam = %4.1f\n",xl,xu,D,gam));
  for(it in 1:nout){
    cat(sprintf("\n         t        x        u(x,t)\n"));
    iv=seq(from=1,to=n,by=5);
    for(i in iv){
      cat(sprintf("%7.3f%7.3f%10.6f\n",t[it],x[i],
                  u[i,it]));
    }
    cat(sprintf("\n"));
  }
  cat(sprintf(" ncall = %4d\n",ncall));
```

Every fifth value of $u(x, t)$ is displayed to conserve space. The number of calls to pde_1a is displayed at the end of the solution.
- The equilibrium solution, Eq. (12.4), is computed and displayed (ue).

```
#
# Equilibrium solution
  if(ncase>1){
    ue=rep(0,n);
    cat(sprintf("\n        Equilibrium solution\n"));
    for(i in 1:n){
      ue[i]=tanh(x[i]/sqrt(2*gam));
    }
    iv=seq(from=1,to=n,by=5);
    for(i in iv){
      cat(sprintf("\n      %7.3f%10.6f",x[i],ue[i]));
    }
  }
```

Every fifth value of ue is displayed to conserve space.

- The numerical solution is plotted with `matplot` and the steady-state solution of Eq. (12.4) is plotted with `matpoints`.

```
#
# Plot 2D numerical solution
    matplot(x,u,type="l",lwd=2,col="black",lty=1,
      xlab="x",ylab="u(x,t)",main="Cahn-Hilliard");
    if(ncase>1){
    matpoints(x,ue,pch="o",col="black");
    }
#
# Plot 3D numerical solution
    persp(x,t,u,theta=0,phi=55,xlim=c(xl,xu),
          ylim=c(t0,t[nout]),xlab="x",ylab="t",
          zlab="u(x,t)");
```

The numerical solution is also plotted in 3D with `persp`.

The ODE/MOL routine pde_1a called by `lsodes` in the main program of Listing 12.1 is considered next.

12.3 ODE Routine

pde_1a for Eq. (12.2) follows.

```
  pde_1a=function(t,u,parm){
#
# Function pde_1a computes the t derivative vector
# for u(x,t)
#
# BCs
  u[1]=-1;
  u[n]= 1;
#
# uxx
  tablexx=splinefun(x,u);
  uxx=tablexx(x,deriv=2);
#
# BCs
  uxx[1]=0;
  uxx[n]=0;
#
# u^3-u-gam*uxx
  ug=rep(0,n);
```

```
  for(i in 1:n){
    ug[i]=u[i]^3-u[i]-gam*uxx[i]
  }
#
# (u^3-u-gam*uxx)xx
  tablegxx=splinefun(x,ug);
  ugxx=tablegxx(x,deriv=2);
#
# PDE
  ut=rep(0,n);
  for(i in 2:(n-1)){
    ut[i]=D*ugxx[i];
  }
  ut[1]=0;ut[n]=0;
#
# Increment calls to pde_1a
  ncall <<- ncall+1;
#
# Return derivative vector
  return(list(c(ut)));
  }
```

Listing 12.2 ODE/MOL routine pde_1a for Eq. (12.2).

We can note the following details about pde_1a:

- The function is defined.

```
  pde_1a=function(t,u,parm){
#
# Function pde_1a computes the t derivative vector
# for u(x,t)
```

t is the current value of t in Eq. (12.2). u is the 51-element dependent variable vector. parm is an argument to pass parameters to pde_1 (unused, but required in the argument list). The arguments must be listed in the order stated to properly interface with lsodes called in the main program. The derivative vector of the LHS of Eq. (12.2) is calculated below and returned to lsodes.

- Nonhomogeneous Dirichlet BCs (12.3b,c) are implemented.

```
#
# BCs
  u[1]=-1;
  u[n]= 1;
```

- The second derivative $\dfrac{\partial^2 u}{\partial x^2}$ in Eq. (12.2) is computed with `splinefun`.

```
# uxx
  tablexx=splinefun(x,u);
  uxx=tablexx(x,deriv=2);
```

- The second-derivative BCs (12.3d,e) are implemented (at $x = -5, 5$) with subscripts `1`, `n`).

```
#
# BCs
  uxx[1]=0;
  uxx[n]=0;
```

- The group $\left(u^3 - u - \gamma \dfrac{\partial^2 u}{\partial x^2} \right)$ in Eq. (12.2) is computed.

```
#
# u^3-u-gam*uxx
  ug=rep(0,n);
  for(i in 1:n){
    ug[i]=u[i]^3-u[i]-gam*uxx[i]
  }
```

- The RHS term $\dfrac{\partial^2 \left(u^3 - u - \gamma \dfrac{\partial^2 u}{\partial x^2} \right)}{\partial x^2}$ in Eq. (12.2) is then differentiated.

```
#
# (u^3-u-gam*uxx)xx
  tablegxx=splinefun(x,ug);
  ugxx=tablegxx(x,deriv=2);
```

- The LHS derivative of Eq. (12.2), $\dfrac{\partial u(x,t)}{\partial t}$, is computed. The 51-derivative vector is placed in `ut`.

```
#
# PDE
  ut=rep(0,n);
  for(i in 2:(n-1)){
    ut[i]=D*ugxx[i];
  }
  ut[1]=0;ut[n]=0;
```

The derivatives $\dfrac{\partial u(x = -5, t)}{\partial t}$, $\dfrac{\partial u(x = 5, t)}{\partial t}$ are set to zero in accordance with BCs (12.3b,c) (so that lsodes does not move the boundary values $u(x = -5, t) = -1, u(x = 5, t) = 1$ from their prescribed values).

- The counter for the calls to pde_1a is returned to the main program of Listing 12.1 by the <<- operator.

```
#
# Increment calls to pde_1a
   ncall <<- ncall+1;
```

- The derivative in ut is returned to lsodes.

```
#
# Return derivative vector
   return(list(c(ut)));
   }
```

Note that a list is used as required by lsodes. The final } concludes pde_1a.

This concludes the coding of Eqs. (12.2)–(12.4). The numerical and graphical output is considered next.

12.4 Model Output

Execution of the preceding main program and ODE routine pde_1a produces the following abbreviated numerical output in Table 12.1 for ncase=1 (set in Listing 12.1).

Table 12.1 Abbreviated output for Eqs. (12.2)–(12.4), ncase=1.

```
[1]  6

[1]  52

xl  =  -5.0      xu  =    5.0
 D  =   1.0     gam  =    0.5

      t        x        u(x,t)
  0.000  -5.000    0.000000
  0.000  -4.000    0.000000
  0.000  -3.000    0.000000
  0.000  -2.000    0.000000
```

(Continued)

Table 12.1 (Continued)

```
0.000 -1.000   0.000000
0.000  0.000   0.000000
0.000  1.000   0.000000
0.000  2.000   0.000000
0.000  3.000   0.000000
0.000  4.000   0.000000
0.000  5.000   0.000000

   t      x      u(x,t)
1.000 -5.000   0.000000
1.000 -4.000  -0.582856
1.000 -3.000   0.015768
1.000 -2.000   0.234439
1.000 -1.000   0.089249
1.000  0.000  -0.000000
1.000  1.000  -0.089249
1.000  2.000  -0.234439
1.000  3.000  -0.015768
1.000  4.000   0.582856
1.000  5.000   0.000000
  .             .
  .             .
  .             .

Output for t = 2,3,4
      removed

  .             .
  .             .
  .             .

   t      x      u(x,t)
5.000 -5.000   0.000000
5.000 -4.000  -0.892626
5.000 -3.000  -0.484573
5.000 -2.000   0.311507
5.000 -1.000   0.555308
5.000  0.000  -0.000000
5.000  1.000  -0.555308
5.000  2.000  -0.311507
5.000  3.000   0.484573
5.000  4.000   0.892626
5.000  5.000   0.000000

ncall =  643
```

We can note the following details about this output.

- The solution array `out` has the dimensions 6 x 52 for 6 output points and 51 ODEs, with *t* included as an additional element, that is, 51 + 1 = 52.
- The interval in *x* is $((5-(-5))/(51-1) = 0.2$ (with every fifth value in Table 12.1).
- IC (12.2a) for `ncase=1` is confirmed.
- The solution at the boundaries at $x = -5, 5$ reflects BCs (12.3b,c).
- The discontinuty between IC (12.2a) and BCs (12.2b,c) is clear in Figure 12.1a.
- The computational effort is modest, `ncall` = 643.

The 3D plot of the solution from `persp` is in Figure 12.1b.

The complexity of the solution for `ncase=1` brings into the question the accuracy of the solution. This question is answered to some extent by `ncase=2` for which the steady state solution Eq. (12.4) is compared with

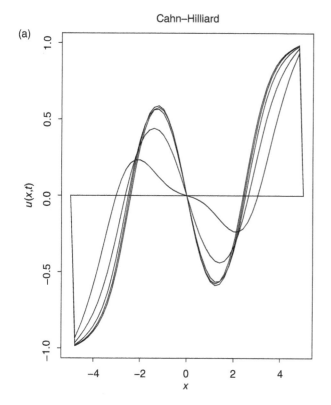

Figure 12.1 (a) Numerical $u(x, t)$ for Eqs. (12.2) and (12.3), `ncase=1,2D`. (b) Numerical $u(x, t)$ for Eqs. (12.2) and (12.3), `ncase=1,3D`.

(b)

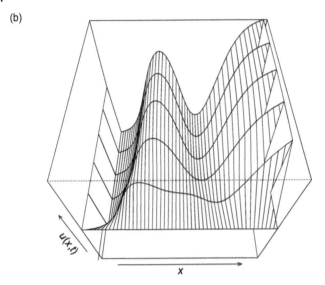

Figure 12.1 (*Continued*)

the steady state numerical solution (see the ncase=2 IC (12.2a) in Listing Listing 12.1).

Abbreviated output in Table 12.2 for ncase=2 follows.

Table 12.2 Abbreviated output for Eqs. (12.2)–(12.4), ncase=2.

```
[1]  6

[1]  52

       t       x      u(x,t)
   0.000 -5.000 -0.999909
   0.000 -4.000 -0.999329
   0.000 -3.000 -0.995055
   0.000 -2.000 -0.964028
   0.000 -1.000 -0.761594
   0.000  0.000  0.000000
```

Table 12.2 (Continued)

```
0.000   1.000   0.761594
0.000   2.000   0.964028
0.000   3.000   0.995055
0.000   4.000   0.999329
0.000   5.000   0.999909

  t        x       u(x,t)
1.000  -5.000  -0.999909
1.000  -4.000  -0.999423
1.000  -3.000  -0.995249
1.000  -2.000  -0.964236
1.000  -1.000  -0.760321
1.000   0.000   0.000000
1.000   1.000   0.760321
1.000   2.000   0.964236
1.000   3.000   0.995249
1.000   4.000   0.999423
1.000   5.000   0.999909
   .               .
   .               .
   .               .

 Output for t = 2,3,4
         removed

   .               .
   .               .
   .               .

  t        x       u(x,t)
5.000  -5.000  -0.999909
5.000  -4.000  -0.999363
5.000  -3.000  -0.995131
5.000  -2.000  -0.964071
5.000  -1.000  -0.760138
5.000   0.000   0.000000
5.000   1.000   0.760139
5.000   2.000   0.964071
5.000   3.000   0.995131
```

(Continued)

Table 12.2 (Continued)

```
5.000   4.000   0.999363
5.000   5.000   0.999909

ncall =   332

      Equilibrium solution

         -5.000 -0.999909
         -4.000 -0.999329
         -3.000 -0.995055
         -2.000 -0.964028
         -1.000 -0.761594
          0.000  0.000000
          1.000  0.761594
          2.000  0.964028
          3.000  0.995055
          4.000  0.999329
          5.000  0.999909
```

Since the IC for `ncase=2` is the steady state solution of Eq. (12.4), the numerical solution remains at the IC as we would expect. Although this may seem like a trivial case, it is worthwhile since if the numerical solution changed from the IC, a programming error would be indicated.

We can note the close agreement between the steady state numerical and analytical solution, e.g., at `t=5`,

`Numerical`

```
5.000   1.000   0.760139
```

`Analytical`

```
        1.000   0.761594
```

The computational effort is modest, `ncall = 332`, as expected since the solution does not change.

The 3D plot of the solution from `persp` is in Figure 12.2b.

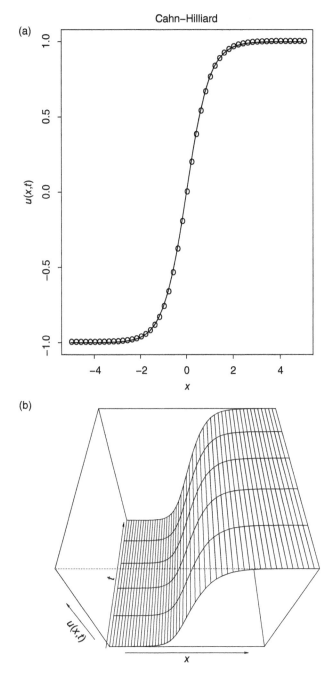

Figure 12.2 (a) Numerical $u(x, t)$ for Eqs. (12.2) and (12.3), ncase=2 num – line, equilibrium – points. (b) Numerical $u(x, t)$ for Eqs. (12.2) and (12.3), ncase=2.

For ncase=3 (set in Listing 12.1), the abbreviated output is (Table 12.3).

Table 12.3 Abbreviated output for
Eqs. (12.2)–(12.4), ncase=3.

```
[1]  6

[1]  52

        t        x       u(x,t)
   0.000  -5.000  -1.000000
   0.000  -4.000  -1.000000
   0.000  -3.000  -1.000000
   0.000  -2.000  -0.666667
   0.000  -1.000  -0.333333
   0.000   0.000   0.000000
   0.000   1.000   0.333333
   0.000   2.000   0.666667
   0.000   3.000   1.000000
   0.000   4.000   1.000000
   0.000   5.000   1.000000

        t        x       u(x,t)
   1.000  -5.000  -1.000000
   1.000  -4.000  -0.954207
   1.000  -3.000  -0.900601
   1.000  -2.000  -0.820670
   1.000  -1.000  -0.596963
   1.000   0.000   0.000000
   1.000   1.000   0.596963
   1.000   2.000   0.820670
   1.000   3.000   0.900601
   1.000   4.000   0.954207
   1.000   5.000   1.000000
        .                .
        .                .
        .                .
    Output for t = 2,3,4
           removed
        .                .
        .                .
        .                .
```

Table 12.3 (Continued)

```
     t       x      u(x,t)
 5.000  -5.000  -1.000000
 5.000  -4.000  -0.994883
 5.000  -3.000  -0.987377
 5.000  -2.000  -0.954654
 5.000  -1.000  -0.750718
 5.000   0.000   0.000000
 5.000   1.000   0.750718
 5.000   2.000   0.954654
 5.000   3.000   0.987377
 5.000   4.000   0.994883
 5.000   5.000   1.000000

ncall =   497

        Equilibrium solution

        -5.000  -0.999909
        -4.000  -0.999329
        -3.000  -0.995055
        -2.000  -0.964028
        -1.000  -0.761594
         0.000   0.000000
         1.000   0.761594
         2.000   0.964028
         3.000   0.995055
         4.000   0.999329
         5.000   0.999909
```

We can note the following details about this output:

- The IC is linear with respect to x for $-3 \leq x \leq 3$, which follows from the programming in Listing 12.1 for ncase=3.
- The solution at the boundaries at $x = -5, 5$ reflects BCs (12.3b,c).
- The solution at $t = 5$ has not quite reached the steady state solution of Eq. (12.4).

Numerical

```
 5.000   1.000   0.750718
```

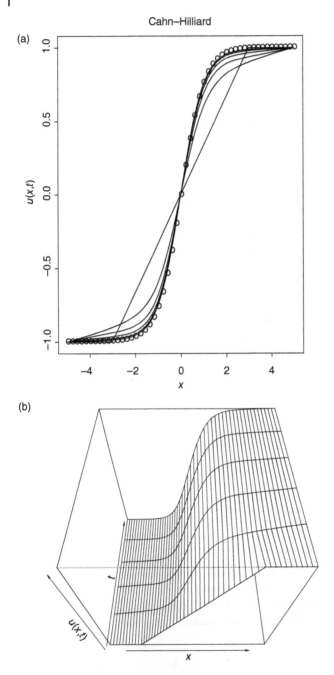

Figure 12.3 (a) Numerical $u(x, t)$ for Eqs. (12.2) and (12.3), ncase=3 num – lines, equilibrium – points. (b) Numerical $u(x, t)$ for Eqs. (12.2) and (12.3), ncase=3.

```
Analytical

      1.000   0.761594
```

- The computational effort is modest, ncall = 497.

The transition of the solution $u(x, t)$ from a linear function in x initially to the steady state of Eq. (12.4) (at $t = 5$) is clear in Figure 12.2a.

This completes the programming and discussion of the solutions to Eqs. (12.2)–(12.4).

12.5 Summary and Conclusions

The programming of the CH equation numerical solution of Eqs. (12.2)–(12.4) within the SCMOL framework is straightforward, and the numerical solution agrees with the steady-state analytical solution of Eq. (12.4). Inclusion of the nonlinear flux term $D\dfrac{\partial^2\left(u^3 - u - \gamma\dfrac{\partial^2 u}{\partial x^2}\right)}{\partial x^2}$ of Eq. (12.2) (in pde_1a of Listing 12.2) is particularly noteworthy.

The linear IC for ncase=3 was used rather than, for example, a finite step (discontinuity) at $x = 0$, which would be undefined at this point and could therefore create numerical difficulties, such as oscillations. Even with this smoother (linear) IC, $\dfrac{\partial u}{\partial x}$ to $\dfrac{\partial^4 u}{\partial x^4}$ are discontinuous at $x = -3, 3$, but the spline approximations of splinefun can accommodate these discontinuities (Figure 12.3a,b).

References

1 Cahn, J.W., and J.E. Hilliard (1958), Free energy of a nonuniform system. I. Interfacial free energy, *Journal Chemical Physics*, **28**, 258.
2 https://en.wikipedia.org/wiki/Cahn-Hilliard_equation (accessed 21 December 2016).

13

Camassa–Holm Equation

The Camassa–Holm (CMH) equation is third order in space and has a mixed partial derivative. The CMH equation has application, for example, to shallow wave propagation and cardiac blood flow.

13.1 PDE Model

The CMH equation to be considered is [1]

$$\frac{\partial u}{\partial t} + 2\kappa \frac{\partial u}{\partial x} - \frac{\partial^3 u}{\partial x^2 \partial t} + 3u\frac{\partial u}{\partial x} = 2\frac{\partial u}{\partial x}\frac{\partial^2 u}{\partial x^2} + u\frac{\partial^3 u}{\partial x^3} \tag{13.1}$$

where

Variable	Interpretation
u	dependent variable[1]
x	Cartesian spatial coordinate
t	time
κ	constant related to the critical shallow wave speed.

Since Eq. (13.1) is first order in t, one IC is required

$$u(x, t = 0) = f(x) \tag{13.2}$$

where $f(x)$ is a function to be specified.

To complete the specification of the problem, we would logically consider the three required BCs for Eq. (13.1) (since it is third order in x). However, if Eq. (13.1) is analyzed over an essentially infinite domain, $-\infty \leq x \leq \infty$,

---

1 $u(x, t)$ might represent fluid velocity of a wave in the x direction or the height of a fluid-free surface above a flat bottom [1].

Spline Collocation Methods for Partial Differential Equations: With Applications in R, First Edition.
William E. Schiesser.
© 2017 John Wiley & Sons, Inc. Published 2017 by John Wiley & Sons, Inc.
Companion website: www.wiley.com/go/Spline_Collocation

and if changes in the solution occur only over a finite interval in x, then BCs at infinity have no effect and are therefore not required. This situation will be clarified through the following discussion (see also Chapter 5 for an example with a comparison of a numerical solution and an analytical solution).

Consequently, Eqs. (13.1) and (13.2) constitute the complete PDE problem.[2] Of particular interest is the approximation of the mixed partial derivative $\dfrac{\partial^3 u}{\partial x^2 \partial t}$.

The spline collocation method of lines (SCMOL) numerical solution of Eqs. (13.1) and (13.2) is considered next.

13.2 Main Program

A main program for Eqs. (13.1) and (13.2) follows.

```
#
# Camassa-Holm
#
# Delete previous workspaces
  rm(list=ls(all=TRUE))
#
# Access ODE integrator
  library("deSolve");
#
# Access functions for numerical solution
  setwd("f:/collocation/chap13");
  source("pde_1.R");
#
# Grid in x
  n=119;xl=-5;xu=7;dx=(xu-xl)/(n+1);
  x=seq(from=(xl+dx),to=(xu-dx),by=dx);
#
# Parameters
  c=0.5;k=1;
#
# Independent variable for ODE integration
  nout=6;t0=0;tf=0.5;
  tout=seq(from=t0,to=tf,by=(tf-t0)/(nout-1));
#
```

---

2 Equations (13.1) and (13.2) constitute a Cauchy problem, that is, an initial value problem (rather than an initial-boundary value problem).

```
# IC
  u0=rep(0,n);
  for(i in 1:n){
    u0[i]=exp(-c*x[i]^2);
  }
  ncall=0;
#
# Coefficient matrix
  cm=matrix(0,nrow=n,ncol=n);
  for(i in 1:n){
  for(j in 1:n){
    if(i==j){cm[i,j]=(1+2/dx^2)};
    if(abs(i-j)==1){cm[i,j]=-1/dx^2};
    if(abs(i-j)>1){cm[i,j]=0};
  }
  }
#
# ODE integration
# out=lsodes(y=u0,times=tout,func=pde_1,
#      sparsetype="sparseint",rtol=1e-6,atol=1e-6,
#      maxord=5);
  out=ode(y=u0,times=tout,func=pde_1);
  nrow(out)
  ncol(out)
#
# Arrays for numerical solution
  u=matrix(0,nrow=n,ncol=nout);
  t=rep(0,nout);
  for(it in 1:nout){
  for(i  in 1:n){
    u[i,it]=out[it,i+1];
      t[it]=out[it,1];
  }
  }
#
# Display selected output
  cat(sprintf("    xl = %4.1f    xu = %4.1f\
    c = %4.1f    k = %4.1f\n",xl,xu,c,k));
  for(it in 1:nout){
    cat(sprintf("\n    t        x      u(x,t)\n"));
    iv=seq(from=1,to=n,by=10);
    for(i in iv){
      cat(sprintf("%7.2f%8.2f%10.5f\n",
```

```
                    t[it],x[i],u[i,it])));
    }
    cat(sprintf("\n"));
  }
  cat(sprintf(" ncall = %4d\n",ncall));
#
# Arrays for plotting
  up=matrix(0,nrow=(n+2),ncol=nout);
  xp=rep(0,(n+2));xp[1]=-5;xp[n+2]=7;
  for(i  in 1:n){xp[i+1]=x[i]};
  for(it in 1:nout){
    up[1,it]=0;up[n+2,it]=0;
    for(i in 1:n){up[i+1,it]=u[i,it]};
  }
#
# Plot numerical solution
  matplot(xp,up,type="l",lwd=2,col="black",lty=1,
    xlab="x",ylab="u(x,t)",main="Camassa-Holm");
```

Listing 13.1 Main program for Eqs. (13.1) and (13.2).

We can note the following details about Listing 13.1:

- Previous workspaces are removed. Then the ODE integrator library deSolve is accessed. Note that the setwd (set working directory) uses / rather than the usual \. setwd requires editing on the local computer (to specify the working directory with the routines for Eqs. (13.1) and (13.2)).

```
#
# Camassa-Holm
#
# Delete previous workspaces
  rm(list=ls(all=TRUE))
#
# Access ODE integrator
  library("deSolve");
#
# Access functions for numerical solution
  setwd("f:/collocation/chap13");
  source("pde_1.R");
```

pde_1 is the routine for the MOL/ODEs (discussed subsequently).

- A uniform grid in x with 119 points is defined with the `seq` utility for the interval $x_l = -5 \le x \le x_u = 7$. Therefore, the grid spacing is $(7 - (-5)/(119 + 1) = 0.1$ and the vector x has the values $x = -5, -4.9, \ldots, 7$.

```
#
# Grid in x
  n=119;xl=-5;xu=7;dx=(xu-xl)/(n+1);
  x=seq(from=(xl+dx),to=(xu-dx),by=dx);
```

n=119 specifies the number of interior grid points and does not include the boundary points corresponding to $x = x_l = -5$, $x = x_u = 7$. This choice of the number of grid points is made to accommodate the mixed partial $\dfrac{\partial^3 u}{\partial x^2 \partial t}$ in Eq. (13.1) as explained subsequently. Also, the solution is a pulse traveling left to right, so the right boundary is $x = 7$ (rather than $x = 5$) to provide an additional length for the traveling pulse. This point is illustrated in the output that follows.
- The model parameters c, κ are defined numerically.

```
#
# Parameters
  c=0.5;k=1;
```

c is the constant in the IC Gaussian pulse explained next. κ is the constant in Eq. (13.1).
- A uniform grid in t of 6 output points is defined with the `seq` utility for the interval $0 \le t \le t_f = 0.5$. Therefore, the vector `tout` has the values $t = 0, 0.1, \ldots, 0.5$.

```
#
# Independent variable for ODE integration
  nout=6;t0=0;tf=0.5;
  tout=seq(from=t0,to=tf,by=(tf-t0)/(nout-1));
```

At this point, the intervals of x and t in Eq. (13.1) are defined.
- The IC Eq. (13.2) is a Gaussian pulse in x (with c defined previously).

```
#
# IC
  u0=rep(0,n);
  for(i in 1:n){
    u0[i]=exp(-c*x[i]^2);
  }
  ncall=0;
```

The counter for the number of calls to pde_1 is initialized.

- An $n \times n$ matrix, cm, is defined for the numerical approximation of the mixed partial derivative in Eq. (13.1).

```
#
# Coefficient matrix
  cm=matrix(0,nrow=n,ncol=n);
  for(i in 1:n){
  for(j in 1:n){
    if(i==j){cm[i,j]=(1+2/dx^2)};
    if(abs(i-j)==1){cm[i,j]=-1/dx^2};
    if(abs(i-j)>1){cm[i,j]=0};
  }
  }
```

To explain this matrix, Eq. (13.1) is approximated at point i of the grid in x as

$$\frac{\partial u_i}{\partial t} - \frac{\partial\left(\dfrac{u_{i+1} - 2u_i + u_{i-1}}{\Delta x^2}\right)}{\partial t} = -2\kappa\frac{\partial u_i}{\partial x} - 3u_i\frac{\partial u_i}{\partial x} + 2\frac{\partial u_i}{\partial x}\frac{\partial^2 u_i}{\partial x^2} + u_i\frac{\partial^3 u_i}{\partial x^3}$$

or

$$\frac{\partial\left(-\dfrac{1}{\Delta x^2}u_{i+1} + (1+\dfrac{2}{\Delta x^2})u_i - \dfrac{1}{\Delta x^2}u_{i-1}\right)}{\partial t}$$
$$= -2\kappa\frac{\partial u_i}{\partial x} - 3u_i\frac{\partial u_i}{\partial x} + 2\frac{\partial u_i}{\partial x}\frac{\partial^2 u_i}{\partial x^2} + u_i\frac{\partial^3 u_i}{\partial x^3} \qquad (13.3a)$$

Here we have made use of the second-order finite difference (FD) approximation for a second-order derivative.

$$\frac{\partial^2 u_i}{\partial x^2} \approx \frac{u_{i+1} - 2u_i + u_{i-1}}{\Delta x^2} + O(\Delta x^2) \qquad (13.3b)$$

The FD approximation of Eq. (13.3b) applies only to interior points in the spatial grid, which is the reason for not including the boundary points ($x = -5, 7$) by using a grid of 119 points rather than 121.

After the derivatives in x in Eq. (13.3a) are approximated with splines, Eq. (13.3a) reduces to a system of n coupled ODEs in t for $\dfrac{du_{i+1}}{dt}, \dfrac{du_i}{dt}, \dfrac{du_{i-1}}{dt}$. This tridiagonal system can be uncoupled by multiplying Eq. (13.3a) by the inverse of the constant coefficient coupling matrix with the terms $-\dfrac{1}{\Delta x^2}$, $1 + \dfrac{2}{\Delta x^2}, -\dfrac{1}{\Delta x^2}$, centered around the diagonal point i. These are the coefficients defined for the matrix cm in the previous coding. The inverse of cm is computed in the ODE/MOL routine pde_1 discussed subsequently.

- The system of 119 ODEs is integrated by the library integrator ode (available in deSolve with the default integrator lsoda). lsodes produced an error

message before attempting the integration (the real work array is of insuf-
ficient length), and rather than go into lsodes to attempt to correct the
problem, ode was used instead. The arguments of ode are the ODE function, pde_1, the IC vector u0, and the vector of output values of t, tout.
The length of u0 (119) informs ode how many ODEs are to be integrated.
func, y, times are reserved names.

```
#
# ODE integration
# out=lsodes(y=u0,times=tout,func=pde_1,
#        sparsetype="sparseint",rtol=1e-6,atol=1e-6,
#        maxord=5);
   out=ode(y=u0,times=tout,func=pde_1);
   nrow(out)
   ncol(out)
```

The numerical solution to the ODEs is returned in matrix out. In this case,
out has the dimensions 6×120 (displayed in the numerical output). The
offset $119 + 1 = 120$ is required since the first element of each column has
the output t (also in tout), and the 2, ... , 120 column elements have the 119
ODE solutions. This indexing of out is used next.

- The numerical solution for $u(x, t)$ is taken from the solution matrix out
returned by ode. The intervals in t and x are covered by two nested fors
(with indices it and i for t and x, respectively). The offset i+1 accounts
for t returned in the first column position of out and placed in the vector
t[it].

```
#
# Arrays for numerical solution
   u=matrix(0,nrow=n,ncol=nout);
   t=rep(0,nout);
   for(it in 1:nout){
   for(i  in 1:n){
     u[i,it]=out[it,i+1];
       t[it]=out[it,1];
   }
   }
```

- The parameters x_l, x_u, c, κ are displayed, followed by the numerical solution
to Eq. (13.1), $u(x, t)$.

```
#
# Display selected output
   cat(sprintf("    xl = %4.1f    xu = %4.1f\
       c = %4.1f    k = %4.1f\n",xl,xu,c,k));
```

```
    for(it in 1:nout){
        cat(sprintf("\n          t          x      u(x,t)\n"));
        iv=seq(from=1,to=n,by=10);
        for(i in iv){
            cat(sprintf("%7.2f%8.2f%10.5f\n",
                        t[it],x[i],u[i,it]));
        }
        cat(sprintf("\n"));
    }
    cat(sprintf(" ncall = %4d\n",ncall));
```

Every tenth value of $u(x,t)$ is displayed to conserve space. The number of calls to pde_1 is displayed at the end of the solution.

- Arrays up, xp are defined with the dimension 119 changed to 121 to include the boundary points $x = x_l = -5, x = x_u = 7$.

```
#
# Arrays for plotting
  up=matrix(0,nrow=(n+2),ncol=nout);
  xp=rep(0,(n+2));xp[1]=-5;xp[n+2]=7;
  for(i  in 1:n){xp[i+1]=x[i]};
  for(it in 1:nout){
    up[1,it]=0;up[n+2,it]=0;
    for(i in 1:n){up[i+1,it]=u[i,it]};
  }
```

- The numerical solution is plotted with matplot.

```
#
# Plot numerical solution
  matplot(xp,up,type="l",lwd=2,col="black",lty=1,
      xlab="x",ylab="u(x,t)",main="Camassa-Holm");
```

The ODE routine pde_1 called by ode in the main program of Listing 13.1 is considered next.

13.3   ODE Routine

pde_1 for Eq. (13.1) follows.

```
  pde_1=function(t,u,parm){
#
# Function pde_1 computes the t derivative vector
# for u(x,t)
#
```

```
# ux
  tablex=splinefun(x,u);
  ux=tablex(x,deriv=1);
#
# uxx
  tablexx=splinefun(x,u);
  uxx=tablexx(x,deriv=2);
#
# uxxx
  tablexxx=splinefun(x,ux);
  uxxx=tablexxx(x,deriv=2);
#
# RHS group
  ug=rep(0,n);
  for(i in 1:n){
    ug[i]=-2*k*ux[i]-3*u[i]*ux[i]+
          2*ux[i]*uxx[i]+u[i]*uxxx[i];
  }
#
# PDE
  ut=rep(0,n);
  ut=solve(cm)%*%ug;
#
# Increment calls to pde_1
  ncall <<- ncall+1;
#
# Return derivative vector
  return(list(c(ut)));
  }
```

Listing 13.2 ODE/MOL routine pde_1 for Eq. (13.1).

We can note the following details about pde_1:

- The function is defined.

```
  pde_1=function(t,u,parm){
#
# Function pde_1 computes the t derivative vector
# for u(x,t)
```

  t is the current value of t in Eq. (13.1). u is the 119-vector. parm is an argument to pass parameters to pde_1 (unused, but required in the argument list). The arguments must be listed in the order stated to properly interface with ode called in the main program. The derivative vector of the LHS of Eq. (13.1) is calculated below and returned to ode.

- The derivatives $\dfrac{\partial u}{\partial x}, \dfrac{\partial^2 u}{\partial x^2}, \dfrac{\partial^3 u}{\partial x^3}$ are computed by a series of calls to spline-fun.

```
#
# ux
   tablex=splinefun(x,u);
   ux=tablex(x,deriv=1);
#
# uxx
   tablexx=splinefun(x,u);
   uxx=tablexx(x,deriv=2);
#
# uxxx
   tablexxx=splinefun(x,ux);
   uxxx=tablexxx(x,deriv=2);
```

- The RHS group of terms of Eq. (13.1) is computed.

```
#
# RHS group
   ug=rep(0,n);
   for(i in 1:n){
      ug[i]=-2*k*ux[i]-3*u[i]*ux[i]+
             2*ux[i]*uxx[i]+u[i]*uxxx[i];
   }
```

The close correspondence of this coding to Eq. (13.1) is a principal feature of the MOL. In particular, the several nonlinearities are easily handled.

- The utility solve is used to solve the tridiagonal system of implicit ODEs for $\dfrac{du_i}{dt}$ at grid point i (see Eq. (13.3a)).[3]

```
#
# PDE
   ut=rep(0,n);
   ut=solve(cm)%*%ug;
```

---

3 solve can accept the RHS vector b of a linear algebraic system $ax = b$ system as a second input argument. If only one input argument is used, solve will return the inverse of a. Then this inverse is multiplied by b using the matrix multiply %*% to complete the solution of $ax = b$. For the present case, a corresponds to the coefficient matrix cm (Listings 13.1 and 13.2), b corresponds to the vector of RHS terms of Eq. (13.1) in ug (Listing 13.2), and x corresponds to the vector $\dfrac{du_1}{dt}, \ldots, \dfrac{du_i}{dt}, \ldots, \dfrac{du_n}{dt}$.

- The counter for the calls to pde_1 is returned to the main program of Listing 13.1 by the <<- operator.

```
#
# Increment calls to pde_1
  ncall <<- ncall+1;
```

- The derivative in ut is returned to ode.

```
#
# Return derivative vector
  return(list(c(ut)));
  }
```

Note that a list is used as required by ode. The final } concludes pde_1.

This concludes the coding of Eqs. (13.1)–(13.3). The numerical and graphical output is considered next.

13.4 Model Output

Execution of the preceding main program and ODE routine pde_1 produces the following abbreviated numerical output in Table 13.1.

Table 13.1 Abbreviated output for Eqs. (13.1)–(13.3).

```
[1] 6

[1] 120
```

t	x	u(x,t)
0.00	-4.90	0.00001
0.00	-3.90	0.00050
0.00	-2.90	0.01492
0.00	-1.90	0.16447
0.00	-0.90	0.66698
0.00	0.10	0.99501
0.00	1.10	0.54607
0.00	2.10	0.11025
0.00	3.10	0.00819
0.00	4.10	0.00022

(Continued)

Table 13.1 (Continued)

```
0.00      5.10     0.00000
0.00      6.10     0.00000
  .                  .
  .                  .
  .                  .

Output for t = 0.1 to 0.4
           removed

  .                  .
  .                  .
  .                  .

    t          x        u(x,t)
  0.50      -4.90     -0.00117
  0.50      -3.90     -0.02242
  0.50      -2.90     -0.07597
  0.50      -1.90     -0.09221
  0.50      -0.90      0.15951
  0.50       0.10      0.67321
  0.50       1.10      0.99850
  0.50       2.10      0.51687
  0.50       3.10      0.20486
  0.50       4.10      0.08916
  0.50       5.10      0.03882
  0.50       6.10      0.01465

  ncall =    44
```

We can note the following details about this output:

- The solution array out has the dimensions 6 x 120 for 6 output points and 119 ODEs, with t included as an additional element, that is, 119 + 1 = 120.
- The interval in x is $((7 - (-5))/(119 + 1) = 0.1$ (with every tenth value in Table 13.1).
- The computational effort is modest, ncall = 44 so that ode is quite efficient. The graphical output is shown in Figure 13.1.

Figure 13.1 indicates that the solution remains at zero at $x = -5, 7$ so that circumventing the use of BCs appears to be justified as discussed

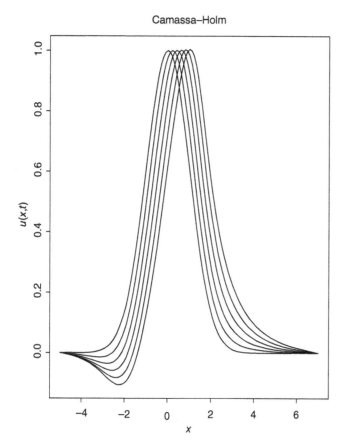

Figure 13.1 Numerical $u(x, t)$ for Eqs. (13.1)–(13.3).

previously. Also, the extension of the right boundary to $x = 7$ is clearly required.

Figure 13.1 also indicates that the solution is a pulse moving left to right with a peak that appears to have a constant height. This peak can be further investigated by changing the output in the interval in x from every tenth value with

```
iv=seq(from=1,to=n,by=10);
```

to every value with

```
iv=seq(from=1,to=n,by=1);
```

The initial peak value (at $t = 0$) can then be identified at each successive value of t. The peak values are listed in Table 13.2.

Table 13.2 Peak values of the solution $u(x, t)$.

t	x	u(x,t)
0.00	0.00	1.00000
0.10	0.20	1.00026
0.20	0.40	1.00094
0.30	0.60	1.00193
0.40	0.80	1.00299
0.50	1.00	1.00380

The peaks move left to right in x with an apparent velocity $0.2/0.1 = 2$, which corresponds to the term $2\kappa\dfrac{\partial u}{\partial x}$ in Eq. (13.1) with $\kappa = 1$ (Listing 13.1).

This completes the programming and discussion of the solutions to Eqs. (13.1)–(13.3).

13.5 Summary and Conclusions

The programming of the CMH equation numerical solution of Eqs. (13.1) and (13.2) within the SCMOL framework is straightforward. In particular, the correspondence of the coding in pde_1 of Listing 13.2 with Eq. (13.1) is clear, which is a principal advantage of the SCMOL.

For additional analysis, computing and displaying the individual terms of Eq. (13.1) would be of interest, which is straightforward by using the code in pde_1. In this way, the relative contributions of the terms can be assessed and a more complete understanding of the solution would be available.

Finally, the numerical approximation of the mixed partial derivative in Eq. (13.1) as implemented via the coefficient matrix cm is tested in the chapter appendix with a second example PDE that has an analytical solution for evaluation of the numerical example.

13.6 Appendix A13.1: Second Example of a PDE with a Mixed Partial Derivative

To test the efficacy of the approximation for the mixed partial derivative based on Eqs. (13.3), we consider briefly a second PDE with an analytical solution that can be compared with the numerical solution.

$$\frac{\partial u}{\partial t} + \frac{\partial^3 u}{\partial x^2 \partial t} = u \tag{A13.1}$$

The IC for Eq. (A13.1) is

$$u(x, t = 0) = \sin(\pi x/L) \tag{A13.2a}$$

and the BCs (homogeneous Dirichlet) are

$$u(x = 0, t) = u(x = L, t) = 0 \tag{A13.2b,c}$$

The analytical solution to Eqs. (A13.1) and (A13.2) is

$$u(x, t) = \sin(\pi x/L)e^{at} \tag{A13.3}$$

with where L is an arbitrary constant and $a = \dfrac{1}{1 - (\pi/L)^2}$.

R routines for the SCMOL solution of Eqs. (A13.1)–(A13.3) follow. They are similar to the R routines in Listings 13.1 and 13.2, and, therefore, the following discussion is essentially limited to the differences from the preceding routines.

The main program follows.

13.7 Main Program

```
#
# Test PDE with mixed partial derivative
#
# Delete previous workspaces
  rm(list=ls(all=TRUE))
#
# Access ODE integrator
  library("deSolve");
#
# Access functions for numerical solution
  setwd("f:/collocation/chap13");
```

```
  source("pde_1a.R");
#
# Grid in x
  n=39;xl=0;xu=1;dx=(xu-xl)/(n+1);
  x=seq(from=(xl+dx),to=(xu-dx),by=dx);
#
# Parameters
  a=1/(1-(pi/(xu-xl))^2);
#
# Independent variable for ODE integration
  nout=6;t0=0;tf=20;
  tout=seq(from=t0,to=tf,by=(tf-t0)/(nout-1));
#
# IC
  u0=rep(0,n);
  for(i in 1:n){
    u0[i]=sin(pi*x[i]/(xu-xl));
  }
  ncall=0;
#
# Coefficient matrix
  cm=matrix(0,nrow=n,ncol=n);
  for(i in 1:n){
  for(j in 1:n){
    if(i==j){cm[i,j]=(1-2/dx^2)};
    if(abs(i-j)==1){cm[i,j]=1/dx^2};
    if(abs(i-j)>1){cm[i,j]=0};
  }
  }
#
# ODE integration
  out=lsodes(y=u0,times=tout,func=pde_1a,
      sparsetype="sparseint",rtol=1e-6,atol=1e-6,
      maxord=5);
  nrow(out)
  ncol(out)
#
# Arrays for numerical solution
  u=matrix(0,nrow=n,ncol=nout);
  t=rep(0,nout);
  for(it in 1:nout){
  for(i  in 1:n){
    u[i,it]=out[it,i+1];
```

```
      t[it]=out[it,1];
   }
   }
#
# Arrays for analytical solution, errors in
# numerical solution
   ua=matrix(0,nrow=n,ncol=nout);
  err=matrix(0,nrow=n,ncol=nout);
  for(it in 1:nout){
    u[1,it]=sin(pi*x[1]/(xu-xl))*exp(a*t[it]);
    u[n,it]=sin(pi*x[n]/(xu-xl))*exp(a*t[it]);
    for(i in 1:n){
      ua[i,it]=sin(pi*x[i]/(xu-xl))*exp(a*t[it]);
      err[i,it]=u[i,it]-ua[i,it];
    }
  }
#
# Display selected output
   cat(sprintf(" xl = %4.1f    xu = %4.1f   a = %4.1f\n",
              xl,xu,a));
   for(it in 1:nout){
     cat(sprintf("\n       t        x      u(x,t)     ua(x,t)
                 err(x,t)\n"));
     iv=seq(from=1,to=n,by=1);
     for(i in iv){
       cat(sprintf("%6.1f%7.3f%10.6f%10.6f%10.6f\n",
               t[it],x[i],u[i,it],ua[i,it],err[i,it]));
     }
     cat(sprintf("\n"));
   }
   cat(sprintf(" ncall = %4d\n",ncall));
#
# Arrays for plotting
    up=matrix(0,nrow=(n+2),ncol=nout);
   uap=matrix(0,nrow=(n+2),ncol=nout);
   xp=rep(0,(n+2));xp[1]=0;xp[n+2]=1;
    for(i in 1:n){xp[i+1]=x[i]};
   for(it in 1:nout){
     up[1,it]=0;up[n+2,it]=0;
     for(i in 1:n){up[i+1,it]=u[i,it]};
     uap[1,it]=0;uap[n+2,it]=0;
     for(i in 1:n){uap[i+1,it]=ua[i,it]};
   }
```

```
#
# Plot 2D numerical, analytical solutions
    matplot(xp,up,type="l",lwd=2,col="black",lty=1,
       xlab="x",ylab="u(x,t)",main="Mixed partial");
    matpoints(xp,uap,pch="o",col="black");
```

Listing A13.1 Main program for Eqs. (A13.1)–(A13.3).

We can note the following details about Listing A13.1:

- The ODE/MOL routine is pde_1a.
- The interval in x is $0 \le x \le 1$ defined on a grid of 39 interior points $x = 0, 0.025, \ldots, 1$ (with $L = x_u - x_l = 1$).
- The interval in t is $0 \le t \le 20$ defined on 6 output points, $t = 0, 4, \ldots, 20$.
- The analytical solution of Eq. (A13.3) is used for the IC (with $t = 0$).
- The coupling matrix cm is slightly different than in Listing 13.1 because in Eq. (13.1) the mixed partial derivative is minus, while in Eq. (A13.1) it is plus. Otherwise, the cm matrices in Listings 13.1 and A13.1 are similar.
- lsodes is the ODE integrator. The numerical solution returned in array out is placed in two arrays, u, t, as in Listing 13.1.
- The analytical solution (Eq. (A13.3)) is placed in an array ua, and the difference between the numerical and analytical solutions is placed in array err.
- The boundary points $x = 0, 1$ are included for both the numerical and analytical solutions for plotting (in arrays xp, up, uap). Homogeneous Dirichlet BCs (A13.2b,c) are included in the expansion to include the boundary points.
- The numerical solution is plotted with matplot and the analytical solution is superimposed with matpoints. The resulting plot is Figure A13.1 discussed subsequently.

The ODE routine follows.

13.8 ODE Routine

The ODE routine differs from pde_1 in Listing 13.2 to reflect the differences between Eq. (13.1) and (A13.1).

```
  pde_1a=function(t,u,parm){
#
# Function pde_1a computes the t derivative vector for
# a PDE with a mixed partial derivative
```

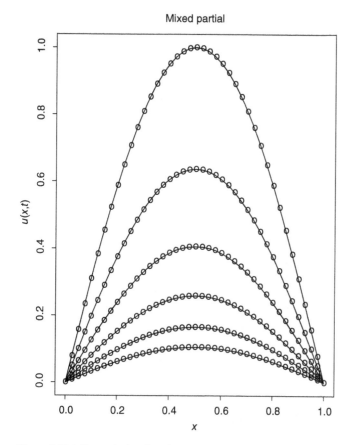

Figure A13.1 Numerical and analytical $u(x, t)$ for Eqs. (A13.1)–(A13.3).

```
#
# BCs
  u[1]=sin(pi*x[1]/(xu-xl))*exp(a*t);
  u[n]=sin(pi*x[n]/(xu-xl))*exp(a*t);
#
# PDE
  ut=rep(0,n);
  ut=solve(cm)%*%u;
  ut[1]=0;ut[n]=0;
#
# Increment calls to pde_1a
  ncall <<- ncall+1;
#
```

```
# Return derivative vector
  return(list(c(ut)));
  }
```

Listing A13.2 ODE/MOL routine for Eqs. (A13.1)–(A13.3).

We can note the following details about Listing A13.2:

- Dirichlet BCs are included based on the analytical solution of Eq. (A13.3).

```
#
# BCs
  u[1]=sin(pi*x[1]/(xu-xl))*exp(a*t);
  u[n]=sin(pi*x[n]/(xu-xl))*exp(a*t);
```

The boundary values are not necessarily homogeneous (zero) since $x[1]$, $x[n]$ do not equal $x = 0, 1$.
- Derivatives in x are not computed with splinefun since they do not appear in Eq. (A13.1). However, the purpose of this PDE problem is to test the implementation of a mixed partial derivative via the coupling matrix cm.
- solve is used as in Listing 13.2.

```
#
# PDE
  ut=rep(0,n);
  ut=solve(cm)%*%u;
  ut[1]=0;ut[n]=0;
```

Note the use of Eq. (A13.1) in ut=solve(cm)%*%u. The boundary derivatives in t are zeroed to preserve the boundary condition values of u set previously.
- The 39-vector of derivatives in t is returned to lsodes.

The output from these routines is discussed briefly.

13.9 Model Output

Abbreviated numerical output from the R routines of Listings A13.1 and A13.2 follows.
We can note the following details about Table A13.1:

- For $t = 0$, the numerical and analytical solutions are the same, as expected, since the analytical solution is used in Listing A13.1 to define the numerical IC.
- The agreement between the numerical and analytical solutions is satisfactory, including $t = 20$.

Table A13.1 Abbreviated output from the R routines of Listings A13.1 and A13.2.

```
[1]  6

[1]  40

xl  =   0.0    xu  =   1.0   a  =  -0.1
        t        x      u(x,t)      ua(x,t)    err(x,t)
      0.0    0.025    0.078459    0.078459    0.000000
      0.0    0.050    0.156434    0.156434    0.000000
      0.0    0.075    0.233445    0.233445    0.000000
      0.0    0.100    0.309017    0.309017    0.000000
                 .                                .
                 .                                .
                 .                                .

     Output for x = 0.125 to 0.375 removed
                 .                                .
                 .                                .
                 .                                .

      0.0    0.400    0.951057    0.951057    0.000000
      0.0    0.425    0.972370    0.972370    0.000000
      0.0    0.450    0.987688    0.987688    0.000000
      0.0    0.475    0.996917    0.996917    0.000000
      0.0    0.500    1.000000    1.000000    0.000000
      0.0    0.525    0.996917    0.996917    0.000000
      0.0    0.550    0.987688    0.987688    0.000000
      0.0    0.575    0.972370    0.972370    0.000000
      0.0    0.600    0.951057    0.951057    0.000000
                 .                                .
                 .                                .
                 .                                .

     Output for x = 0.625 to 0.875 removed
                 .                                .
                 .                                .
                 .                                .
```

(Continued)

Table A13.1 (Continued)

```
0.0   0.900   0.309017   0.309017   0.000000
0.0   0.925   0.233445   0.233445   0.000000
0.0   0.950   0.156434   0.156434   0.000000
0.0   0.975   0.078459   0.078459   0.000000
                .                      .
                .                      .
                .                      .
        Output for t = 4 to 16 removed
                .                      .
                .                      .
                .                      .
   t      x      u(x,t)     ua(x,t)    err(x,t)
20.0   0.025   0.008229   0.008229   0.000000
20.0   0.050   0.016387   0.016408  -0.000021
20.0   0.075   0.024454   0.024485  -0.000031
20.0   0.100   0.032370   0.032411  -0.000041
                .                      .
                .                      .
                .                      .
     Output for x = 0.125 to 0.375 removed
                .                      .
                .                      .
                .                      .
20.0   0.400   0.099625   0.099751  -0.000126
20.0   0.425   0.101858   0.101987  -0.000129
20.0   0.450   0.103463   0.103594  -0.000131
20.0   0.475   0.104429   0.104562  -0.000132
20.0   0.500   0.104752   0.104885  -0.000133
20.0   0.525   0.104429   0.104562  -0.000132
20.0   0.550   0.103463   0.103594  -0.000131
20.0   0.575   0.101858   0.101987  -0.000129
20.0   0.600   0.099625   0.099751  -0.000126
                .                      .
                .                      .
                .                      .
```

Table A13.1 (Continued)

```
Output for x = 0.625 to 0.875 removed
            .                        .
            .                        .
            .                        .
20.0   0.900   0.032370   0.032411  -0.000041
20.0   0.925   0.024454   0.024485  -0.000031
20.0   0.950   0.016387   0.016408  -0.000021
20.0   0.975   0.008229   0.008229   0.000000

ncall =   120
```

- The computational effort is modest with 120 calls to pde_1a.

The agreement between the numerical and analytical solutions is clear in Figure A13.1.

In summary, the test problem of Eqs. (A13.1)–(A13.3) indicates the use of the coupling matrix cm for the mixed partial derivative in Eqs. (13.1) and (A13.1) is valid. Of course, this example does not constitute a general proof but rather provides confidence that the numerical procedure based on cm is flexible and accurate.

Reference

1 R. Camassa, and D.D. Holm (1993), An integrable shallow water equation with peaked solitons, *Physical Review Letters*, **71**, no. 11, 1661–1664.

14

Burgers–Huxley Equation

The Burgers–Huxley (BH) equation is a traveling wave nonlinear PDE ([1], Chapter 6) with a known analytical solution [2] that is used to verify the numerical solution. The following spline collocation method of lines (SCMOL) solutions are for Dirichlet and Neumann boundary conditions (BCs) (two cases).

14.1 PDE Model

The BH equation in 1D is

$$\frac{\partial u}{\partial t} + u^2 \frac{\partial u}{\partial x} - \frac{\partial^2 u}{\partial x^2} = \frac{2}{3}u^3(1 - u^2) \tag{14.1}$$

Equation (14.1) is first order in t and therefore requires one initial condition (IC), which is based on the analytical solution (discussed as follows).

$$u(x, t = 0) = \left[\frac{1}{2} + \frac{1}{2}\tanh\left(\frac{1}{3}x\right)\right]^{1/2} \tag{14.2}$$

Equation (14.1) is second order in x and therefore requires two BCs. Two cases are considered:

ncase=1: Dirichlet BCs (based on the analytical solution).

$$u(x = x_l, t) = \left[\frac{1}{2} + \frac{1}{2}\tanh\left(\frac{1}{9}(3x_l + t)\right)\right]^{1/2} \tag{14.3a}$$

$$u(x = x_u, t) = \left[\frac{1}{2} + \frac{1}{2}\tanh\left(\frac{1}{9}(3x_u + t)\right)\right]^{1/2} \tag{14.3b}$$

with $x_l = -20, x_u = 10$.

ncase=2: Neumann BCs (homogeneous).

$$\frac{\partial u(x = x_l, t)}{\partial x} = \frac{\partial u(x = x_u, t)}{\partial x} = 0 \tag{14.3c,d}$$

Spline Collocation Methods for Partial Differential Equations: With Applications in R, First Edition.
William E. Schiesser.
© 2017 John Wiley & Sons, Inc. Published 2017 by John Wiley & Sons, Inc.
Companion website: www.wiley.com/go/Spline_Collocation

The analytical solution is

$$u_a(x, t) = \left[\frac{1}{2} + \frac{1}{2} \tanh\left(\frac{1}{9}(3x + t) \right) \right]^{1/2} \tag{14.4}$$

$u_a(x, t)$ is a traveling wave solution with Lagrangian variable $k(x - vt) = \frac{1}{9}(3x + t) = \frac{1}{3}(x - (-1/3)t)$ or $v = -1/3$, so the solution travels right to left.

Equations (14.1)–(14.3) constitute the PDE model with analytical solution (14.4) to verify the numerical solution.

14.2 Main Program

A main program for Eqs. (14.1)–(14.3) follows.

```
#
# Burgers-Huxley
#
# Delete previous workspaces
  rm(list=ls(all=TRUE))
#
# Access ODE integrator
  library("deSolve");
#
# Access functions for numerical, analytical solutions
  setwd("f:/collocation/chap14");
  source("pde_1.R");source("ua_1.R");
#
# Grid in x
  n=51;xl=-20;xu=10;
  x=seq(from=xl,to=xu,by=(xu-xl)/(n-1));
#
# Independent variable for ODE integration
  nout=4;t0=0;tf=15;
  tout=seq(from=t0,to=tf,by=(tf-t0)/(nout-1));
#
# Initial condition
  u0=rep(0,n);
  for(i in 1:n){
    u0[i]=ua_1(x[i],t0);
  }
  ncall=0;
#
```

```
# ODE integration
  out=lsodes(y=u0,times=tout,func=pde_1,
      sparsetype="sparseint",rtol=1e-6,atol=1e-6,
      maxord=5);
  nrow(out)
  ncol(out)
#
# Arrays for numerical solution
   u=matrix(0,nrow=n,ncol=nout);
  t=rep(0,nout);
  for(it in 1:nout){
  for(i  in 1:n){
    u[i,it]=out[it,i+1];
    t[it]=out[it,1];
  }
  }
#
# Arrays for analytical solution, errors in
# numerical solution
   ua=matrix(0,nrow=n,ncol=nout);
  err=matrix(0,nrow=n,ncol=nout);
  for(it in 1:nout){
    u[1,it]=ua_1(x[1],t[it]);
    u[n,it]=ua_1(x[n],t[it]);
    for(i in 1:n){
       ua[i,it]=ua_1(x[i],t[it]);
      err[i,it]=u[i,it]-ua[i,it];
    }
  }
#
# Display selected output
  cat(sprintf("\n n = %3d,  xl = %4.2f,
              xu = %4.2f\n",n,xl,xu));
  for(it in 1:nout){
    cat(sprintf("\n      t        x         u(x,t)
                ua(x,t)      err(x,t)\n"));
    iv=seq(from=1,to=n,by=5);
    for(i in iv){
      cat(sprintf("%6.2f%8.3f%12.6f%12.6f
                  %12.6f\n",t[it],x[i],u[i,it],
                  ua[i,it],err[i,it]));
    }
```

```
    cat(sprintf("\n"));
  }
  cat(sprintf(" ncall = %4d\n",ncall));
#
# Graphical output
    matplot(x,u,type="l",lwd=2,col="black",lty=1,
      xlab="x",ylab="u(x,t)",main="Burgers-Huxley");
    matpoints(x,ua,pch="o",col="black");
```

Listing 14.1 Main program for Eqs. (14.1–14.3).

We can note the following details about Listing 14.1:

- Previous workspaces are removed. Then the ODE integrator library deSolve is accessed. Note that the setwd (set working directory) uses / rather than the usual \. setwd requires editing for the local computer (to specify the folder with the MOL routines).

```
#
# Burgers-Huxley
#
# Delete previous workspaces
  rm(list=ls(all=TRUE))
#
# Access ODE integrator
  library("deSolve");
#
# Arrays for analytical solution, errors in
# numerical solution
  setwd("f:/collocation/chap14");
  source("pde_1.R");source("ua_1.R");
```

pde_1 is the routine for the MOL/ODEs (discussed subsequently). ua_1 implements Eq. (14.4).

- A uniform grid in x with 51 points is defined with the seq utility for the interval $x_l = -20 \leq x \leq x_u = 10$. Therefore, the vector x has the values $x = -20, -19.4, \dots, 10$.

```
#
# Grid in x
  n=51;xl=-20;xu=10;
  x=seq(from=xl,to=xu,by=(xu-xl)/(n-1));
```

- A uniform grid in t of 4 output points is defined with the `seq` utility for the interval $0 \leq t \leq t_f = 15$. Therefore, the vector `tout` has the values $t = 0, 5, 10, 15$.

```
#
# Independent variable for ODE integration
  nout=4;t0=0;tf=15;
  tout=seq(from=t0,to=tf,by=(tf-t0)/(nout-1));
```

At this point, the intervals of x and t in Eq. (14.1) are defined.
- Equation (14.2) is the IC for Eq. (14.1) (ua_1 is called with $t = t_0 = 0$).

```
#
# Initial condition
  u0=rep(0,n);
  for(i in 1:n){
    u0[i]=ua_1(x[i],t0);
  }
  ncall=0;
```

ua_1 for Eqs. (14.2) and (14.4) is discussed subsequently. The counter for the calls to the ODE routine pde_1 (discussed subsequently) is initialized.
- The system of 51 ODEs is integrated by the library integrator lsodes (available in deSolve) with the sparse matrix option specified. As expected, the inputs to lsodes are the ODE function, pde_1, the IC vector u0, and the vector of output values of t, tout. The length of u0 (51) informs lsodes how many ODEs are to be integrated. func,y,times are reserved names.

```
#
# ODE integration
  out=lsodes(y=u0,times=tout,func=pde_1,
      sparsetype="sparseint",rtol=1e-6,atol=1e-6,
      maxord=5);
  nrow(out)
  ncol(out)
```

The numerical solution to the ODEs is returned in matrix out. In this case, out has the dimensions 4×52 (displayed in the numerical output). The offset $51 + 1 = 52$ is required since the first element of each column has the output t (also in tout), and the $2, \ldots, 52$ column elements have the 51 ODE solutions. This indexing of out is used next.
- The numerical solution for $u(x, t)$ is taken from the solution matrix out returned by lsodes. The intervals in t and x are covered by two nested fors (with indices it and i for t and x, respectively). The offset i+1 accounts

for *t* returned in the first column position of out and placed in the vector
t[it].

```
#
# Arrays for numerical solution
   u=matrix(0,nrow=n,ncol=nout);
   t=rep(0,nout);
   for(it in 1:nout){
   for(i   in 1:n){
     u[i,it]=out[it,i+1];
       t[it]=out[it,1];
   }
   }
```

- The numerical and analytical solutions from Eqs. (14.1) and (14.4), and the
  difference in these solutions (err), are placed in arrays.

```
#
# Arrays for analytical solution, errors in
# numerical solution
   ua=matrix(0,nrow=n,ncol=nout);
   err=matrix(0,nrow=n,ncol=nout);
   for(it in 1:nout){
     u[1,it]=ua_1(x[1],t[it]);
     u[n,it]=ua_1(x[n],t[it]);
     for(i in 1:n){
        ua[i,it]=ua_1(x[i],t[it]);
        err[i,it]=u[i,it]-ua[i,it];
     }
   }
```

The numerical solution of Eq. (14.1) at $x = -20, 10$, which is also the analytical solution of Eq. (14.4) from the BCs for Eq. (14.1), is placed in u[1,it],
u[n,it] since these values are not returned from the ODE routine pde_1
in array out.
- The parameters of the numerical and analytical solutions are displayed,
  followed by every fifth value of the solutions and their difference.

```
#
# Display selected output
   cat(sprintf("\n n = %3d,   xl = %4.2f,
             xu = %4.2f\n",n,xl,xu));
   for(it in 1:nout){
     cat(sprintf("\n      t       x       u(x,t)
               ua(x,t)     err(x,t)\n"));
```

```
    iv=seq(from=1,to=n,by=5);
    for(i in iv){
      cat(sprintf("%6.2f%8.3f%12.6f%12.6f
                  %12.6f\n",t[it],x[i],u[i,it],
                  ua[i,it],err[i,it]));
    }
    cat(sprintf("\n"));
  }
  cat(sprintf(" ncall = %4d\n",ncall));
```

The number of calls to pde_1 is displayed at the end of the solution.

- The numerical and analytical solutions are plotted with matplot and matpoints, respectively.

```
#
# Graphical output
    matplot(x,u,type="l",lwd=2,col="black",lty=1,
      xlab="x",ylab="u(x,t)",main="Burgers-Huxley");
    matpoints(x,ua,pch="o",col="black");
```

The ODE routine pde_1 called by lsodes in the main program of Listing 14.1 is considered next.

14.3 ODE Routine

pde_1 for Eq. (14.1) follows.

```
  pde_1=function(t,u,parms){
#
# Function pde_1 computes the t derivative vector
# of u(x,t)
#
# BCs
  u[1]=ua_1(x[1],t);
  u[n]=ua_1(x[n],t);
#
# ux
  tablex=splinefun(x,u);
  ux=tablex(x,deriv=1);
#
# uxx
  tablexx=splinefun(x,ux);
  uxx=tablexx(x,deriv=1);
#
```

```
# PDE
  ut=rep(0,n);
  for(i in 2:(n-1)){
    ut[i]=-(u[i]^2)*ux[i]+uxx[i]+
          (2/3)*u[i]^3*(1-u[i]^2);
  }
  ut[1]=0;ut[n]=0;
#
# Increment calls to pde_1
  ncall <<- ncall+1;
#
# Return derivative vector
  return(list(c(ut)));
  }
```

Listing 14.2a ODE/MOL routine pde_1 for Eq. (14.1), Dirichlet BCs.

We can note the following details about pde_1:

- The function is defined.

```
    pde_1=function(t,u,parms){
#
# Function pde_1 computes the t derivative vector
# of u(x,t)
```

t is the current value of t in Eq. (14.1) that is used in the BCs (below). u is the 51-vector of dependent variables. parm is an argument to pass parameters to pde_1 (unused, but required in the argument list). The arguments must be listed in the order stated to properly interface with lsodes called in the main program. The derivative vector of the LHS of Eq. (14.1) is calculated next and returned to lsodes.

- The analytical solution of Eq. (14.4) is used for Dirichlet BCs (14.3a,b) for Eq. (14.1).

```
#
# BCs
  u[1]=ua_1(x[1],t);
  u[n]=ua_1(x[n],t);
```

Note the subscripts 1, n corresponding to $x = -20, 10$.

- The second derivative $\dfrac{\partial^2 u}{\partial x^2}$ in Eq. (14.1) is computed with splinefun.

```
#
# ux
```

```
   tablex=splinefun(x,u);
   ux=tablex(x,deriv=1);
#
# uxx
   tablexx=splinefun(x,ux);
   uxx=tablexx(x,deriv=1);
```

splinefun is called twice for the stagewise differentiation u to ux to uxx.

- The LHS derivative of Eq. (14.1), $\frac{\partial u}{\partial t}$, is computed.

```
#
# PDE
   ut=rep(0,n);
   for(i in 2:(n-1)){
     ut[i]=-(u[i]^2)*ux[i]+uxx[i]+
           (2/3)*u[i]^3*(1-u[i]^2);
   }
   ut[1]=0;ut[n]=0;
```

Since the values $u(x = -20, t), u(x = 10, t)$ are set as BCs, the boundary derivatives in t are set to zero so that the ODE integration by lsodes does not move the boundary values away from their prescribed values. Also, these boundary values are not returned to the calling program, so they are reset in the main program of Listing 14.1 (discussed previously) for the purpose of displaying the complete numerical solution (only ODE-dependent variables computed by ODEs are returned by pde_1). The straightforward programming of Eq. (14.1) is particularly noteworthy.

- The counter for the calls to pde_1 is returned to the main program of Listing 14.1 by the <<- operator.

```
#
# Increment calls to pde_1
   ncall <<- ncall+1;
```

- The derivative in ut is returned to lsodes (the ODE integrator called in the main program of Listing 14.1).

```
#
# Return derivative vector
   return(list(c(ut)));
   }
```

Note that a list is used as required by lsodes. The final } concludes pde_1.

A variant of pde_1 is listed next, pde_1a, which is similar to pde_1, but with Neumann BCs (14.3c,d) in place of Dirichlet BCs (14.3a,b).

```
  pde_1a=function(t,u,parms){
#
# Function pde_1a computes the t derivative vector
# of u(x,t)
#
# ux
  tablex=splinefun(x,u);
  ux=tablex(x,deriv=1);
#
# BCs
  ux[1]=0;ux[n]=0;
#
# uxx
  tablexx=splinefun(x,ux);
  uxx=tablexx(x,deriv=1);
#
# PDE
  ut=rep(0,n);
  for(i in 1:n){
    ut[i]=-(u[i]^2)*ux[i]+uxx[i]+
           (2/3)*u[i]^3*(1-u[i]^2);
  }
#
# Increment calls to pde_1a
  ncall <<- ncall+1;
#
# Return derivative vector
  return(list(c(ut)));
  }
```

Listing 14.2b ODE/MOL routine pde_1a for Eq. (14.1), Neumann BC.

We can note the following details about pde_1a:

- ux is computed first.

```
#
# ux
```

```
   tablex=splinefun(x,u);
   ux=tablex(x,deriv=1);
```

- Homogeneous Neumann BCs (14.3c,d) are then applied.

```
#
# BCs
   ux[1]=0;ux[n]=0;
```

- With the boundary values ux [1] =0 ; ux [n] =0 defined, uxx is computed.

```
#
# uxx
   tablexx=splinefun(x,ux);
   uxx=tablexx(x,deriv=1);
```

- Equation (14.1) is programmed for for (i in 1:n) rather than for (i in 2:(n-1)) as in pde_1 (Listing 14.2a).

```
#
# PDE
   ut=rep(0,n);
   for(i in 1:n){
     ut[i]=-(u[i]^2)*ux[i]+uxx[i]+
           (2/3)*u[i]^3*(1-u[i]^2);
   }
```

In this case, the BCs are defined as derivatives in x (Neumann BCs), so the boundary values vary with t as defined by Eq. (14.1).

The solutions from pde_1 (Listing 14.2a) and pde_1a (Listing 14.2b) are compared subsequently.

Also, the main program of Listing 14.1 is modified in a few places to use pde_1a (rather than pde_1), and BCs (14.3c,d) (rather than BCs (14.3a,b)).

```
                         .
                         .
                         .

#
# Access functions for numerical, analytical solutions
   setwd("f:/collocation/chap14");
   source("pde_1a.R");source("ua_1.R");
                         .
                         .
                         .

#
```

```
# ODE integration
  out=lsodes(y=u0,times=tout,func=pde_1a,
      sparsetype="sparseint",rtol=1e-6,atol=1e-6,
      maxord=5);
  nrow(out)
  ncol(out)
              .
              .
              .


#
# Arrays for analytical solution, errors in
# numerical solution
  ua=matrix(0,nrow=n,ncol=nout);
  err=matrix(0,nrow=n,ncol=nout);
  for(it in 1:nout){
    for(i in 1:n){
        ua[i,it]=ua_1(x[i],t[it]);
        err[i,it]=u[i,it]-ua[i,it];
    }
  }
              .
              .
              .
```

The subordinate routine ua_1 for the analytical solution of Eq. (14.4) follows.

14.4 Subordinate Routine

```
  ua_1=function(x,t){
#
# Function ua_1 computes the exact solution of the
# Burgers-Huxley equation for comparison with the
# numerical solution
#
# Analytical solution
  expp=exp( (1/3)*x+(1/9)*t);
  expm=exp(-(1/3)*x-(1/9)*t);
  ua=((1/2)*(1+(expp-expm)/(expp+expm)))^0.5;
#
# Return solution
```

```
return(c(ua));
    }
```

Listing 14.3 Function ua_1 for the analytical solution of Eq. (14.4).

ua_1 is a straightforward implementation of Eq. (14.4). A numerical value ua is returned (not a list as in Listings 14.2a and 14.2b).

This completes the programming of Eq. (14.1). The numerical and graphical output is considered next.

14.5 Model Output

Execution of the preceding main program (Listing 14.1) and subordinate routines pde_1, ua_1 produces the following abbreviated numerical output in Table 14.1.

We can note the following details about this output:

- As expected, the solution array out has the dimensions 4 x 52 for 4 output points and 51 ODEs, with t included as an additional element, that is, $51 + 1 = 52$.
- The interval in x is $((10 - (-20))/(51 - 1) = 0.6$ (with every fifth value in Table 14.1).
- The numerical and analytical ICs at $t = 0$ agree exactly since the analytical solution (Eq. (14.4)) is used in the main program of Listing 14.1 to define the numerical IC.
- The numerical and analytical solutions at $x = -5, 10$ agree exactly from the BCs set in pde_1.
- The agreement between the numerical and analytical solutions is quite acceptable given that only 51 points were used in x.
- The computational effort is modest, ncall = 144.

The solution is displayed graphically in Figure 14.1, which again demonstrates the agreement between the numerical and analytical solutions. The solution at $x = -20, 10$ varies (only slightly) with t as defined by the BCs in pde_1.

The traveling wave characteristic (dependent on only the Lagrangian variable x-ct = x+(1/3)t) is clear (the use of "characteristic" has a mathematical interpretation since x-ct is the characteristic of Eq. (14.1)).

Also, the slope of the solution at $x = -20, 10$ is close to zero, which suggests that the homogeneous Neumann BCs (14.3c,d) could be used in place of the Dirichlet BCs (14.3a,b). This is confirmed by comparing the solutions from

Table 14.1 Abbreviated output for Eqs. (14.1), Dirichlet BCs.

```
[1]  4

[1]  52

n =   51,   xl = -20.00,   xu = 10.00

     t        x       u(x,t)       ua(x,t)       err(x,t)
  0.00 -20.000      0.001273      0.001273      0.000000
  0.00 -17.000      0.003459      0.003459      0.000000
  0.00 -14.000      0.009403      0.009403      0.000000
  0.00 -11.000      0.025553      0.025553      0.000000
  0.00  -8.000      0.069316      0.069316      0.000000
  0.00  -5.000      0.185594      0.185594      0.000000
  0.00  -2.000      0.456737      0.456737      0.000000
  0.00   1.000      0.812869      0.812869      0.000000
  0.00   4.000      0.966970      0.966970      0.000000
  0.00   7.000      0.995331      0.995331      0.000000
  0.00  10.000      0.999364      0.999364      0.000000
              .                          .
              .                          .
              .                          .
          Output for t = 5, 10 removed
              .                          .
              .                          .
              .                          .

     t        x       u(x,t)       ua(x,t)       err(x,t)
 15.00 -20.000      0.006738      0.006738       0.000000
 15.00 -17.000      0.018312      0.018313      -0.000000
 15.00 -14.000      0.049726      0.049725       0.000001
 15.00 -11.000      0.134113      0.134113       0.000000
 15.00  -8.000      0.345239      0.345258      -0.000019
 15.00  -5.000      0.707096      0.707107      -0.000011
 15.00  -2.000      0.938502      0.938508      -0.000006
 15.00   1.000      0.990966      0.990966      -0.000001
 15.00   4.000      0.998763      0.998763       0.000000
 15.00   7.000      0.999832      0.999832       0.000000
 15.00  10.000      0.999977      0.999977       0.000000

 ncall =   144
```

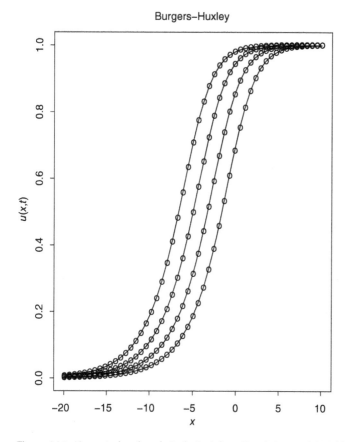

Figure 14.1 Numerical and analytical $u(x, t)$ from Eqs. (14.1–14.4) Dirichlet BCs, lines – num, points – anal.

pde_1a (Listing 14.2b) and pde_1 (Listing 14.2a). Abbreviated numerical output from pde_1a follows.

We can note the following details about the numerical output:

- The agreement of the numerical and analytical solutions at $x = x_l$ in Table 14.2 is not as good as in Table 14.1.

```
Table 14.1

     t        x        u(x,t)       ua(x,t)      err(x,t)
 15.00 -20.000      0.006738      0.006738      0.000000
```

Table 14.2 Abbreviated output for Eqs. (14.1), Neumann BCs.

```
[1]  4

[1]  52

n =   51,  xl = -20.00,  xu = 10.00

     t        x        u(x,t)       ua(x,t)      err(x,t)
  0.00 -20.000     0.001273      0.001273      0.000000
  0.00 -17.000     0.003459      0.003459      0.000000
  0.00 -14.000     0.009403      0.009403      0.000000
  0.00 -11.000     0.025553      0.025553      0.000000
  0.00  -8.000     0.069316      0.069316      0.000000
  0.00  -5.000     0.185594      0.185594      0.000000
  0.00  -2.000     0.456737      0.456737      0.000000
  0.00   1.000     0.812869      0.812869      0.000000
  0.00   4.000     0.966970      0.966970      0.000000
  0.00   7.000     0.995331      0.995331      0.000000
  0.00  10.000     0.999364      0.999364      0.000000
                      .                            .
                      .                            .
                      .                            .
           Output for t = 5, 10 removed
                      .                            .
                      .                            .
                      .                            .
     t        x        u(x,t)       ua(x,t)      err(x,t)
 15.00 -20.000     0.013019      0.006738      0.006282
 15.00 -17.000     0.020385      0.018313      0.002072
 15.00 -14.000     0.050346      0.049725      0.000621
 15.00 -11.000     0.134281      0.134113      0.000168
 15.00  -8.000     0.345285      0.345258      0.000027
 15.00  -5.000     0.707110      0.707107      0.000003
 15.00  -2.000     0.938503      0.938508     -0.000005
 15.00   1.000     0.990965      0.990966     -0.000001
 15.00   4.000     0.998763      0.998763      0.000000
 15.00   7.000     0.999832      0.999832     -0.000000
 15.00  10.000     0.999970      0.999977     -0.000008

ncall =   157
```

Table 14.2

t	x	u(x,t)	ua(x,t)	err(x,t)
15.00	-20.000	0.013019	0.006738	0.006282

This can be explained by a nonzero slope at $x = x_l$, which is captured by the analytical solution (Eq. (14.4)) used for the Dirichlet BCs Eqs. (14.3a,b) (Table 14.1), but not by the approximate homogeneous Neumann BCs Eqs. (14.3c,d) (Table 14.2).

- Generally, except near the boundary $x = x_l$, the agreement between the numerical and analytical solutions is better than 0.1%, including where the solutions are changing most rapidly (Table 14.2). The differences in the graphical output for the second case (homogeneous Neumann BCs) are indistinguishable from the first case and are not presented here.
- The computational effort is modest, ncall = 157.

Three variations can be considered to test the idea that the slope is not zero at the left boundary $x = x_l$.

- The left boundary can be extended. For example, if $x = x_l = -25$ (in Listing 14.1), the agreement of the numerical and analytical solutions is better.

xl = -20 (Table 14.2)

t	x	u(x,t)	ua(x,t)	err(x,t)
15.00	-20.000	0.013019	0.006738	0.006282
15.00	-17.000	0.020385	0.018313	0.002072
15.00	-14.000	0.050346	0.049725	0.000621

xl = -25

t	x	u(x,t)	ua(x,t)	err(x,t)
15.00	-25.000	0.002458	0.001273	0.001186
15.00	-21.500	0.004408	0.004087	0.000321
15.00	-18.000	0.013198	0.013123	0.000075

- The number of grid points can be increased above $n = 51$ (not discussed here).
- The analytical slope (first derivative of $u_a(x, t)$ with respect to x from Eq. (14.4)) can be used (not discussed here).

These results for the left-end BC suggest that knowledge of the solution can lead to approximate BCs that do not require an analytical solution. For example, the spatial domain can be extended, as illustrated by using xl=20, then xl=25, so that the solution does not change with x and homogeneous Neumann BCs can be used. This idea can also be employed, for example, if infinite BCs ($x = \pm\infty$) are specified in the original PDE problem.

14.6 Summary and Conclusions

The programming of the BH equation (14.1) numerical solution and the analytical solution Eq. (14.4) within the SCMOL framework is straightforward, and the two solutions agree with acceptable accuracy and computational effort. The solutions for Dirichlet BCs (14.3a,b) and Neumann BCs (14.3c,d) are essentially the same.

References

1 Griffiths, G.W., and W.E. Schiesser (2012), *Traveling Wave Analysis of Partial Differential Equations*, Elsevier, Burligton, MA.

2 Wang, X.Y. (1990), Solitary wave solutions of the generalized Burgers-Huxley equation, *Journal Physics A: Mathematics General*, **23**, 271–274.

15

Gierer–Meinhardt Equations

The Gierer–Meinhardt equations (GM) are a system of nonlinear diffusion–reaction PDEs with application to the diffusion and reaction of an attractant and an inhibitor that can contribute to pattern formation. We start with a coordinate-free version of the GM equations [1].

15.1   PDE Model

The 2 × 2 PDE system (two equations in two unknowns) discussed in ([1], eqs. (1)) is

$$\frac{\partial u_1}{\partial t} = r\left(1 + \frac{u_1^2}{u_2}\right) - \mu u_1 + D_1 \nabla^2 u_1 \tag{15.1a}$$

$$\frac{\partial u_2}{\partial t} = r u_1^2 - v u_2 + D_2 \nabla^2 u_2 \tag{15.1b}$$

where

u_1	attractant concentration
u_2	inhibitor concentration
∇	spatial differential operator
$\nabla^2 = \nabla \cdot \nabla$	Laplacian operator
t	time
D_1, D_2	diffusivities for u_1, u_2
r, μ, v	rate constants.

Spline Collocation Methods for Partial Differential Equations: With Applications in R, First Edition.
William E. Schiesser.
© 2017 John Wiley & Sons, Inc. Published 2017 by John Wiley & Sons, Inc.
Companion website: www.wiley.com/go/Spline_Collocation

Equations (15.1) are then specialized to 1D in Cartesian coordinates to provide the PDE system that is studied next.

$$\frac{\partial u_1}{\partial t} = r\left(1 + \frac{u_1^2}{u_2}\right) - \mu u_1 + D_1 \frac{\partial^2 u_1}{\partial x^2} \tag{15.2a}$$

$$\frac{\partial u_2}{\partial t} = r u_1^2 - v u_2 + D_2 \frac{\partial^2 u_2}{\partial x^2} \tag{15.2b}$$

Equations (15.2) are first order in t and second order in x, and, therefore, each requires one initial condition (IC) and two boundary conditions (BCs).

$$u_1(x, t = 0) = f_1(x); \quad u_2(x, t = 0) = f_2(x) \tag{15.3a,b}$$

$$\frac{\partial u_1(x = x_l, t)}{\partial x} = \frac{\partial u_1(x = x_u, t)}{\partial x} = 0 \tag{15.4a,b}$$

$$\frac{\partial u_2(x = x_l, t)}{\partial x} = \frac{\partial u_2(x = x_u, t)}{\partial x} = 0 \tag{15.4c,d}$$

15.2 Main Program

A main program for Eqs. (15.2)–(15.4) follows.

```
#
# Gierer-Meinhardt
#
# Delete previous workspaces
  rm(list=ls(all=TRUE))
#
# Access ODE integrator
  library("deSolve");
#
# Access functions for numerical solution
  setwd("f:/collocation/chap15");
  source("pde_1.R");
#
# Grid in x
  n=101;xl=0;xu=1;
  x=seq(from=xl,to=xu,by=(xu-xl)/(n-1));
#
# Parameters
  ncase=1;
  D1=1;D2=1;c=10;
  mu=1;nu=1;
  if(ncase==1){r=1 ;}
```

```
  if(ncase==2){r=10;}
#
# Independent variable for ODE integration
  nout=11;t0=0;tf=0.1;
  tout=seq(from=t0,to=tf,by=(tf-t0)/(nout-1));
#
# ICs
  u0=rep(0,2*n);
  for(i in 1:n){
    u0[i]   =exp(-c*x[i]^2);
    u0[i+n]=exp(-c*x[i]^2);
  }
  ncall=0;
#
# ODE integration
  out=lsodes(y=u0,times=tout,func=pde_1,
      sparsetype="sparseint",rtol=1e-6,atol=1e-6,
      maxord=5);
  nrow(out)
  ncol(out)
#
# Arrays for numerical solution
  u1=matrix(0,nrow=n,ncol=nout);
  u2=matrix(0,nrow=n,ncol=nout);
  t=rep(0,nout);
  for(it in 1:nout){
  for(i  in 1:n){
    u1[i,it]=out[it,i+1];
    u2[i,it]=out[it,i+1+n];
      t[it]=out[it,1];
  }
  }
#
# Display selected output
  cat(sprintf("\n r = %3.1f   mu = %3.1f   nu =
              %3.1f\n",r,mu,nu));
  for(it in 1:nout){
    cat(sprintf("\n     t           x     u1(x,t)
                u2(x,t)\n"));
    iv=seq(from=1,to=n,by=10);
    for(i in iv){
      cat(sprintf("%6.2f%9.3f%10.6f%10.6f\n",
```

```
                   t[it],x[i],u1[i,it],u2[i,it])) ;
    }
  cat(sprintf("\n")) ;
  }
  cat(sprintf(" ncall = %4d\n",ncall)) ;
#
# Plot 2D numerical solution
    matplot(x,u1,type="l",lwd=2,col="black",
            lty=1,xlab="x",ylab="u1(x,t)",
            main="Gierer-Meinhardt") ;
    matplot(x,u2,type="l",lwd=2,col="black",
            lty=1,xlab="x",ylab="u2(x,t)",
            main="Gierer-Meinhardt") ;
```

Listing 15.1 Main program for Eqs. (15.2–15.4).

We can note the following details about Listing 15.1:

- Previous workspaces are removed. Then the ODE integrator library deS-olve is accessed. Note that the setwd (set working directory) uses / rather than the usual \. The setwd requires editing for the local computer (to specify the directory with the MOL routines).

```
#
# Gierer-Meinhardt
#
# Delete previous workspaces
  rm(list=ls(all=TRUE))
#
# Access ODE integrator
  library("deSolve") ;
#
# Access functions for numerical solution
  setwd("f:/collocation/chap15") ;
  source("pde_1.R") ;
```

pde_1 is the routine for the MOL/ODEs (discussed subsequently).

- A uniform grid in x with 101 points is defined with the seq utility for the interval $x_l = 0 \le x \le x_u = 1$. Therefore, the vector x has the values $x = 0, 0.01, \ldots, 1$.

```
#
# Grid in x
  n=101;xl=0;xu=1;
  x=seq(from=xl,to=xu,by=(xu-xl)/(n-1)) ;
```

- The model parameters (constants) in Eqs. (15.2) are defined numerically.

```
#
# Parameters
  ncase=1;
  D1=1;D2=1;c=10;
  mu=1;nu=1;
  if(ncase==1){r=1 ;}
  if(ncase==2){r=10;}
```

Two cases are programmed for a variation in r, which is a sensitive parameter as discussed subsequently. c is the constant used in the Gaussian ICs discussed next.

- A uniform grid in t of 11 output points is defined with the seq utility for the interval $0 \le t \le t_f = 0.1$. Therefore, the vector tout has the values $t = 0, 0.01, \ldots, 0.1$.

```
#
# Independent variable for ODE integration
  nout=11;t0=0;tf=0.1;
  tout=seq(from=t0,to=tf,by=(tf-t0)/(nout-1));
```

At this point, the intervals of x and t in Eqs. (15.1) are defined.

- The ICs for Eqs. (15.2) (Eqs. (15.3)) are placed in a vector u0 of length $2(101) = 202$.

```
#
# ICs
  u0=rep(0,2*n);
  for(i in 1:n){
    u0[i]   =exp(-c*x[i]^2);
    u0[i+n] =exp(-c*x[i]^2);
  }
  ncall=0;
```

Equations (15.2a,b) have the same IC, as well as BCs (programmed in the ODE routine pde_1), so that the only difference between the two PDEs is their RHSs. A comparison of the solutions (to follow) therefore gives a direction indication of the contributions of the RHSs. The counter for the number of calls to pde_1 is initialized.

- The system of 202 ODEs is integrated by the library integrator lsodes (available in deSolve) with the sparse matrix option specified. As expected, the inputs to lsodes are the ODE function, pde_1, the IC vector u0, and the vector of output values of t, tout. The length of u0 (202) informs lsodes how many ODEs are to be integrated. func,y,times are reserved names.

```
#
# ODE integration
  out=lsodes(y=u0,times=tout,func=pde_1,
      sparsetype="sparseint",rtol=1e-6,atol=1e-6,
      maxord=5);
  nrow(out)
  ncol(out)
```

The numerical solution to the ODEs is returned in matrix out. In this case, out has the dimensions 11×203 (displayed in the numerical output). The offset $202 + 1 = 203$ is required since the first element of each column has the output t (also in tout), and the 2, ... , 203 column elements have the 202 ODE solutions. This indexing of out is used next.

- The solutions to Eqs. (15.2), $u_1(x, t)$, $u_2(x, t)$, are placed in arrays u1, u2 for subsequent numerical and graphical output.

```
#
# Arrays for numerical solution
  u1=matrix(0,nrow=n,ncol=nout);
  u2=matrix(0,nrow=n,ncol=nout);
  t=rep(0,nout);
  for(it in 1:nout){
  for(i  in 1:n){
    u1[i,it]=out[it,i+1];
    u2[i,it]=out[it,i+1+n];
      t[it]=out[it,1];
  }
  }
```

- The parameters of the numerical solutions are displayed, followed by every tenth value of the solutions.

```
#
# Display selected output
  cat(sprintf("\n r = %3.1f   mu = %3.1f   nu =
              %3.1f\n",r,mu,nu));
  for(it in 1:nout){
    cat(sprintf("\n      t            x    u1(x,t)
                u2(x,t)\n"));
    iv=seq(from=1,to=n,by=10);
    for(i in iv){
      cat(sprintf("%6.2f%9.3f%10.6f%10.6f\n",
                t[it],x[i],u1[i,it],u2[i,it]));
    }
```

```
    cat(sprintf("\n"));
    }
  cat(sprintf(" ncall = %4d\n",ncall));
```

The number of calls to pde_1 is displayed at the end of the solution.
- The numerical solutions are plotted with matplot.

```
  #
  # Plot 2D numerical solution
      matplot(x,u1,type="l",lwd=2,col="black",
              lty=1,xlab="x",ylab="u1(x,t)",
              main="Gierer-Meinhardt");
      matplot(x,u2,type="l",lwd=2,col="black",
              lty=1,xlab="x",ylab="u2(x,t)",
              main="Gierer-Meinhardt");
```

The ODE routine pde_1 called by lsodes in the main program of Listing 15.1 is considered next.

15.3   ODE Routine

pde_1 for Eqs. (15.2) follows.

```
  pde_1=function(t,u,parm){
#
# Function pde_1 computes the t derivative vector
# for u1(x,t), u2(2,t)
#
# One vector to two vectors
  u1=rep(0,n);u2=rep(0,n);
  for(i  in 1:n){
    u1[i]=u[i];
    u2[i]=u[i+n];
  }
#
# u1x
  tablex=splinefun(x,u1);
  u1x=tablex(x,deriv=1);
#
# u2x
  tablex=splinefun(x,u2);
  u2x=tablex(x,deriv=1);
#
# BCs
```

```
  u1x[1]=0;u1x[n]=0;
  u2x[1]=0;u2x[n]=0;
#
# u1xx
  tablexx=splinefun(x,u1x);
  u1xx=tablexx(x,deriv=1);
#
# u2xx
  tablexx=splinefun(x,u2x);
  u2xx=tablexx(x,deriv=1);
#
# PDE
  u1t=rep(0,n);u2t=rep(0,n);
  for(i in 1:n){
    u1t[i]=r*(1+u1[i]^2/u2[i])-mu*u1[i]+D1*u1xx[i];
    u2t[i]=r*u1[i]^2-nu*u2[i]+D2*u2xx[i];
  }
#
# Two vectors to one vector
  ut=rep(0,2*n);
  for(i in 1:n){
    ut[i]  =u1t[i];
    ut[i+n]=u2t[i];
  }
#
# Increment calls to pde_1
  ncall <<- ncall+1;
#
# Return derivative vector
  return(list(c(ut)));
  }
```

Listing 15.2 ODE/MOL routine pde_1 for Eqs. (15.2).

We can note the following details about pde_1:

- The function is defined.

```
  pde_1=function(t,u,parm){
#
# Function pde_1 computes the t derivative vector
# for u1(x,t), u2(x,t)
```

t is the current value of *t* in Eqs. (15.2). u is the dependent variable 202-vector. parm is an argument to pass parameters to pde_1 (unused, but required in the argument list). The arguments must be listed in the order stated to properly interface with lsodes called in the main program. The derivative vector of the LHS of Eqs. (15.2) is calculated next and returned to lsodes (below).

- u is placed in two 101-vectors, u1, u2, to facilitate the programming of Eqs. (15.2).

```
#
# One vector to two vectors
  u1=rep(0,n);u2=rep(0,n);
  for(i  in 1:n){
    u1[i]=u[i];
    u2[i]=u[i+n];
  }
```

- The first derivatives $\frac{\partial u_1}{\partial x}, \frac{\partial u_2}{\partial x}$, in Eqs. (15.2) are computed with splinefun.

```
#
# u1x
  tablex=splinefun(x,u1);
  u1x=tablex(x,deriv=1);
#
# u2x
  tablex=splinefun(x,u2);
  u2x=tablex(x,deriv=1);
```

- Neumann BCs Eqs. (15.4) are implemented.

```
#
# BCs
  u1x[1]=0;u1x[n]=0;
  u2x[1]=0;u2x[n]=0;
```

Subscripts 1, n correspond to $x = x_l = 0, x = x_u = 1$

- The second derivatives $\frac{\partial^2 u_1}{\partial x^2}, \frac{\partial^2 u_2}{\partial x^2}$, in Eqs. (15.2) are computed from the first derivatives (stagewise differentiation).

```
#
# u1xx
  tablexx=splinefun(x,u1x);
  u1xx=tablexx(x,deriv=1);
#
# u2xx
```

```
tablexx=splinefun(x,u2x);
u2xx=tablexx(x,deriv=1);
```

- The LHS derivatives of Eqs. (15.2) $\dfrac{\partial u_1(x,t)}{\partial t}$, $\dfrac{\partial u_2(x,t)}{\partial t}$ are computed in a for over the interval in *x*. The close correspondence of the coding with the PDEs is noteworthy. The two 101-derivative vectors are placed in u1t, u2t.

```
#
# PDE
  u1t=rep(0,n);u2t=rep(0,n);
  for(i in 1:n){
    u1t[i]=r*(1+u1[i]^2/u2[i])-mu*u1[i]+D1*u1xx[i];
    u2t[i]=r*u1[i]^2-nu*u2[i]+D2*u2xx[i];
  }
```

- u1t, u2t are placed in a 202-vector ut.

```
#
# Two vectors to one vector
  ut=rep(0,2*n);
  for(i in 1:n){
    ut[i]   =u1t[i];
    ut[i+n] =u2t[i];
  }
```

- The counter for the calls to pde_1 is returned to the main program of Listing 15.1 by the <<- operator.

```
#
# Increment calls to pde_1
  ncall <<- ncall+1;
```

- The derivative in ut is returned to lsodes.

```
#
# Return derivative vector
  return(list(c(ut)));
  }
```

Note that a list is used as required by lsodes. The final } concludes pde_1.

15.4 Model Output

Execution of the preceding main program and ODE routine pde_1 produces the following abbreviated numerical output in Table 15.1.

Table 15.1 Abbreviated output for Eqs. (15.2)–(15.4), ncase=1.

---

[1] 11

[1] 203

r = 1.0   mu = 1.0   nu = 1.0

t	x	u1(x,t)	u2(x,t)
0.00	0.000	1.00000	1.00000
0.00	0.100	0.90484	0.90484
0.00	0.200	0.67032	0.67032
0.00	0.300	0.40657	0.40657
0.00	0.400	0.20190	0.20190
0.00	0.500	0.08208	0.08208
0.00	0.600	0.02732	0.02732
0.00	0.700	0.00745	0.00745
0.00	0.800	0.00166	0.00166
0.00	0.900	0.00030	0.00030
0.00	1.000	0.00005	0.00005

.
.
.

Output for t = 0.01 to 0.09 removed

.
.
.

t	x	u1(x,t)	u2(x,t)
0.10	0.000	0.55483	0.43181
0.10	0.100	0.54610	0.42317
0.10	0.200	0.52101	0.39835
0.10	0.300	0.48262	0.36042
0.10	0.400	0.43541	0.31390
0.10	0.500	0.38463	0.26398
0.10	0.600	0.33550	0.21577
0.10	0.700	0.29261	0.17372
0.10	0.800	0.25956	0.14133
0.10	0.900	0.23881	0.12098
0.10	1.000	0.23175	0.11405

ncall = 356

We can note the following details about this output:

- As expected, the solution array `out` has the dimensions `11 x 203` for 11 output points and 202 ODEs, with t included as an additional element, that is, $202 + 1 = 203$.
- The interval in x is $(1-0)/(101-1) = 0.01$ (with every tenth value in Table 15.1).
- The equality of the ICs of Eqs. (15.2) (Eqs. (15.3)) as programmed in Listing 15.1 is confirmed (for $t = 0$, $u_1(x,t) = u_2(x,t)$).
- $u_1(x,t)$ and $u_2(x,t)$ are still changing with t at $t = 0.1$, but appear to be approaching a uniform steady state (equilibrium solution). This is confirmed in Figure 15.1a,b.
- The computational effort is modest, `ncall = 356`.

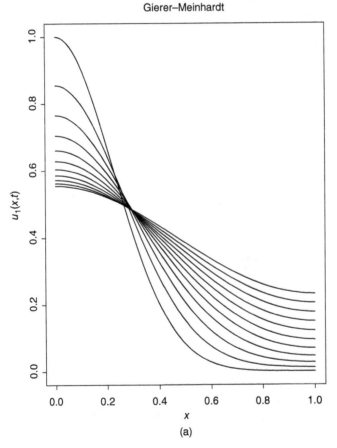

Gierer–Meinhardt

(a)

Figure 15.1 (a) $u_1(x,t)$ from Eqs. (15.2)–(15.4), `ncase=1`. (b) $u_2(x,t)$ from Eqs. (15.2)–(15.4), `ncase=1`.

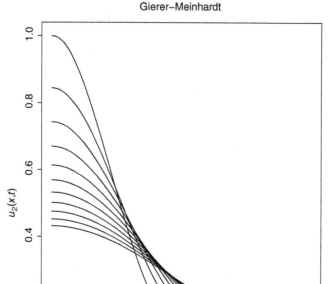

Gierer–Meinhardt

(b)

Figure 15.1 (*Continued*)

Figure 15.1a,b are similar suggesting a small difference in the RHSs of Eqs. (15.2). This would occur since (if, when) $D_1 = D_2$ and $\mu = \nu$ (with the same ICs) and the coupling terms with r are small. To investigate this explanation, ncase=2 corresponds to a larger value of r (Listing 15.1) so that the coupling terms have a greater effect. Abbreviated numerical output and graphical output follows in Table 15.2.

We can note the following details about this output:

- $u_1(x, t)$ and $u_2(x, t)$ are substantially different (at $t = 0.1$), with the term in r having a greater effect for Eq. (15.2b).
- $u_1(x, t)$ and $u_2(x, t)$ do not appear to be reaching a stable steady state.
- The computational effort is still modest, ncall = 1206.

These features of the solutions also appear in Figure 15.2a,b.

Table 15.2 Abbreviated output for Eqs. (15.2)–(15.4), ncase=2.

```
[1] 11

[1] 203

r = 1.0   mu = 1.0   nu = 1.0

       t          x      u1(x,t)      u2(x,t)
    0.00      0.000     1.00000      1.00000
    0.00      0.100     0.90484      0.90484
    0.00      0.200     0.67032      0.67032
    0.00      0.300     0.40657      0.40657
    0.00      0.400     0.20190      0.20190
    0.00      0.500     0.08208      0.08208
    0.00      0.600     0.02732      0.02732
    0.00      0.700     0.00745      0.00745
    0.00      0.800     0.00166      0.00166
    0.00      0.900     0.00030      0.00030
    0.00      1.000     0.00005      0.00005
                  .                     .
                  .                     .
                  .                     .
  Output for t = 0.01 to 0.09 removed
                  .                     .
                  .                     .
                  .                     .
       t          x      u1(x,t)      u2(x,t)
    0.10      0.000     4.02781      6.98454
    0.10      0.100     4.06812      7.26663
    0.10      0.200     4.16475      7.91453
    0.10      0.300     4.32614      8.99918
    0.10      0.400     4.54792     10.50139
    0.10      0.500     4.81717     12.35090
    0.10      0.600     5.11051     14.40358
    0.10      0.700     5.39544     16.43803
    0.10      0.800     5.63511     18.18119
    0.10      0.900     5.79525     19.36233
    0.10      1.000     5.84146     19.70614

ncall = 1206
```

15.5 Summary and Conclusions

The programming of the GM equations of Eqs. (15.2)–(15.4) within the SCMOL framework is straightforward. Extensions of the preceding cases could be considered, for example,

- Since ICs (15.3) are even (Gaussian) functions of x (symmetric with respect to the vertical axes), and there are no explicit functions of x in the RHSs of Eqs. (15.2), we can expect that the solutions will remain even. This could be checked by using a spatial interval $-1 \leq x \leq 1$ rather than $0 \leq x \leq 1$. In other words, Eqs. (15.3a,c) can be considered as symmetry BCs (so the solutions for $x < 0$ are mirror images of the solutions for $x > 0$). The advantage of using

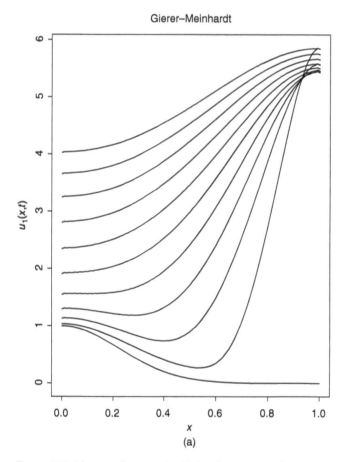

Gierer–Meinhardt

Figure 15.2 (a) $u_1(x, t)$ from Eqs. (15.2)–(15.4), ncase=2. (b) $u_2(x, t)$ from Eqs. (15.2)–(15.4), ncase=2.

Gierer–Meinhardt

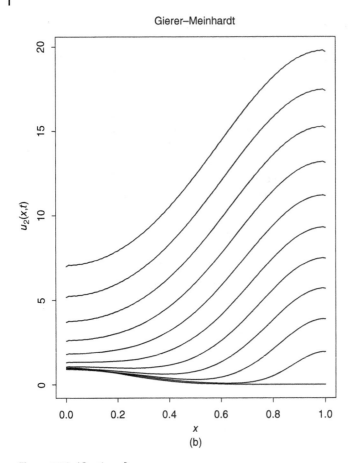

Figure 15.2 (*Continued*)

the interval $0 \le x \le 1$ is that the additional computation required to produce the solutions for $-1 \le x \le 0$ can be avoided.

- The features of the numerical solutions could be investigated by computing and displaying the LHSs of Eqs. (15.2) (derivatives in t) and the various RHS terms to demonstrate the relative contributions of these terms. This additional analysis is easily implemented by using the available numerical solutions $u_1(x, t), u_2(x, t)$.
- Various forms of the diffusion terms can be studied, for example, linear (Fickian) diffusion replaced by chemotaxis.
- Additional components could be included (by adding the associated PDEs).

These extensions would elucidate the features of the GM equations.

Table 15.3 Abbreviated output for Eqs. (15.2)–(15.4), `ncase=1, n=101, 201,`
$t = 0.1$.

```
n=101

[1]  11

[1]  203

 r = 1.0   mu = 1.0   nu = 1.0

      t          x      u1(x,t)    u2(x,t)
    0.10      0.000     0.55483    0.43181
    0.10      0.100     0.54610    0.42317
    0.10      0.200     0.52101    0.39835
    0.10      0.300     0.48262    0.36042
    0.10      0.400     0.43541    0.31390
    0.10      0.500     0.38463    0.26398
    0.10      0.600     0.33550    0.21577
    0.10      0.700     0.29261    0.17372
    0.10      0.800     0.25956    0.14133
    0.10      0.900     0.23881    0.12098
    0.10      1.000     0.23175    0.11405

 ncall =   356

n=201

[1]  11

[1]  403

 r = 1.0   mu = 1.0   nu = 1.0

      t          x      u1(x,t)    u2(x,t)
    0.10      0.000     0.55483    0.43181
    0.10      0.100     0.54610    0.42317
    0.10      0.200     0.52101    0.39835
    0.10      0.300     0.48262    0.36042
    0.10      0.400     0.43541    0.31391
```

Table 15.3 (Continued)

0.10	0.500	0.38463	0.26398
0.10	0.600	0.33550	0.21577
0.10	0.700	0.29261	0.17372
0.10	0.800	0.25956	0.14133
0.10	0.900	0.23881	0.12098
0.10	1.000	0.23175	0.11405

ncall = 567

An analytical solution to verify the numerical solutions may not be available (the usual situation for applications based on simultaneous, nonlinear PDEs). The accuracy of the numerical solutions can be inferred by changing the number of spatial grid points and observing the effect on the solutions, that is, h refinement. The following Table 15.3 gives the numerical output at $t = 0.1$ for n=101, 201 (with iv=seq(from=1,to=n,by=10) changed to iv=seq(from=1,to=n,by=20) in the main program of Listing 15.1).

The solutions for $n = 101, 201$ are essentially identical to five figures.

Of course, this numerical analysis (based on h refinement) does not constitute a proof of the validity (accuracy) of the numerical solution. But this method, which does not require an analytical solution, gives some assurance that the numerical solution is accurate to the indicated number of significant figures.

The caveat that should be kept in mind is that both solutions were computed with splinefun and both could be incorrect. To address this question, calculation of SCMOL solutions with splines of higher order than splinefun could be considered (a form of p refinement). In fact, the development of routines analogous to splinefun, but with higher order splines, would be an important and useful next step.

Reference

1 Page, K.M., P.K. Maini, and N.A.M. Monk (2005), Complex pattern formation in reaction-diffusion systems with spatially varying patterns, *Physica D*, **202**, no. 1-2, 95–115.

16

Keller–Segel Equations

The Keller–Segel (KES) equations [2] are an extension of the linear diffusion equation with a chemotaxis[1] (nonlinear, non-Fickian) diffusion flux. We consider here a 2×2 (two equations in two unknowns) chemotaxis PDE system with a known analytical solution ([3] p68) that is used to verify a numerical solution computed by a spline collocation method of lines (SCMOL).

16.1 PDE Model

The KES chemotaxis equations are

$$\frac{\partial u_1}{\partial t} = -ku_2 \tag{16.1a}$$

$$\frac{\partial u_2}{\partial t} = D\frac{\partial}{\partial x}\left[\frac{\partial u_2}{\partial x} - 2\frac{u_2}{u_1}\frac{\partial u_1}{\partial x}\right] \tag{16.1b}$$

where k, D are positive (assigned) constants.

The RHS diffusion term in Eq. (16.1b) consists of two parts:

1)
$$D\frac{\partial}{\partial x}\left[\frac{\partial u_2}{\partial x}\right] = D\frac{\partial^2 u_2}{\partial x^2}$$

is the usual linear (Fickian) diffusion with diffusivity D.

2)
$$-D\frac{\partial}{\partial x}\left[2\frac{u_2}{u_1}\frac{\partial u_1}{\partial x}\right]$$

---

1 *Chemotaxis* is an established term that denotes movement (taxis) of, for example, biochemicals, biological cells, viruses, under the influence of chemical gradients (chemo).

Spline Collocation Methods for Partial Differential Equations: With Applications in R, First Edition.
William E. Schiesser.
© 2017 John Wiley & Sons, Inc. Published 2017 by John Wiley & Sons, Inc.
Companion website: www.wiley.com/go/Spline_Collocation

Note in particular that the rate of diffusion (flux) of u_2 is proportional to the gradient (derivative) of u_1, $\dfrac{\partial u_1}{\partial x}$, and because of the minus, the flux is in the direction of increasing u_1.[2]

Also, the ratio u_2/u_1 is a multiplying factor in determining the flux. This ratio causes the rate of transfer to increase with increasing u_2 (the LHS-dependent variable of Eq. (16.1b)), and to increase with a decrease in u_1 (the LHS-dependent variable of Eq. (16.1a)).[3]

In a biological context [2], the variables and parameters in Eqs. (16.1) are

u_1	attractant concentration
u_2	microorganism concentration
x	spatial coordinate
t	time
k	rate constant
D	diffusivity.

Since the second RHS term has a rather unconventional form, we would expect that it will introduce unusual features in the solution when compared with the usual diffusion modeled by linear (Fickian) diffusion. These features will be discussed when the output from the following R routines for Eqs. (16.1) is discussed subsequently.

Equations (16.1) are first order in t, and, therefore, each requires an initial condition (IC).

$$u_1(x, t = 0) = f_1(x); \quad u_2(x, t = 0) = f_2(x) \tag{16.2a,b}$$

where f_1, f_2 are functions to be specified.

Equation (16.2b) is second order in x and therefore requires two boundary conditions (BCs).

$$\frac{\partial u_2(x \to -\infty, t)}{\partial x} = \frac{\partial u_2(x \to \infty, t)}{\partial x} = 0 \tag{16.3a,b}$$

Since $u_1(x, t)$ is a function of x, and it appears in a second derivative term in Eq. (16.2b), we will also assign it two BCs.

$$\frac{\partial u_1(x \to -\infty, t)}{\partial x} = \frac{\partial u_1(x \to \infty, t)}{\partial x} = 0 \tag{16.3c,d}$$

---

2 For example, bacteria might move in the direction of increasing oxygen or nutrient (food). This effect is clearly a feature of an animate (living) system, such as bacteria, rather than an inanimate system.
3 For example, bacteria might move faster with increasing concentration (density, population) and decreasing oxygen or nutrient (food).

An analytical solution to Eqs. (16.1)–(16.3) is ([3], p68)

$$u_1(z) = [1 + e^{-cz/D}]^{-1} \tag{16.4a}$$

$$u_2(z) = \frac{c^2}{kD} e^{-cz/D} [1 + e^{-cz/D}]^{-2} \tag{16.4b}$$

where $z = x - ct$; c is a constant to be specified (a velocity). Note that $u_1(z), u_2(z)$ are a function of the single Lagrangian variable z. In other words, these solutions are invariant for a given value of z, regardless of how x and t may vary. A solution with this property is termed a *traveling wave*. This property is discussed further when the numerical solution to Eqs. (16.1)–(16.3) is considered subsequently.

Equations (16.4) are used for ICs (16.2) with $t = 0, z = x$.

$$f_1(x) = u_1(x, t = 0) = [1 + e^{-cx/D}]^{-1} \tag{16.5a}$$

$$f_2(x) = u_2(x, t = 0) = \frac{c^2}{kD} e^{-cx/D} [1 + e^{-cx/D}]^{-2} \tag{16.5b}$$

The analytical solution, Eqs. (16.4), is used to evaluate the numerical solution. The R routines for Eqs. (16.1)–(16.5) follow.

16.2 Main Program

A main program for Eqs. (16.1)–(16.5) follows next.

```
#
# Keller-Segel
#
# Delete previous workspaces
  rm(list=ls(all=TRUE))
#
# Access ODE integrator
  library("deSolve");
#
# Access functions for numerical solution
  setwd("f:/collocation/chap16");
  source("pde_1.R");source("u1a.R");
  source("u2a.R");
#
# Grid in x
  n=101;xl=-10;xu=15;
  x=seq(from=xl,to=xu,by=(xu-xl)/(n-1));
#
# Parameters
  k=1;D=1;c=1;
```

```
   cat (sprintf ("\n\n k = %5.2f    D = %5.2f
               c = %5.2f\n",k,D,c));
#
# Independent variable for ODE integration
   nout=6;t0=0;tf=5;
   tout=seq(from=t0,to=tf,by=(tf-t0)/(nout-1));
#
# Initial condition from analytical solutions (t=0)
   u0=rep(0,2*n);
   for(i in 1:n){
     u0[i]   =u1a(x[i],0);
     u0[i+n]=u2a(x[i],0);
   }
   ncall=0;
#
# ODE integration
   out=lsodes(y=u0,times=tout,func=pde_1,
       sparsetype="sparseint",rtol=1e-6,atol=1e-6,
       maxord=5);
   nrow(out)
   ncol(out)
#
# Arrays for plotting numerical,
# analytical solutions
   u1=matrix(0,nrow=n,ncol=nout);
   u2=matrix(0,nrow=n,ncol=nout);
  ua1=matrix(0,nrow=n,ncol=nout);
  ua2=matrix(0,nrow=n,ncol=nout);
    t=rep(0,nout);
    for(it in 1:nout){
      t[it]=out[it,1];
      for(i in 1:n){
         u1[i,it]=out[it,i+1];
         u2[i,it]=out[it,i+1+n];
        ua1[i,it]=u1a(x[i],t[it]);
        ua2[i,it]=u2a(x[i],t[it]);
      }
    }
#
# Display selected output
#
# Step through t
```

```
  for(it in 1:nout){
    cat(sprintf("\n        t          x    u1(x,t)
                 ua1(x,t)         diff1\n"));
    cat(sprintf("    "      t          x    u2(x,t)
                 ua2(x,t)         diff2\n"));
#
#    Step through x
    iv=seq(from=1,to=n,by=10);
    for(i in iv){
      diff1=u1[i,it]-ua1[i,it];
      cat(sprintf("%6.2f%8.3f%10.6f%12.6f%12.6f\n",
              t[it],x[i],u1[i,it],ua1[i,it],diff1));
      diff2=u2[i,it]-ua2[i,it];
      cat(sprintf("%6.2f%8.3f%10.6f%12.6f%12.6f\n",
              t[it],x[i],u2[i,it],ua2[i,it],diff2));
    cat(sprintf("\n"));
    }
  }
  cat(sprintf(" ncall = %4d\n",ncall));
#
# Plot 2D numerical solutions
    matplot(x,u1,type="l",lwd=2,col="black",lty=1,
      xlab="x",ylab="u1(x,t)",main="Keller-Segel");
    matpoints(x,ua1,pch="o",col="black");
    matplot(x,u2,type="l",lwd=2,col="black",lty=1,
      xlab="x",ylab="u2(x,t)",main="Keller-Segel");
    matpoints(x,ua2,pch="o",col="black");
```

Listing 16.1 Main program for Eqs. (16.1)–(16.4).

We can note the following details about Listing 16.1:

- Previous workspaces are removed. Then the ODE integrator library deSolve is accessed. Note that the setwd (set working directory) uses / rather than the usual \. The setwd requires editing for the local computer (to specify the directory for the R routines).

```
#
# Keller-Segel
#
# Delete previous workspaces
  rm(list=ls(all=TRUE))
#
# Access ODE integrator
```

```
library("deSolve");
#
# Access functions for numerical solution
setwd("f:/collocation/chap16");
source("pde_1.R");source("u1a.R");
source("u2a.R");
```

pde_1 is the routine for the MOL/ODEs (discussed subsequently). u1a, u2a are the functions that implement the analytical solution of Eqs. (16.4).

- A uniform grid in x with 101 points is defined with the seq utility for the interval $x_l = -10 \leq x \leq x_u = 15$. Therefore, the vector x has the values $x = -10, -9.75, \ldots, 15$.

```
#
# Grid in x
n=101;xl=-10;xu=15;
x=seq(from=xl,to=xu,by=(xu-xl)/(n-1));
```

- The parameters (constants) in Eqs. (16.1) and (16.4) are defined numerically.

```
#
# Parameters
k=1;D=1;c=1;
cat(sprintf("\n\n k = %5.2f   D = %5.2f
              c = %5.2f\n",k,D,c));
```

- A uniform grid in t of 6 output points is defined with the seq utility for the interval $0 \leq t \leq t_f = 5$. Therefore, the vector tout has the values $t = 0, 1, \ldots, 5$.

```
#
# Independent variable for ODE integration
nout=6;t0=0;tf=5;
tout=seq(from=t0,to=tf,by=(tf-t0)/(nout-1));
```

At this point, the intervals of x and t in Eqs. (16.1) are defined. The homogeneous Neumann BCs of Eqs. (16.3) at $\pm\infty$ are replaced with BCs at finite x, $-10 \leq x \leq 15$. This equivalence (of infinite and finite BCs) will be observed in the numerical solution.

- The ICs for Eqs. (16.1), Eqs. (16.5), are placed in a vector u0 of length $2(101) = 202$.

```
#
# Initial condition from analytical solutions (t=0)
u0=rep(0,2*n);
for(i in 1:n){
  u0[i]   =u1a(x[i],0);
```

```
        u0[i+n]=u2a(x[i],0);
    }
    ncall=0;
```

The functions for the analytical u_1, u_2 of Eqs. (16.4) are used with $t = 0$ according to Eqs. (16.5). The counter for the number of calls to pde_1 is initialized.

- The system of 202 ODEs is integrated by the library integrator lsodes (available in deSolve) with the sparse matrix option specified. As expected, the inputs to lsodes are the ODE function, pde_1, the IC vector u0, and the vector of output values of t, tout. The length of u0 (202) informs lsodes how many ODEs are to be integrated. func,y,times are reserved names.

```
#
# ODE integration
  out=lsodes(y=u0,times=tout,func=pde_1,
      sparsetype="sparseint",rtol=1e-6,atol=1e-6,
      maxord=5);
  nrow(out)
  ncol(out)
```

The numerical solution to the ODEs is returned in matrix out. In this case, out has the dimensions 6×203 (displayed in the numerical output). The offset $202 + 1 = 203$ is required since the first element of each column has the output t (also in tout), and the 2, ... , 203 column elements have the 202 ODE solutions. This indexing of out is used next.

- The numerical solution for $u_1(x, t)$, $u_2(x, t)$ is taken from the solution matrix out returned by lsodes. The intervals in t and x are covered by two nested fors (with indices it and i for t and x, respectively). The offset i+1 accounts for t returned in the first column position of out and placed in the vector t[it].

```
#
# Arrays for plotting numerical,
# analytical solutions
  u1=matrix(0,nrow=n,ncol=nout);
  u2=matrix(0,nrow=n,ncol=nout);
 ua1=matrix(0,nrow=n,ncol=nout);
 ua2=matrix(0,nrow=n,ncol=nout);
   t=rep(0,nout);
   for(it in 1:nout){
     t[it]=out[it,1];
     for(i in 1:n){
        u1[i,it]=out[it,i+1];
```

```
        u2[i,it]=out[it,i+1+n];
        ua1[i,it]=u1a(x[i],t[it]);
        ua2[i,it]=u2a(x[i],t[it]);
    }
  }
```

The analytical solution of Eqs. (16.4) is placed in two arrays ua1, ua2.

- A heading for the numerical and analytical solutions and their difference is displayed at each value of t. The numerical and analytical solutions and the difference are displayed for every 10th value of x.

```
#
# Display selected output
#
# Step through t
  for(it in 1:nout){
    cat(sprintf("\n      t           x    u1(x,t)
                    ua1(x,t)    diff1\n"));
    cat(sprintf(  "      t           x    u2(x,t)
                    ua2(x,t)    diff2\n"));
#
#     Step through x
    iv=seq(from=1,to=n,by=10);
    for(i in iv){
      diff1=u1[i,it]-ua1[i,it];
      cat(sprintf("%6.2f%8.3f%10.6f%12.6f%12.6f\n",
              t[it],x[i],u1[i,it],ua1[i,it],diff1));
      diff2=u2[i,it]-ua2[i,it];
      cat(sprintf("%6.2f%8.3f%10.6f%12.6f%12.6f\n",
              t[it],x[i],u2[i,it],ua2[i,it],diff2));
      cat(sprintf("\n"));
      }
    }
  cat(sprintf("  ncall = %4d\n",ncall));
```

The number of calls to pde_1 is displayed at the end of the solution.

- The numerical and analytical solutions are plotted with matplot and matpoints, respectively.

```
#
# Plot 2D numerical solutions
    matplot(x,u1,type="l",lwd=2,col="black",lty=1,
      xlab="x",ylab="u1(x,t)",main="Keller-Segel");
    matpoints(x,ua1,pch="o",col="black");
    matplot(x,u2,type="l",lwd=2,col="black",lty=1,
```

```
      xlab="x",ylab="u2(x,t)",main="Keller-Segel");
    matpoints(x,ua2,pch="o",col="black");
```

The ODE routine pde_1 called by lsodes in the main program of Listing 16.1 is considered next.

16.3 ODE Routine

pde_1 for Eqs. (16.1) follows.

```
  pde_1=function(t,u,parm){
#
# Function pde_1 computes the t derivative vector
# for u1(x,t) u2(x,t)
#
# One vector to two vectors
  u1=rep(0,n);u2=rep(0,n);
  for(i in 1:n){
    u1[i]=u[i];
    u2[i]=u[i+n];
  }
#
# u1x
  tablex=splinefun(x,u1);
  u1x=tablex(x,deriv=1);
#
# u2x
  tablex=splinefun(x,u2);
  u2x=tablex(x,deriv=1);
#
# BCs
  u1x[1]=0;u1x[n]=0;
  u2x[1]=0;u2x[n]=0;
#
# Nonlinear term
  u1u2x=rep(0,n);
  for(i in 1:n){
    u1u2x[i]=2*u2[i]/u1[i]*u1x[i];
  }
#
# u1u2xx
  tablex=splinefun(x,u1u2x);
  u1u2xx=tablex(x,deriv=1);
```

```
#
# u2xx
  tablexx=splinefun(x,u2x);
  u2xx=tablexx(x,deriv=1);
#
# PDEs
  u1t=rep(0,n);u2t=rep(0,n);
  for(i in 1:n){
    u1t[i]=-k*u2[i];
    u2t[i]=D*(u2xx[i]-u1u2xx[i]);
  }
#
# Two vectors to one vector
  ut=rep(0,2*n);
  for(i in 1:n){
    ut[i]   =u1t[i];
    ut[i+n]=u2t[i];
  }
#
# Increment calls to pde_1
  ncall <<- ncall+1;
#
# Return derivative vector
  return(list(c(ut)));
  }
```

Listing 16.2 ODE/MOL routine pde_1 for Eqs. (16.1).

We can note the following details about pde_1:

- The function is defined.

```
  pde_1=function(t,u,parm){
#
# Function pde_1 computes the t derivative vector
# for u1(x,t),  u2(x,t)
```

t is the current value of *t* in Eqs. (16.1). u is the dependent variable 202-vector. parm is an argument to pass parameters to pde_1 (unused, but required in the argument list). The arguments must be listed in the order stated to properly interface with lsodes called in the main program. The derivative vector of the LHS of Eqs. (16.1) is calculated next and returned to lsodes.

- u is placed in two 101-vectors, u1, u2, to facilitate the programming of Eqs. (16.1).

```
#
# One vector to two vectors
  u1=rep(0,n);u2=rep(0,n);
  for(i   in 1:n){
    u1[i]=u[i];
    u2[i]=u[i+n];
  }
```

- $\dfrac{\partial u_1}{\partial x}, \dfrac{\partial u_2}{\partial x}$ are computed with splinefun.

```
#
# u1x
  tablex=splinefun(x,u1);
  u1x=tablex(x,deriv=1);
#
# u2x
  tablex=splinefun(x,u2);
  u2x=tablex(x,deriv=1);
```

- Homogeneous Neumann BCs (16.3a–d) are implemented. The subscripts 1, n correspond to $x = -10, 15$.

```
#
# BCs
  u1x[1]=0;u1x[n]=0;
  u2x[1]=0;u2x[n]=0;
```

- $2\dfrac{u_2}{u_1}\dfrac{\partial u_1}{\partial x}$ in Eq. (16.1b) is computed and placed in u1u2x.

```
#
# Nonlinear term
  u1u2x=rep(0,n);
  for(i in 1:n){
    u1u2x[i]=2*u2[i]/u1[i]*u1x[i];
  }
```

- The derivative of $2\dfrac{u_2}{u_1}\dfrac{\partial u_1}{\partial x}$ is computed with splinefun.

```
#
# u1u2xx
  tablex=splinefun(x,u1u2x);
  u1u2xx=tablex(x,deriv=1);
```

- The second derivative $\dfrac{\partial^2 u_2}{\partial x^2}$ in Eq. (16.1b) is computed by differentiating the first derivative (i.e., stagewise differentiation).

```
#
# u2xx
  tablexx=splinefun(x,u2x);
  u2xx=tablexx(x,deriv=1);
```

- The LHS derivatives of Eqs. (16.1) are computed and the two 101-derivative vectors are placed in u1t, u2t.

```
#
# PDEs
  u1t=rep(0,n);u2t=rep(0,n);
  for(i in 1:n){
    u1t[i]=-k*u2[i];
    u2t[i]=D*(u2xx[i]-u1u2xx[i]);
  }
```

The resemblance of this code to Eqs. (16.1) and the accommodation of the nonlinear flux are particularly noteworthy.

- u1t, u2t are placed in a 202-vector ut.

```
#
# Two vectors to one vector
  ut=rep(0,2*n);
  for(i in 1:n){
    ut[i]   =u1t[i];
    ut[i+n] =u2t[i];
  }
```

- The counter for the calls to pde_1 is returned to the main program of Listing 16.1 by the <<- operator.

```
#
# Increment calls to pde_1
  ncall <<- ncall+1;
```

- The derivative in ut is returned to lsodes.

```
#
# Return derivative vector
  return(list(c(ut)));
  }
```

note that a list is used as required by lsodes. The final } concludes pde_1.

16.4 Subordinate Routines

Functions u1a, u2a for the analytical solutions of Eqs. (16.4) are listed next.

```
u1a=function (x,t) {
#
# Function u1a computes the analytical solution
# for u1(x,t) of the Keller-Segel equations
z=x-c*t;
u1a=1/(1+exp(-c*z/D));
#
# Return solution
return(c(u1a));
}
```

Listing 16.3 Routine u1a for analytical solution (16.4a).

```
u2a=function (x,t) {
#
# Function u2a computes the analytical solution
# for u2(x,t) of the Keller-Segel equations
#
z=x-c*t;
u2a=(c^2/(k*D))*exp(-c*z/D)/(1+exp(-c*z/D))^2;
#
# Return solution
return(c(u2a));
}
```

Listing 16.4 Routine u2a for analytical solution (16.4b).

Functions u1a, u2a are straightforward implementations of the analytical solutions, Eqs. (16.4).

16.5 Model Output

Execution of the preceding main program, ODE routine pde_1 and subordinate routines u1a, u2a produces the following abbreviated numerical output in Table 16.1.

We can note the following details about this output:

- As expected, the solution array out has the dimensions 6 x 203 for 6 output points and 202 ODEs, with t included as an additional element, that is, $202 + 1 = 203$.

Table 16.1 Abbreviated output for Eqs. (16.1).

```
k =  1.00   D =  1.00   c =  1.00

[1] 6

[1] 203
```

t	x	u1(x,t)	ua1(x,t)	diff1
t	x	u2(x,t)	ua2(x,t)	diff2
0.00	-10.000	0.000045	0.000045	0.000000
0.00	-10.000	0.000045	0.000045	0.000000
0.00	-7.500	0.000553	0.000553	0.000000
0.00	-7.500	0.000552	0.000552	0.000000
0.00	-5.000	0.006693	0.006693	0.000000
0.00	-5.000	0.006648	0.006648	0.000000
0.00	-2.500	0.075858	0.075858	0.000000
0.00	-2.500	0.070104	0.070104	0.000000
0.00	0.000	0.500000	0.500000	0.000000
0.00	0.000	0.250000	0.250000	0.000000
0.00	2.500	0.924142	0.924142	0.000000
0.00	2.500	0.070104	0.070104	0.000000
0.00	5.000	0.993307	0.993307	0.000000
0.00	5.000	0.006648	0.006648	0.000000
0.00	7.500	0.999447	0.999447	0.000000
0.00	7.500	0.000552	0.000552	0.000000
0.00	10.000	0.999955	0.999955	0.000000
0.00	10.000	0.000045	0.000045	0.000000
0.00	12.500	0.999996	0.999996	0.000000
0.00	12.500	0.000004	0.000004	0.000000
0.00	15.000	1.000000	1.000000	0.000000

Table 16.1 (Continued)

t	x	u1(x,t)	ua1(x,t)	diff1

```
0.00   15.000   0.000000      0.000000      0.000000
                    .             .
                    .             .
                    .             .

         Output for t = 1 to t = 4 removed

                    .             .
                    .             .
                    .             .
```

t	x	u1(x,t)	ua1(x,t)	diff1
t	x	u2(x,t)	ua2(x,t)	diff2
5.00	-10.000	0.000002	0.000000	0.000002
5.00	-10.000	0.000003	0.000000	0.000003
5.00	-7.500	0.000005	0.000004	0.000002
5.00	-7.500	0.000007	0.000004	0.000003
5.00	-5.000	0.000050	0.000045	0.000004
5.00	-5.000	0.000052	0.000045	0.000007
5.00	-2.500	0.000565	0.000553	0.000012
5.00	-2.500	0.000562	0.000552	0.000010
5.00	0.000	0.006716	0.006693	0.000023
5.00	0.000	0.006649	0.006648	0.000001
5.00	2.500	0.075881	0.075858	0.000023
5.00	2.500	0.070085	0.070104	-0.000019
5.00	5.000	0.500008	0.500000	0.000008
5.00	5.000	0.249985	0.250000	-0.000015
5.00	7.500	0.924140	0.924142	-0.000001
5.00	7.500	0.070104	0.070104	-0.000000
5.00	10.000	0.993307	0.993307	-0.000000
5.00	10.000	0.006648	0.006648	0.000000
5.00	12.500	0.999444	0.999447	-0.000003
5.00	12.500	0.000556	0.000552	0.000004

(Continued)

Table 16.1 (Continued)

5.00	15.000	0.999910	0.999955	-0.000045
5.00	15.000	0.000091	0.000045	0.000045

ncall = 390

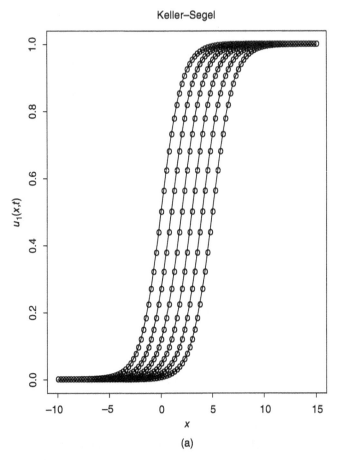

Keller–Segel

(a)

Figure 16.1 (a) Numerical and analytical $u_1(x, t)$ from Eqs. (16.1) lines – num, points – anal. (b) Numerical and analytical $u_2(x, t)$ from Eqs. (16.1) lines – num, points – anal.

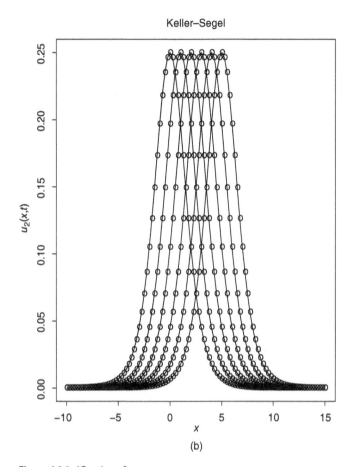

(b)

Figure 16.1 (*Continued*)

- The interval in x is $((15 - (-10))/(101 - 1) = 0.25$ (with every tenth value in Table 16.1).
- The numerical and analytical ICs at $t = 0$ agree exactly since the analytical solution is used in the main program of Listing 16.1 to define the numerical IC.
- The good agreement between the numerical and analytical solutions, including the points of rapid change in x, for example, $u_1(x = 5, t = 5) = 0.5$, $u_2(x = 5, t = 5) = 0.25$.
- The computational effort is modest, `ncall = 390`.

The solution is displayed graphically in Figure 16.1a,b, which again demonstrate the agreement between the numerical and analytical solutions. The traveling wave solution is evident, that is, with a velocity $c = 1$, the solution travels according to the Lagrangian variable $z = x - ct = x - t$. For example, the peak values of $u_2(x = 0, t = 0) = 0.25$, $u_2(x = 5, t = 5) = 0.25$ (Table 16.1) demonstrate the accurate resolution in x and t.

Also, Figure 16.1a,b indicate the equivalence of the solutions for the infinite BCs of Eqs. (16.3a–d) with the finite BCs at $x = -10, 15$.

16.6 Summary and Conclusions

The programming of the numerical solutions of the KES equations (16.1) and the analytical solution, Eqs. (16.4), within the SCMOL framework is straightforward, and the two solutions agree with acceptable accuracy and computational effort.

Additionally,

- The traveling wave solutions are closely approximated numerically.
- The finite domain $-10 \leq x \leq 15$ is equivalent to the infinite domain $-\infty \leq x \leq \infty$ for the velocity $c = 1$ and the interval $0 \leq t \leq 5$.
- The nonlinear chemotaxis diffusion is readily accommodated numerically.

Modification and extension of the chemotaxis model can be made numerically, even for cases for which an analytical solution is not available (generally the case in applications). Alternate forms of the diffusion flux are listed in Appendix A16.1.

Appendix A16.1. Diffusion Models

Several diffusion models with one and two components are briefly reviewed here. The two component models can be classified as chemotaxis in the sense that the diffusion of one of the components is determined in part by the second component.

The one-component models are listed in the following table, which gives the flux of component 1, q_{x,u_1}, as a function of the component, u_1, for a 1D, Cartesian coordinate system with spatial variable x.

One component, u_1

No.	Flux	Comments
(1)	$q_{x,u_1} = -D\dfrac{\partial u_1}{\partial x}$	Fick's first law
(2)	$q_{x,u_1} = -Du_1\dfrac{\partial u_1}{\partial x}$	nonlinear diffusion
(3)	$q_{x,u_1} = -f(u_1)\dfrac{\partial u_1}{\partial x}$	arbitrary function $f(u_1)$

u_1 is the concentration of the (one) component. Model (1) is the conventional, linear Fickian model with diffusivity D (a constant). Models (2),(3) are nonlinear extensions of (1). $f(u_1)$ is a positive function that can be used to define the diffusivity as a function of the concentration u_1.

The fluxes of (1) – (3) can be used in a conservation PDE,

$$\frac{\partial u_1}{\partial t} = -\frac{\partial q_{x,u_1}}{\partial x} \tag{A16.1}$$

For example, using (1), we have

$$\frac{\partial u_1}{\partial t} = -\frac{\partial \left(-D\dfrac{\partial u_1}{\partial x}\right)}{\partial x} = D\frac{\partial^2 u_1}{\partial x^2} \tag{A16.2}$$

which is Fick's second law or the diffusion equation (also known as Fourier's second law in heat transfer).

Similar substitutions of (2),(3) in Eq. (A16.1) give nonlinear versions of the diffusion equation. Analytical solutions for these nonlinear diffusion equations may not be available, but in principle, they can be integrated numerically.

The two-component models are listed in the following table, which gives the flux of component 1, q_{x,u_1}, as a function of the components, u_1, u_2, for a 1D, Cartesian coordinate system with spatial variable x.

Two components, u_1, u_2

No.	Flux, u_1	Comments
(4)	$q_{x,u_1} = -f_1(u_2)\dfrac{\partial u_1}{\partial x} + u_1 f_2(u_2)\dfrac{\partial u_2}{\partial x}$	chemotaxis 1 [2]
(5)	$q_{x,u_1} = -D_1\dfrac{\partial u_1}{\partial x} + D_2 u_1\dfrac{\partial u_2}{\partial x}$	chemotaxis 2 [3]
(6)	$q_{x,u_1} = -D_1\dfrac{\partial u_1}{\partial x} + D_2\left(\dfrac{u_1}{u_2}\right)\dfrac{\partial u_2}{\partial x}$	chemotaxis 3 [3]

Chemotaxis has been discussed in an extensive literature. A survey of PDE chemotaxis models is provided in [1].

References

1 Hillen, T., and K.J. Painter (2009), A user's guide to PDE models for chemotaxis, *Journal of Mathematical Biology*, **58**, no. 1-2, 183–217.
2 Keller, E.F., and L.A. Segel (1971), Traveling bands of chemotactic bacteria: a theoretical analysis, *Journal of Theoretical Biology*, **30**, no. 2, 235–248.
3 Murray, J.D. (2003), *Mathematical Biology, II: Spatial Models and Biomedical Applications*, 3rd ed., Springer-Verlag, New York.

17

Fitzhugh–Nagumo Equations

The Fitzhugh–Nagumo (FN) equations were originally reported as a 2×2 (two equations in two unknowns) ODE system to model the membrane potential of a nerve axon [1]. A diffusion term was then added for the conduction process of the action potential along a nerve, so that a 2×2 PDE system resulted. This PDE system can then be reduced to a 1×1 special case PDE for which an analytical solution is available, which is the starting point for the following spline collocation method of lines (SCMOL) solution.

17.1 PDE Model

The FN is

$$\frac{\partial u}{\partial t} = D\frac{\partial^2 u}{\partial x^2} + f(u); \quad f(u) = u(u-1)(a-u) \tag{17.1a,b}$$

with arbitrary constants a, D.

Equation (17.1a) is first order in t and second order in x. It therefore requires one initial condition (IC) and two boundary conditions (BCs).

$$u(x, t = 0) = g_1(x) \tag{17.2a}$$

$$u(x = x_l, t) = g_2(t); \quad u(x = x_u, t) = g_3(t) \tag{17.2b,c}$$

where $g_1(x), g_2(t), g_3(t)$ are prescribed functions.

An analytical solution is available for Eqs. (17.1) and (17.2) [1].

$$u(x, t) = \cfrac{1}{1 + \exp\left[\cfrac{x}{\sqrt{2D}} + (a - \frac{1}{2})t\right]} \tag{17.3}$$

Equation (17.3) can be used to define the IC and BCs.

$$u(x, t = 0) = g_1(x) = \cfrac{1}{1 + \exp\left[\cfrac{x}{\sqrt{2D}}\right]} \tag{17.2d}$$

Spline Collocation Methods for Partial Differential Equations: With Applications in R, First Edition.
William E. Schiesser.
© 2017 John Wiley & Sons, Inc. Published 2017 by John Wiley & Sons, Inc.
Companion website: www.wiley.com/go/Spline_Collocation

$$u(x = x_l, t) = g_2(t) = \cfrac{1}{1 + \exp\left[\cfrac{x_l}{\sqrt{2D}} + (a - \cfrac{1}{2})t\right]} \tag{17.2e}$$

$$u(x = x_u, t) = g_3(t) = \cfrac{1}{1 + \exp\left[\cfrac{x_u}{\sqrt{2D}} + (a - \cfrac{1}{2})t\right]} \tag{17.2f}$$

Equations (17.1)–(17.3) constitute the PDE model and are programmed in the following routines.

17.2 Main Program

A main program for Eqs. (17.1)–(17.3) follows.

```
#
# Fitzhugh-Nagumo
#
# Delete previous workspaces
  rm(list=ls(all=TRUE))
#
# Access ODE integrator
  library("deSolve");
#
# Access functions for numerical solution
  setwd("f:/collocation/chap17");
  source("pde_1.R");
#
# Grid in x
  n=101;xl=-60;xu=20;
  x=seq(from=xl,to=xu,by=(xu-xl)/(n-1));
#
# Parameters
  ncase=1;
  if(ncase==1){a=1;D=1  ;}
  if(ncase==2){a=1;D=0.1;}
#
# Independent variable for ODE integration
  nout=4;t0=0;tf=60;
  tout=seq(from=t0,to=tf,by=(tf-t0)/(nout-1));
#
# Initial condition
```

```
  u0=rep(0,n);sr=1/sqrt(2*D);
  for(i in 1:n){
    u0[i]=1/(1+exp(sr*x[i]));
  }
  ncall=0;
#
# ODE integration
  out=lsodes(y=u0,times=tout,func=pde_1,
      sparsetype="sparseint",rtol=1e-6,atol=1e-6,
      maxord=5);
  nrow(out)
  ncol(out)
#
# Arrays for numerical solution
   u=matrix(0,nrow=n,ncol=nout);
  ua=matrix(0,nrow=n,ncol=nout);
  t=rep(0,nout);
  for(it in 1:nout){
  for(i  in 1:n){
     t[it]=out[it,1];
    u[i,it]=out[it,i+1];
   ua[i,it]=1/(1+exp(sr*x[i]+(a-0.5)*t[it]));
  }
  }
#
# Display selected output
  cat(sprintf("\n a = %4.2f,  D = %4.2f\n",a,D));
  for(it in 1:nout){
    cat(sprintf("\n      t          x        u(x,t)
                  ua(x,t)           diff\n"));
    u[1,it]=1;u[n,it]=0;
    iv=seq(from=1,to=n,by=10);
    for(i in iv){
      diff=u[i,it]-ua[i,it];
      cat(sprintf("%6.2f%9.3f%12.6f%12.6f%12.6f\n",
              t[it],x[i],u[i,it],ua[i,it],diff));
    }
    cat(sprintf("\n"));
  }
  cat(sprintf(" ncall = %4d\n",ncall));
#
# Graphical output
#
```

```
# u, ua
    matplot(x,u,type="l",lwd=2,col="black",lty=1,
      xlab="x",ylab="u(x,t)",main="Fitzhugh-Nagumo");
    matpoints(x,ua,pch="o",col="black");
```

Listing 17.1 Main program for Eqs. (17.1)–(17.3).

We can note the following details about Listing 17.1:

- Previous workspaces are removed. Then the ODE integrator library deS-olve is accessed. Note that the setwd (set working directory) uses / rather than the usual \. The setwd requires editing for the local computer (to specify the directory with the R routines).

```
#
# Fitzhugh-Nagumo
#
# Delete previous workspaces
  rm(list=ls(all=TRUE))
#
# Access ODE integrator
  library("deSolve");
#
# Access functions for numerical solution
  setwd("f:/collocation/chap17");
  source("pde_1.R");
```

pde_1 is the routine for the MOL/ODEs (discussed subsequently).

- A uniform grid in x with 101 points is defined with the seq utility for the interval $x_l = -60 \le x \le x_u = 20$. Therefore, the vector x has the values $x = -60, -59.2, \dots, 20$.

```
#
# Grid in x
  n=101;xl=-60;xu=20;
  x=seq(from=xl,to=xu,by=(xu-xl)/(n-1));
```

The x interval is extended to the left since the solution of Eq. (17.3) is a wave traveling right to left. That is, the solution (Eq. (17.3)) is a function of the Lagrangian variable $\lambda = \dfrac{x}{\sqrt{2D}} + (a - \dfrac{1}{2})t$, and as t increases, x must decrease for λ to remain constant (for $a > 1/2$). For example, for $\lambda = 0$, from Eq. (17.3),

$$u(x,t) = \cfrac{1}{1 + \exp\left[\dfrac{x}{\sqrt{2D}} + (a - \dfrac{1}{2})t\right]} = \frac{1}{1 + \exp[0]} = \frac{1}{2}$$

and the solution $u(x, t) = 1/2$ requires x to decrease as t increases. This feature of the analytical and numerical solutions is evident in the graphical output that follows.

- The model parameters (constants) are defined numerically for two cases.

```
#
# Parameters
  ncase=1;
  if(ncase==1){a=1;D=1  ;}
  if(ncase==2){a=1;D=0.1;}
```

- A uniform grid in t of 4 output points is defined with the seq utility for the interval $0 \leq t \leq t_f = 60$. Therefore, the vector tout has the values $t = 0, 20, 40, 60$.

```
#
# Independent variable for ODE integration
  nout=4;t0=0;tf=60;
  tout=seq(from=t0,to=tf,by=(tf-t0)/(nout-1));
```

At this point, the intervals of x and t in Eqs. (17.1) are defined.

- The IC for Eqs. (17.1) is given by Eq. (17.2d) (with $t = 0$).

```
#
# Initial condition
  u0=rep(0,n);sr=1/sqrt(2*D);
  for(i in 1:n){
    u0[i]=1/(1+exp(sr*x[i]));
  }
  ncall=0;
```

The counter for the calls to the ODE routine pde_1 (discussed subsequently) is initialized.

- The system of 101 ODEs is integrated by the library integrator lsodes (available in deSolve) with the sparse matrix option specified. As expected, the inputs to lsodes are the ODE function, pde_1, the IC vector u0, and the vector of output values of t, tout. The length of u0 (101) informs lsodes how many ODEs are to be integrated. func,y,times are reserved names.

```
#
# ODE integration
  out=lsodes(y=u0,times=tout,func=pde_1,
      sparsetype="sparseint",rtol=1e-6,atol=1e-6,
      maxord=5);
  nrow(out)
  ncol(out)
```

The numerical solution to the ODEs is returned in matrix out. In this case, out has the dimensions 4×102 (displayed in the numerical output). The offset $101 + 1 = 102$ is required since the first element of each column has the output t (also in tout), and the 2, ..., 102 column elements have the 101 ODE solutions. This indexing of out is used next.

- The numerical solution for $u(x, t)$ is taken from the solution matrix out returned by lsodes. The intervals in t and x are covered by two nested fors (with indices it and i for t and x, respectively). The offset i+1 accounts for t returned in the first column position of out and placed in the vector t[it].

```
#
# Arrays for numerical solution
   u=matrix(0,nrow=n,ncol=nout);
   ua=matrix(0,nrow=n,ncol=nout);
   t=rep(0,nout);
   for(it in 1:nout){
   for(i   in 1:n){
      t[it]=out[it,1];
     u[i,it]=out[it,i+1];
    ua[i,it]=1/(1+exp(sr*x[i]+(a-0.5)*t[it]));
    }
   }
```

The analytical solution of Eq. (17.3) is also placed in ua.

- The parameters of the numerical and analytical solutions are displayed, followed by every tenth value of the solutions and their difference.

```
#
# Display selected output
   cat(sprintf("\n a = %4.2f,   D = %4.2f\n",a,D));
   for(it in 1:nout){
     cat(sprintf("\n      t              x          u(x,t)
                  ua(x,t)          diff\n"));
    u[1,it]=1;u[n,it]=0;

    iv=seq(from=1,to=n,by=10);
    for(i in iv){
      diff=u[i,it]-ua[i,it];
      cat(sprintf("%6.2f%9.3f%12.6f%12.6f%12.6f\n",
              t[it],x[i],u[i,it],ua[i,it],diff));
     }
    cat(sprintf("\n"));
   }
   cat(sprintf(" ncall = %4d\n",ncall));
```

The solution at $x = x_l = -60, x = x_u = 20$ is set to the boundary values since these values of u (explained subsequently) are not returned from lsodes (only dependent variables computed by integration of ODEs are returned by lsodes). The number of calls to pde_1 is displayed at the end of the solution.

- The numerical and analytical solutions are plotted with matplot and matpoints, respectively.

```
#
# Graphical output
#
# u, ua
    matplot(x,u,type="l",lwd=2,col="black",lty=1,
       xlab="x",ylab="u(x,t)",main="Fitzhugh-Nagumo");
    matpoints(x,ua,pch="o",col="black");
```

The ODE routine pde_1 called by lsodes in the main program of Listing 17.1 is considered next.

17.3 ODE Routine

pde_1 for Eqs. (17.1) and (17.2) follows.

```
    pde_1=function(t,u,parms){
#
# Function pde_1 computes the t derivative vector of
# u(x,t)
#
# BCs
  u[1]=1;
  u[n]=0;
#
# ux
  tablex=splinefun(x,u);
  ux=tablex(x,deriv=1);
#
# uxx
  tablexx=splinefun(x,ux);
  uxx=tablexx(x,deriv=1);
#
# PDE
  ut=rep(0,n);
  for(i in 1:n){
    f=u[i]*(u[i]-1)*(a-u[i]);
```

```
    ut [i] =D*uxx [i] +f;
    }
    ut [1] =0;ut [n] =0;
#
# Increment calls to pde_1
    ncall <<- ncall+1;
#
# Return derivative vector
    return (list (c (ut) ) ) ;
    }
```

Listing 17.2 ODE/MOL routine pde_1 for Eqs. (17.1) and (17.2).

We can note the following details about pde_1:

- The function is defined.

```
    pde_1=function (t,u,parms) {
#
# Function pde_1 computes the t derivative vector of
# u (x,t)
```

t is the current value of t in Eqs. (17.1). u is the 101-vector of ODE-dependent variables. parm is an argument to pass parameters to pde_1 (unused, but required in the argument list). The arguments must be listed in the order stated to properly interface with lsodes called in the main program. The derivative vector of the LHS of Eq. (17.1a) is calculated as follows and returned to lsodes.
- Dirichlet BCs at $x = x_l, x_u$ are used.

```
#
# BCs
    u [1] =1;
    u [n] =0;
```

Note the subscripts 1, n corresponding to $x = -60, 20$. These boundary values follow from Eq. (17.2e,f). For example, from Eq. (17.2e),
$\left[\dfrac{-60}{\sqrt{2D}} + (a - \dfrac{1}{2})t \right]$ is large and negative so the exp is small ($\ll 1$) and

$$u(x = x_l, t) = g_2(t) = \frac{1}{1 + \exp \left[\dfrac{-60}{\sqrt{2D}} + (a - \dfrac{1}{2})t \right]} \approx 1$$

Similarly, for Eq. (17.2f), $\left[\dfrac{20}{\sqrt{2D}} + (a - \dfrac{1}{2})t\right]$ is large and positive so the exp is large ($\gg 1$) and

$$u(x = x_u, t) = g_3(t) = \frac{1}{1 + \exp\left[\dfrac{20}{\sqrt{2D}} + (a - \dfrac{1}{2})t\right]} \approx 0$$

which are the two values boundary values ($u(x = x_l, t) = 1, u(x = x_u, t) = 0$) used in pde_1.

- The second derivative in Eq. (17.1a), $\dfrac{\partial^2 u}{\partial x^2}$, is computed by two first-order differentiations.

```
#
# ux
  tablex=splinefun(x,u);
  ux=tablex(x,deriv=1);
#
# uxx
  tablexx=splinefun(x,ux);
  uxx=tablexx(x,deriv=1);
```

- Equations (17.1) are programmed in a for that steps through x. The cubic nonlinearity of Eq. (17.1b) is computed first (as f).

```
#
# PDE
  ut=rep(0,n);
  for(i in 1:n){
    f=u[i]*(u[i]-1)*(a-u[i]);
    ut[i]=D*uxx[i]+f;
  }
  ut[1]=0;ut[n]=0;
```

Since the values $u(x = -60, t) = 1, u(x = 20, t) = 0$ are set as BCs, the boundary derivatives in t are set to zero so that the ODE integration by lsodes does not move the boundary values away from their prescribed values. Also, these boundary values are not returned to the calling program, so they are reset in the main program of Listing 17.1 (discussed previously) for the purpose of displaying the complete numerical solution (only dependent variables computed by integrating ODEs are returned by pde_1).

- The counter for the calls to pde_1 is returned to the main program of Listing 17.1 by the <<- operator.

```
#
# Increment calls to pde_1
  ncall <<- ncall+1;
```

- The derivative in ut is returned to lsodes.

```
#
# Return derivative vector
  return(list(c(ut)));
  }
```

list is used as required by lsodes. The final } concludes pde_1.

This completes the programming of Eqs. (17.1)–(17.3). The numerical and graphical output is considered next.

17.4 Model Output

Execution of the preceding main program and ODE routine pde_1 produces the following abbreviated numerical output in Table 17.1.

Table 17.1 Abbreviated output for Eqs. (17.1)–(17.3), ncase=1.

```
[1] 4

[1] 102

a = 1.00,   D = 1.00

      t        x        u(x,t)      ua(x,t)         diff
    0.00   -60.000    1.000000     1.000000      0.000000
    0.00   -52.000    1.000000     1.000000      0.000000
    0.00   -44.000    1.000000     1.000000      0.000000
    0.00   -36.000    1.000000     1.000000      0.000000
    0.00   -28.000    1.000000     1.000000      0.000000
    0.00   -20.000    0.999999     0.999999      0.000000
    0.00   -12.000    0.999794     0.999794      0.000000
    0.00    -4.000    0.944193     0.944193      0.000000
    0.00     4.000    0.055807     0.055807      0.000000
    0.00    12.000    0.000206     0.000206      0.000000
    0.00    20.000    0.000000     0.000001     -0.000001
                         .            .
                         .            .
                         .            .
           Output for t = 20, 40 removed
```

Table 17.1 (Continued)

.
.
.

t	x	u(x,t)	ua(x,t)	diff
60.00	-60.000	1.000000	0.999996	0.000004
60.00	-52.000	0.998857	0.998853	0.000004
60.00	-44.000	0.753207	0.752632	0.000575
60.00	-36.000	0.010546	0.010517	0.000029
60.00	-28.000	0.000038	0.000037	0.000000
60.00	-20.000	0.000000	0.000000	-0.000000
60.00	-12.000	0.000000	0.000000	-0.000000
60.00	-4.000	0.000000	0.000000	-0.000000
60.00	4.000	-0.000000	0.000000	-0.000000
60.00	12.000	-0.000000	0.000000	-0.000000
60.00	20.000	0.000000	0.000000	-0.000000

ncall = 575

We can note the following details about this output:

- As expected, the solution array out has the dimensions 4 x 102 for 4 output points and 101 ODEs, with t included as an additional element, that is, 101 + 1 = 102.
- The interval in x is $((20 - (-60))/(101 - 1) = 0.8$ (with every tenth value in Table 17.1).
- The numerical and analytical ICs at $t = 0$ agree exactly since the analytical solution of Eq. (17.3) is used in the main program of Listing 17.1 to define the numerical IC.
- The numerical and analytical solutions at $x = -60, 20$ agree exactly with the BCs set in pde_1.
- The agreement between the numerical and analytical solutions is quite good given the relatively rapid change in the moving front solution.
- The computational effort is modest, ncall = 575.

The agreement of the numerical and analytical solutions is displayed graphically in Figure 17.1.

The traveling wave characteristic (dependent on only the Lagrangian variable $\lambda = \dfrac{x}{\sqrt{2D}} + (a - \dfrac{1}{2})t$) is clear (the use of "characteristic" has a mathematical interpretation since λ is the characteristic of Eqs. (17.1)).

The solution for ncase=2 (in Listing 17.1) follows.

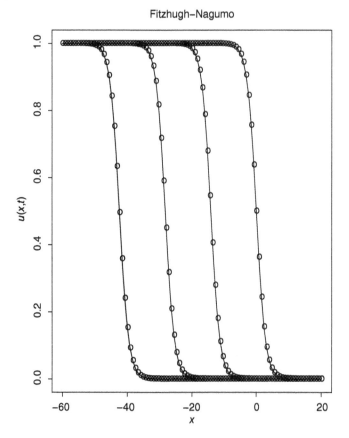

Figure 17.1 Numerical and analytical $u(x, t)$ from Eqs. (17.1)–(17.3) ncase=1, lines – num, points – anal.

We can note the following details about this output:

- With the decrease in the diffusivity from D = 1, ncase = 1 to D = 0.1, ncase = 2, the front of the traveling wave sharpens (see Figure 17.2).
- The sharp front requires $n = 101$ for adequate spatial resolution.
- The errors in the numerical solution appear to be relatively large, for example, from Table 17.2,

t	x	u(x,t)	ua(x,t)	diff
60.00	-12.000	0.052510	0.040419	0.012091

However, these errors occur on the front, which has a rapid variation in $u(x, t)$ over a narrow band (range) in x. Thus, the errors are not as significant as might be anticipated (see Figure 17.2).

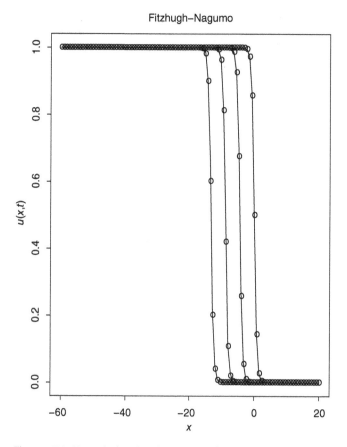

Figure 17.2 Numerical and analytical $u(x, t)$ from Eqs. (17.1)–(17.3) ncase=2, lines – num, points – anal.

In other words, the agreement between the numerical and analytical solutions is quite acceptable given the relatively rapid change in the moving front solution.

- The computational effort is modest, ncall = 700.

A spatial grid with $n = 101$ gives adequate resolution of the sharpening front, but for smaller values of D, additional points would be required, and as the solution approaches a finite discontinuity, the MOL would be expected to fail (so that a nonlinear approximation would be required, e.g., flux limiter).

In general, this is an expected requirement for a parabolic–hyperbolic PDE as the PDE becomes more (strongly) hyperbolic. Alternatively, for a strongly hyperbolic PDE, a small amount of dispersion might be added to stabilize the calculation of the solution, which is the so-called *viscosity method*. In physical terms, most systems have a certain amount of dispersion (few physical systems

Table 17.2 Abbreviated output for Eqs. (17.1)–(17.3), ncase=2.

---

[1] 4

[1] 102

a = 1.00,   D = 0.10

t	x	u(x,t)	ua(x,t)	diff
0.00	-60.000	1.000000	1.000000	0.000000
0.00	-52.000	1.000000	1.000000	0.000000
0.00	-44.000	1.000000	1.000000	0.000000
0.00	-36.000	1.000000	1.000000	0.000000
0.00	-28.000	1.000000	1.000000	0.000000
0.00	-20.000	1.000000	1.000000	0.000000
0.00	-12.000	1.000000	1.000000	0.000000
0.00	-4.000	0.999870	0.999870	0.000000
0.00	4.000	0.000130	0.000130	0.000000
0.00	12.000	0.000000	0.000000	0.000000
0.00	20.000	0.000000	0.000000	-0.000000

.
.
.

Output for t = 20, 40 removed

.
.
.

t	x	u(x,t)	ua(x,t)	diff
60.00	-60.000	1.000000	1.000000	0.000000
60.00	-52.000	1.000000	1.000000	0.000000
60.00	-44.000	1.000001	1.000000	0.000001
60.00	-36.000	1.000008	1.000000	0.000008
60.00	-28.000	1.000077	1.000000	0.000077
60.00	-20.000	1.000479	1.000000	0.000479
60.00	-12.000	0.052510	0.040419	0.012091
60.00	-4.000	-0.000008	0.000000	-0.000008
60.00	4.000	-0.000000	0.000000	-0.000000
60.00	12.000	0.000000	0.000000	0.000000
60..00	20.000	0.000000	0.000000	-0.000000

ncall =   700

---

display discontinuous characteristics), so the error of the viscosity method may be small (relative to the physical dispersion) and therefore the computed solution is valid. Some physical intuition, reasoning, and experimentation may be required to resolve the problem of resolving steep moving fronts.

17.5 Summary and Conclusions

The programming of the FN Eqs. (17.1) within the SCMOL framework is straightforward, with reasonable computational effort required to produce numerical solutions for the particular parameter values (ncase = 1,2). The cubic nonlinearity of Eq. (17.1b) is particularly noteworthy, although in principle, other nonlinear forms can be readily accommodated numerically. The agreement of the numerical and analytical solutions in Tables 17.1 and 17.2 and Figures 17.1 and 17.2 indicates acceptable accuracy for the cases considered.

Reference

1 Griffiths, G.W., and W.E. Schiesser (2012), *Traveling Wave Analysis of Partial Differential Equations*, Chapter 9, Elsevier, Burlington, MA.

18

Euler–Poisson–Darboux Equation

The Euler–Poisson–Darboux (EPD) equation is a variant of the linear wave equation (see Section 2.5 and Chapter 6) with a damping term that is a function of t. Numerous applications of the EPD equation have been reported, such as sound wave propagation. Here we focus on the spline collocation method of lines (SCMOL) solution of the EPD equation to demonstrate some unique features of this equation.

18.1 PDE Model

The EPD equation is [1, 2]

$$\frac{\partial^2 u}{\partial t^2} + \frac{\lambda}{t}\frac{\partial u}{\partial t} = \nabla^2 u \tag{18.1}$$

The variables, parameters, and operators are shown in Table 18.1.

Note in particular the damping term $\dfrac{\lambda}{t}\dfrac{\partial u}{\partial t}$ with the variable coefficient in t.

Equation (18.1) is considered in 1D spherical coordinates.

$$\frac{\partial^2 u}{\partial t^2} + \frac{\lambda}{t}\frac{\partial u}{\partial t} = \frac{1}{r^2}\frac{\partial}{\partial r}\left(r^2\frac{\partial u}{\partial r}\right) = \frac{\partial^2 u}{\partial r^2} + \frac{2}{r}\frac{\partial u}{\partial r} \tag{18.2}$$

with $0 \le r \le r_0$.

Equation (18.2) has a combination of a singularity in t $\left(\dfrac{\lambda}{t}\dfrac{\partial u}{\partial t}\text{ for } t = 0\right)$ and a singularity in r $\left(\dfrac{1}{r^2}\dfrac{\partial}{\partial r}\left(r^2\dfrac{\partial u}{\partial r}\right)\text{ for } r = 0\right)$.

Equation (18.2) is second order in t and second order in r. It therefore requires two initial conditions (ICs) and two boundary conditions (BCs).

$$u(r, t = 0) = f(r); \quad \frac{\partial u(r, t = 0)}{\partial r} = 0 \tag{18.3a,b}$$

where $f(r)$ is a function to be specified.

$$\frac{\partial u(r = 0, t)}{\partial r} = \frac{\partial u(r = r_0, t)}{\partial r} = 0 \tag{18.3c,d}$$

Spline Collocation Methods for Partial Differential Equations: With Applications in R, First Edition.
William E. Schiesser.
© 2017 John Wiley & Sons, Inc. Published 2017 by John Wiley & Sons, Inc.
Companion website: www.wiley.com/go/Spline_Collocation

Table 18.1 Variables, parameters, and operators in Eq. (18.1).

u	dependent variable
∇	coordinate-free spatial differential operator
∇^2	Laplacian operator
t	time
λ	arbitrary constant

Equations (18.3c,d) are homogeneous Neumann BCs.

Equations (18.2) and (18.3) constitute the PDE system. The MOL solution is implemented in the R routines discussed next.

18.2 Main Program

The following main program is for Eqs. (18.2) and (18.3).

```
#
# Euler-Poisson-Darboux
#
# Delete previous workspaces
  rm(list=ls(all=TRUE))
#
# Access ODE integrator, routines
  library(deSolve);
  setwd("f:/collocation/chap18");
  source("pde_1a.R");
#
# Select case
  ncase=1;
#
# Parameters
  if(ncase==1){lam=0;c=5;};
  if(ncase==2){lam=1;c=5;};
#
# Spatial interval
  n=51;r0=1;
  r=seq(from=0,to=r0,by=(r0-0)/(n-1));
#
# Time interval
```

```
  t0=0;tf=0.5;nout=6;
  tout=seq(from=t0,to=tf,by=(tf-t0)/(nout-1));
  ncall=0;
#
# Initial conditions (ICs)
  u0=rep(0,2*n);
  for(i in 1:n){
    u0[i]=exp(-c*r[i]^2);
    u0[i+n]=0;
  }
#
# ODE integration
  out=lsodes(y=u0,times=tout,func=pde_1a,
      sparsetype="sparseint",rtol=1e-6,atol=1e-6,
      maxord=5);
  nrow(out)
  ncol(out)
#
# Store numerical solution for plotting
  u1=matrix(0,nrow=n,ncol=nout);
  u2=matrix(0,nrow=n,ncol=nout);
   t=rep(0,nout);
  for(it in 1:nout){
    t[it]=out[it,1];
  for(i in 1:n){
    u1[i,it]=out[it,i+1];
    u2[i,it]=out[it,i+1+n];
  }
  }
#
# Numerical solution
  for(it in 1:nout){
  cat(sprintf("\n        t         r     u(r,t)"));
  iv=seq(from=1,to=n,by=5);
  for(i in iv){
      cat(sprintf("\n %6.3f%8.3f%9.4f",
              t[it],r[i],u1[i,it]));
  }
    cat(sprintf("\n"));
  }
  cat(sprintf("\n ncall = %4d\n",ncall));
#
# Plot numerical solution
```

```
matplot(r,u1,type="l",lwd=2,col="black",
        lty=1,xlab="r",ylab="u1(r,t)",
        main="Euler-Poisson-Darboux");
```

Listing 18.1 Main program for Eqs. (18.2) and (18.3).

We can note the following details about Listing 18.1:

- Previous workspaces are removed. Then the ODE integrator library deSolve is accessed. Note that the setwd (set working directory) uses / rather than the usual \. The setwd requires editing for the local computer (to specify the directory with the R routines).

```
#
# Euler-Poisson-Darboux
#
# Delete previous workspaces
  rm(list=ls(all=TRUE))
#
# Access ODE integrator, routines
  library(deSolve);
  setwd("f:/collocation/chap18");
  source("pde_1a.R");
```

pde_1a is the routine for the MOL/ODEs (discussed subsequently).

- Two cases are programmed for no damping (lam=0 for the classical linear wave equation to initially test the R implementation) and with damping (lam=1). The choice of the nonzero value of lam is explained subsequently.

```
#
# Select case
  ncase=1;
#
# Parameters
  if(ncase==1){lam=0;c=5;};
  if(ncase==2){lam=1;c=5;};
```

c is the constant in the IC Gaussian function (Eq. (18.3a))

$$f(r) = e^{-cr^2} \tag{18.3e}$$

- A uniform grid in r of 51 points is defined with the seq utility for the interval $r = 0 \le r \le r_0 = 1$. Therefore, the vector r has the values $r = 0, 0.02, \ldots, 1$.

```
#
# Spatial interval
```

```
n=51;r0=1;
r=seq(from=0,to=r0,by=(r0-0)/(n-1));
```

- A uniform grid in t of 6 output points is defined with the seq utility for the interval $0 \le t \le t_f = 0.5$. Therefore, the vector tout has the values $t = 0, 0.1, \dots, 0.5$.

```
#
# Time interval
  t0=0;tf=0.5;nout=6;
  tout=seq(from=t0,to=tf,by=(tf-t0)/(nout-1));
  ncall=0;
```

At this point, the intervals of r and t in Eqs. (18.2) and (18.3) are defined. Also, the counter for the calls to the MOL/ODE routine pde_1a is initialized (and passed to pde_1a without a special designation).

- The ICs, Eqs. (18.3a,b), are defined.

```
#
# Initial conditions (ICs)
  u0=rep(0,2*n);
  for(i in 1:n){
    u0[i]=exp(-c*r[i]^2);
    u0[i+n]=0;
  }
```

These ICs reflect the use of two dependent variables u_1, u_2, defined as

$$u(r, t) = u_1(r, t); \quad \frac{\partial u(r, t)}{\partial t} = u_2(r, t)$$

This approach to PDEs second order in t is discussed in Section 2.5. The total number of first-order ODEs is therefore $2(51) = 102$.

- The system of 102 MOL/ODEs is integrated by the library integrator lsodes (available in deSolve) with the sparse matrix option specified. As expected, the inputs to lsodes are the ODE function, pde_1a, the IC vector u0, and the vector of output values of t, tout. The length of u0 (102) informs lsodes how many ODEs are to be integrated. func, y, times are reserved names.

```
#
# ODE integration
  out=lsodes(y=u0,times=tout,func=pde_1a,
      sparsetype="sparseint",rtol=1e-6,atol=1e-6,
      maxord=5);
  nrow(out)
  ncol(out)
```

The numerical solution to the ODEs is returned in matrix out. In this case, out has the dimensions $nout \times (n + 1) = 6 \times 103$. The offset $n + 1$ is required since the first element of each column has the output t (also in tout), and the $2, \ldots, n + 1 = 2, \ldots, 103$ column elements have the 102 ODE solutions. This indexing of out is used next.

- The numerical u_1, u_2 are taken from the solution matrix out returned by lsodes. The intervals in t and r are covered by two nested fors (with indices it and i for t and r, respectively). The offset i+1 accounts for t placed in the first column position of out (t [it]=out [it,1]).

```
#
# Store numerical solution for plotting
  u1=matrix(0,nrow=n,ncol=nout);
  u2=matrix(0,nrow=n,ncol=nout);
   t=rep(0,nout);
  for(it in 1:nout){
    t[it]=out[it,1];
  for(i in 1:n){
    u1[i,it]=out[it,i+1];
    u2[i,it]=out[it,i+1+n];
  }
  }
```

- The numerical solution for $u = u_1$ ($u(r,t)$ from Eq. (18.2)) is displayed.

```
#
# Numerical solution
  for(it in 1:nout){
  cat(sprintf("\n        t        r    u(r,t)"));
  iv=seq(from=1,to=n,by=5);
  for(i in iv){
      cat(sprintf("\n %6.3f%8.3f%9.4f",
              t[it],r[i],u1[i,it]));
  }
    cat(sprintf("\n"));
  }
  cat(sprintf("\n ncall = %4d\n",ncall));
```

To conserve space, only every fifth value in r of the solution is displayed numerically (using the subscript iv). Thus, the displayed spacing in r is $5(0.02) = 0.1$. Finally, the number of calls to the ODE/MOL routine pde_1a (considered next) is displayed as a measure of the computational effort required for computing the numerical solution.

- The numerical and analytical solutions for $u = u_1$ are plotted with the `matplot` utility.

```
#
# Plot numerical solution
  matplot(r,u1,type="l",lwd=2,col="black",
          lty=1,xlab="r",ylab="u1(r,t)",
          main="Euler-Poisson-Darboux");
```

This concludes the discussion of the main program in Listing 18.1. The ODE/MOL routine `pde_1a` called by `lsodes` in the main program of Listing 18.1 is considered next.

18.3 ODE Routine

`pde_1a` for Eq. (18.2) follows.

```
  pde_1a=function(t,u,parms){
#
# Function pde_1a computes the t derivative vectors
# for u1(r,t), u2(r,t)
#
# One vector to two vectors
  u1=rep(0,n);u2=rep(0,n);
  for(i in 1:n){
   u1[i]=u[i];
   u2[i]=u[i+n];
   }
#
# u1r
  table1=splinefun(r,u1);
  u1r=table1(r,deriv=1);
#
# Neumann BCs
  u1r[1]=0;u1r[n]=0;
#
# u1rr
  table2=splinefun(r,u1r);
  u1rr=table2(r,deriv=1);
#
# PDE
  u1t=rep(0,n);u2t=rep(0,n);
```

```
for(i in 1:n){
  if(t< 1.0e-04){
    if(i==1){
      u1t[i]=u2[i];
      u2t[i]=1/(1+lam)*3*u1rr[i];
    }
    if(i> 1){
      u1t[i]=u2[i];
      u2t[i]=1/(1+lam)*(u1rr[i]+(2/r[i])*
             u1r[i]);
    }
  }
  if(t>=1.0e-04){
    if(i==1){
      u1t[i]=u2[i];
      u2t[i]=-lam/t*u2[i]+3*u1rr[i];
    }
    if(i> 1){
      u1t[i]=u2[i];
      u2t[i]=-lam/t*u2[i]+(u1rr[i]+(2/r[i])*
             u1r[i]);
    }
  }
}
#
# Two vectors to one vector
  ut=rep(0,2*n);
  for(i in 1:n){
      ut[i]=u1t[i];
    ut[i+n]=u2t[i];
  }
#
# Increment calls to pde_1a
  ncall <<- ncall+1;
#
# Return derivative vector
  return(list(c(ut)))
}
```

Listing 18.2 ODE/MOL routine pde_1a for Eq. (18.2).

We can note the following details about pde_1a:

- The function is defined.

```
  pde_1a=function(t,u,parms){
#
# Function pde_1a computes the t derivative vectors
# for u1(r,t), u2(r,t)
```

t is the current value of t in Eq. (18.2). u is the ODE-dependent variable $2(51) = 102$-vector. parm is an argument to pass parameters to pde_1a (unused, but required in the argument list). The arguments must be listed in the order stated to properly interface with lsodes called in the main program.

- u is placed in two vectors, u1, u2 in accordance with $u_1(r, t) = u(r, t)$,

$$u_2(r, t) = \frac{\partial u_1(r, t)}{\partial t} = \frac{\partial u(r, t)}{\partial t}.$$

```
#
# One vector to two vectors
  u1=rep(0,n);u2=rep(0,n);
  for(i in 1:n){
    u1[i]=u[i];
    u2[i]=u[i+n];
    }
```

- $\dfrac{\partial u_1}{\partial r}$ in Eqs. (18.2) is computed with splinefun.

```
#
# u1r
  table1=splinefun(r,u1);
  u1r=table1(r,deriv=1);
```

- BCs (18.3c,d) are used.

```
#
# Neumann BCs
  u1r[1]=0;u1r[n]=0;
```

- $\dfrac{\partial^2 u_1}{\partial r^2}$ in Eq. (18.2) is computed with splinefun by differentiating $\dfrac{\partial u_1}{\partial r}$ (stagewise differentiation).

```
#
# u1rr
  table2=splinefun(r,u1r);
  u1rr=table2(r,deriv=1);
```

- Equation (18.2) is programmed through a series of steps.
  - A for is used to step through r with index i. The derivative vectors in t result when the for is completed, $\dfrac{\partial u_1}{\partial t} = $ u1t, $\dfrac{\partial u_2}{\partial t} = $ u2t.

```
#
# PDE
  u1t=rep(0,n);u2t=rep(0,n);
  for(i in 1:n){
```

- Two intervals in t are programmed for the variable coefficient term $\dfrac{\lambda}{t}\dfrac{\partial u}{\partial t}$ in Eq. (18.2). For the singularity in t, by l'Hospital's rule and IC (18.3b),

$$\frac{\lambda}{t}\frac{\partial u}{\partial t} = \lambda\frac{\partial^2 u}{\partial t^2}$$

so that

$$\frac{\partial^2 u}{\partial t^2} + \frac{\lambda}{t}\frac{\partial u}{\partial t} = (1+\lambda)\frac{\partial^2 u}{\partial t^2}$$

For the singularity in r, by l'Hospital's rule and BC (18.3c),

$$\frac{2}{r}\frac{\partial u}{\partial r} = 2\frac{\partial^2 u}{\partial r^2}$$

so that

$$\frac{\partial^2 u}{\partial r^2} + \frac{2}{r}\frac{\partial u}{\partial r} = 3\frac{\partial^2 u}{\partial r^2}$$

- The programming of these two special cases is therefore for $t < 10^{-4}$ (corresponding to $t = 0$) and $r = 0$,

```
if(i==1){
   u1t[i]=u2[i];
   u2t[i]=1/(1+lam)*3*u1rr[i];
```

- For $t < 10^{-4}$ (corresponding to $t = 0$) and $r > 0$,

```
if(i> 1){
   u1t[i]=u2[i];
   u2t[i]=1/(1+lam)*(u1rr[i]+(2/r[i])*
       u1r[i]);
}
```

The singularity in t is handled as before and $\dfrac{\partial^2 u}{\partial r^2} + \dfrac{2}{r}\dfrac{\partial u}{\partial r}$ is programmed as (u1rr[i]+(2/r[i])*u1r[i]).
- For $t > 10^{-4}$ (corresponding to $t > 0$) and $r = 0$,

```
if(t>=1.0e-04){
   if(i==1){
      u1t[i]=u2[i];
      u2t[i]=-lam/t*u2[i]+3*u1rr[i];
   }
```

The singularity in r is handled as before and division by t is possible since $t \neq 0$. Also, $\dfrac{\partial u}{\partial t} = u2$.

- For $t > 10^{-4}$ (corresponding to $t > 0$) and $r > 0$,

```
if(i> 1){
    u1t[i]=u2[i];
    u2t[i]=-lam/t*u2[i]+(u1rr[i]+(2/r[i])*
            u1r[i]);
    }
  }
}
```

The programming of Eq. (18.2) does not involve a singularity in t or r. Two final }s are required to complete the `if` for $t = 0, \neq 0$ and the `for` in `i` (over the grid in r).

To conclude this discussion of the programming of Eq. (18.2), we can note the following details:

- For $\lambda = 0$, the coding is for the undamped wave equation in 1D spherical coordinates.
- For $\lambda > 0, t = 0$, damping is added through the RHS factor `1/(1+lam)`.
- For $\lambda > 0, t > 0$, damping is added through the additional RHS term `-lam/t*u2[i]`. The contribution of $\dfrac{\partial u}{\partial t} (= u2)$ in Eq. (18.2) is clear. The damping decreases with increasing t.

- This completes the programming of Eq. (18.2) in terms of u_1, u_2. The derivatives $\dfrac{\partial u_1}{\partial t} = $ `u1t`, $\dfrac{\partial u_2}{\partial t} = $ `u2t` are placed in one 102-vector, `ut`, to return to `lsodes` called in the main program of Listing 18.1.

```
#
# Two vectors to one vector
  ut=rep(0,2*n);
  for(i in 1:n){
      ut[i]=u1t[i];
    ut[i+n]=u2t[i];
  }
```

- The counter for the calls to `pde_1a` is incremented and its value is returned to the calling program (of Listing 18.1) with the «- operator.

```
#
# Increment calls to pde_1a
  ncall <<- ncall+1;
```

- The derivative vector `ut` is returned to `lsodes`, which requires a list. `c` is the R vector utility. The combination of `return`, `list`, `c` gives `lsodes` (the ODE integrator called in the main program of Listing 18.1) the required derivative vector for the next step along the solution.

```
#
# Return derivative vector
  return(list(c(ut)))
}
```

The final } concludes pde_1a.

This completes the programming of Eqs. (18.2) and (18.3). In summary, the switching value $t = 10^{-4}$ is used to distinguish between $t = 0$ and $t > 0$ during the integration (stepping in t) by lsodes. This value is selected arbitrarily as a value less than the time step used by lsodes in moving away from $t = 0$. It is not a sensitive parameter affecting the numerical solution of Eq. (18.2).

The numerical and graphical output from the R routines in Listings 18.1 and 18.2 is considered next.

18.4 Model Output

Abbreviated numerical output in Table 18.2a is for $\lambda = 0$ (no damping in Eq. (18.2)).

Table 18.2a Abbreviated output for Eqs. (18.2) and (18.3), $\lambda = 0$.

[1] 6

[1] 103

t	r	u(r,t)
0.000	0.000	1.0000
0.000	0.100	0.9512
0.000	0.200	0.8187
0.000	0.300	0.6376
0.000	0.400	0.4493
0.000	0.500	0.2865
0.000	0.600	0.1653
0.000	0.700	0.0863
0.000	0.800	0.0408

Table 18.2a (Continued)

t	r	u(r,t)
0.000	0.900	0.0174
0.000	1.000	0.0067

t	r	u(r,t)
0.100	0.000	0.8561
0.100	0.100	0.8187
0.100	0.200	0.7160
0.100	0.300	0.5725
0.100	0.400	0.4182
0.100	0.500	0.2789
0.100	0.600	0.1697
0.100	0.700	0.0941
0.100	0.800	0.0476
0.100	0.900	0.0221
0.100	1.000	0.0164

.
.
.

Output for t = 0.2,
0.3, 0.4 removed

.
.
.

t	r	u(r,t)
0.500	0.000	-0.4298
0.500	0.100	-0.4028
0.500	0.200	-0.3272
0.500	0.300	-0.2186
0.500	0.400	-0.0993
0.500	0.500	0.0074
0.500	0.600	0.0941
0.500	0.700	0.1452
0.500	0.800	0.1683
0.500	0.900	0.1750
0.500	1.000	0.1758

ncall = 418

We can note the following details about this output:

- The solution array `out` has the dimensions `6 x 103` for 6 output points and 102 ODEs, with t included as an additional element, that is, `102 + 1 = 103`.
- The interval in r is $((1 - (-0))/(51 - 1) = 0.02$ (with every fifth value in Table 18.2a).
- The output interval in t is 0.1 as specified in the main program of Listing 18.1.
- The solution continues to oscillate (Eq. (18.2) is the linear wave equation with $\lambda = 0$), and as can be inferred from negative values of $u(r, t)$ in the graphical output of Figure 18.1a (this can be verified numerically by using a larger interval in t).

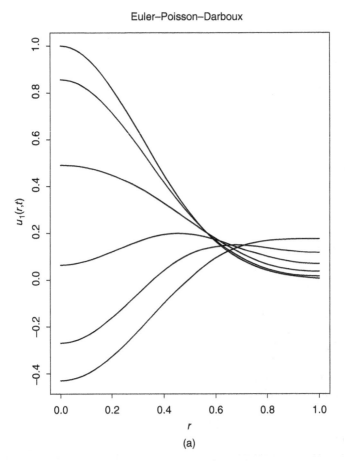

Euler–Poisson–Darboux

Figure 18.1 Numerical solution of Eqs. (18.2) and (18.3), (a) $\lambda = 0$. (b) $\lambda = 1$.

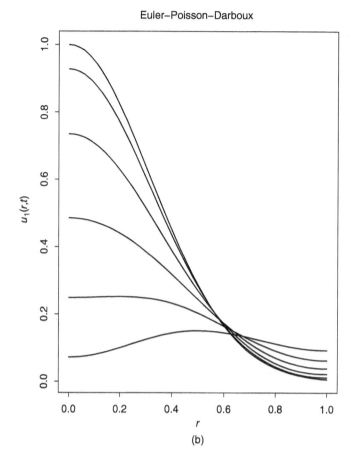

Figure 18.1 (*Continued*)

- The computational effort is modest, ncall = 418.

For $\lambda = 1$, the abbreviated numerical output is displayed in Table 18.2b. We can note the following details about this output:

- The solution is damped, and there are no negative values of $u(r, t)$ in Table 18.2b (as in Table 18.2a). This damping of the solution is apparent in Figure 18.1b (and by comparing Figures 18.2a,b). The long-term solution $u(r, t \rightarrow \infty) = 0$ can be inferred from Figure 18.1b (and can be verified numerically by using a larger interval in t).
- The computational effort is modest, ncall = 340.

Table 18.2b Abbreviated output for Eqs.
(18.2) and (18.3), $\lambda = 1$.

[1] 6

[1] 103

t	r	u(r,t)
0.000	0.000	1.0000
0.000	0.100	0.9512
0.000	0.200	0.8187
0.000	0.300	0.6376
0.000	0.400	0.4493
0.000	0.500	0.2865
0.000	0.600	0.1653
0.000	0.700	0.0863
0.000	0.800	0.0408
0.000	0.900	0.0174
0.000	1.000	0.0067

t	r	u(r,t)
0.100	0.000	0.9273
0.100	0.100	0.8843
0.100	0.200	0.7669
0.100	0.300	0.6048
0.100	0.400	0.4337
0.100	0.500	0.2828
0.100	0.600	0.1676
0.100	0.700	0.0903
0.100	0.800	0.0442
0.100	0.900	0.0197
0.100	1.000	0.0125

.
.
.

Output for t = 0.2,
0.3, 0.4 removed

.
.
.

t	r	u(r,t)
0.500	0.000	0.0723
0.500	0.100	0.0800

Table 18.2b (Continued)

0.500	0.200	0.1001
0.500	0.300	0.1247
0.500	0.400	0.1441
0.500	0.500	0.1508
0.500	0.600	0.1453
0.500	0.700	0.1301
0.500	0.800	0.1119
0.500	0.900	0.0981
0.500	1.000	0.0931

ncall = 340

18.5 Summary and Conclusions

The SCMOL can accommodate the singularities in t and r of the Euler–Poisson –Darboux equation (18.2). For $\lambda = 1$, the solution is clearly damped (compare Figures 18.1a,b). Although Eq. (18.2) is linear, the analytical solutions that have been reported [1, 2] are complicated and therefore are not discussed here.

References

1 Blum, E.K. (1954), The Euler–Poisson–Darboux equation in the exceptional cases, *Proceedings of the American Mathematical Society*, **5**, 511–520.
2 Bresters, D.W. (1973), On the equation of Euler–Poisson–Darboux, *SIAM Journal on Mathematical Analysis*, **4**, no. 1, 31–41.

19

Kuramoto–Sivashinsky Equation

The Kuramoto–Sivashinsky (KS) equation has applications in fluid mechanics, including as a basic description for turbulence [1, 3, 4]. It is also a useful test problem for numerical methods with the following features:

- Higher order (up to $\dfrac{\partial^4 u}{\partial x^4}$ in x)
- Nonlinear (with a convective term $u\dfrac{\partial u}{\partial x}$)
- Dirichlet and Neumann BCs
- Traveling wave solutions
- Available exact (analytical) solutions (to test numerical solutions).

These features are demonstrated in the following discussion.

19.1 PDE Model

The KS is [2]

$$\frac{\partial u}{\partial t} = -u\frac{\partial u}{\partial x} - \alpha\frac{\partial^2 u}{\partial x^2} - \beta\frac{\partial^3 u}{\partial x^3} - \gamma\frac{\partial^4 u}{\partial x^4} \tag{19.1}$$

Equation (19.1) is first order in t and fourth order in x. It therefore requires one initial condition (IC) and four boundary conditions (BCs).

$$u(x, t = 0) = f(x) \tag{19.2}$$

$$u(x = x_l, t) = f_1(t); \quad u(x = x_u, t) = f_2(t) \tag{19.3a,b}$$

$$\frac{\partial u(x = x_l, t)}{\partial x} = f_3(t); \quad \frac{\partial u(x = x_u, t)}{\partial x} = f_4(t) \tag{19.3c,d}$$

where x_l, x_u are the boundary points in x, and $f_1(t), f_2(t), f_3(t), f_4(t)$ are functions to be specified.

Two analytical solutions are available to test the spline collocation method of lines (SCMOL) solution.

Spline Collocation Methods for Partial Differential Equations: With Applications in R, First Edition.
William E. Schiesser.
© 2017 John Wiley & Sons, Inc. Published 2017 by John Wiley & Sons, Inc.
Companion website: www.wiley.com/go/Spline_Collocation

`ncase = 1`:

$$u_a(x, t) = c_0 + c_1 f + c_2 f^2 + c_3 f^3 \tag{19.4a}$$

with

$$f = \frac{k}{1 + e^{-kx - \lambda t}} \tag{19.4b}$$

For Neumann BCs (19.3c,d), the x derivative of the solution is required.

$$\frac{\partial u_a(x, t)}{\partial x} = c_1 \frac{\partial f}{\partial x} + 2 c_2 f \frac{\partial f}{\partial x} + 3 c_3 f^2 \frac{\partial f}{\partial x} \tag{19.4c}$$

with

$$\frac{\partial f}{\partial x} = \frac{k^2 e^{-kx - \lambda t}}{(1 + e^{-kx - \lambda t})^2} \tag{19.4d}$$

$u_a(x, t)$ of Eqs. (19.4) is a traveling wave solution, that is, a function of the Lagrangian variable $-kx - \lambda t$ (a linear combination of x and t).

`ncase = 2`:

The analytical solution is available ([2], p593) for the parameter values $\beta = 0$, $\alpha = \gamma = 1, k = \pm\sqrt{\dfrac{11}{19}}$.

$$u_a(x, t) = \frac{15}{19} k(11 H^3 - 9H + 2); \quad H = \tanh\left(\frac{1}{2} kx - \frac{15}{19} k^2 t\right) \tag{19.5a}$$

For Neumann BCs (19.3c,d), the x derivative of the solution is required.

$$\frac{\partial u}{\partial x} = \frac{15}{19} k\left(33 H^2 \frac{\partial H}{\partial x} - 9 \frac{\partial H}{\partial x}\right) \tag{19.5b}$$

with

$$\frac{\partial H}{\partial x} = \frac{1}{2} k\left(1 - \tanh^2\left(\frac{1}{2} kx - \frac{15}{19} k^2 t\right)\right) \tag{19.5c}$$

$u_a(x, t)$ of Eqs. (19.5) is a traveling wave solution, that is, a function of the Lagrangian variable $\dfrac{1}{2} kx - \dfrac{15}{19} k^2 t$ (a linear combination of x and t).

The two analytical solutions, `ncase=1, 2`, are programmed in the R routines that follow, starting with the main program.

19.2 Main Program

The following main program is for Eqs. (19.1)–(19.5).

```
#
# Kuramoto-Sivashinsky
#
```

```
# Delete previous workspaces
  rm(list=ls(all=TRUE))
#
# Access ODE integrator, routines
  library(deSolve);
  setwd("f:/collocation/chap19");
  source("pde_1a.R");source("ua_1.R");
  source("uax_1.R");
#
# Select case
  ncase=1;
  if(ncase==1){
    alpha=1;gamma=1;c0=1;
    beta=4*(alpha*gamma)^0.5;
    k=(alpha/gamma)^0.5;
    lambda=-c0*k-(3/2)*beta*k^3;
    c1=(15/76)*(16*alpha-beta^2/gamma)+
       15*beta*k+60*gamma*k^2;
    c2=-(15*beta+180*gamma*k);
    c3=60*gamma;
#
#   Spatial grid
    n=51;xl=0;xu=1;
    x=seq(from=xl,to=xu,by=(xu-xl)/(n-1));
#
#   Independent variable for ODE integration
    t0=0;tf=0.15;nout=6;
    tout=seq(from=t0,to=tf,by=(tf-t0)/(nout-1));
    ncall=0;
  }
#
  if(ncase==2){
    alpha=1;beta=0;gamma=1;c0=0;
    k=(11/19)^0.5;
#
#   Spatial grid
    n=101;xl=-10;xu=20;
    x=seq(from=xl,to=xu,by=(xu-xl)/(n-1));
#
#   Independent variable for ODE integration
    t0=0;tf=10;nout=6;
    tout=seq(from=t0,to=tf,by=(tf-t0)/(nout-1));
    ncall=0;
```

```
  }
#
# Initial condition (IC)
  u0=rep(0,n);
  for(i in 1:n){
    u0[i]=ua_1(x[i],0);
  }
#
# ODE integration
  out=lsodes(y=u0,times=tout,func=pde_1a,
      sparsetype="sparseint",rtol=1e-6,atol=1e-6,
      maxord=5);
  nrow(out)
  ncol(out)
#
# Store numerical, analytical solutions for plotting
  u=matrix(0,nrow=n,ncol=nout);
  ua=matrix(0,nrow=n,ncol=nout);
  t=rep(0,nout);
  for(it in 1:nout){
    t[it]=out[it,1];
  for(i in 1:n){
    u[i,it]=out[it,i+1];
    ua[i,it]=ua_1(x[i],t[it]);
  }
  u[1,it]=ua_1(x[1],t[it]);
  u[n,it]=ua_1(x[n],t[it]);
  }
#
# Numerical solution
  for(it in 1:nout){
  cat(sprintf("\n         t         x    u(x,t)
                ua(x,t)       diff"));
  if(ncase==1){iv=seq(from=1,to=n,by=5);}
  if(ncase==2){iv=seq(from=1,to=n,by=10);}
  for(i in iv){
    diff=u[i,it]-ua[i,it];
    cat(sprintf("\n %6.3f%8.3f%9.4f%9.4f%9.4f",
                t[it],x[i],u[i,it],ua[i,it],diff));
  }
    cat(sprintf("\n"));
  }
  cat(sprintf("\n ncall = %4d\n",ncall));
```

```
#
# Plot numerical solution
  matplot(x,u,type="l",lwd=2,col="black",
          lty=1,xlab="x",ylab="u(x,t)",
          main="Kuramoto-Sivashinsky");
   matpoints(x,ua,pch="o",col="black");
#
# Plot 3D numerical solution
    persp(x,t,u,theta=55,phi=45,xlim=c(xl,xu),
          ylim=c(t0,t[nout]),xlab="x",ylab="t",
          zlab="u(x,t)");
```

Listing 19.1 Main program for Eqs. (19.1)–(19.5).

We can note the following details about Listing 19.1:

- Previous workspaces are removed. Then the ODE integrator library deS-olve is accessed. Note that the setwd (set working directory) uses / rather than the usual \. The setwd requires editing for the local computer (to specify the directory with the R routines).

```
#
# Kuramoto-Sivashinsky
#
# Delete previous workspaces
  rm(list=ls(all=TRUE))
#
# Access ODE integrator, routines
  library(deSolve);
  setwd("f:/collocation/chap19");
  source("pde_1a.R");source("ua_1.R");
  source("uax_1.R");
```

pde_1a is the routine for the MOL/ODEs (discussed subsequently). ua_1, uax_1.R are functions for the analytical solutions, Eqs. (19.4) and (19.5).

- Two cases are programmed. For ncase=1, Eqs. (19.4) are programmed.

```
#
# Select case
  ncase=1;
  if(ncase==1){
    alpha=1;gamma=1;c0=1;
    beta=4*(alpha*gamma)^0.5;
    k=(alpha/gamma)^0.5;
```

```
        lambda=-c0*k-(3/2)*beta*k^3;
        c1=(15/76)*(16*alpha-beta^2/gamma)+
            15*beta*k+60*gamma*k^2;
        c2=-(15*beta+180*gamma*k);
        c3=60*gamma;
#
#       Spatial grid
        n=51;xl=0;xu=1;
        x=seq(from=xl,to=xu,by=(xu-xl)/(n-1));
#
#       Independent variable for ODE integration
        t0=0;tf=0.15;nout=6;
        tout=seq(from=t0,to=tf,by=(tf-t0)/(nout-1));
        ncall=0;
      }
```

We can note the following details about this programming:

- The parameters in Eq. (19.1) are defined numerically ($\alpha, \beta, \gamma, \lambda, k$). The constants c_0, c_1, c_2, c_3 in Eqs. (19.4) are also computed. These parameters are available to subordinate routines pde_1a, ua_1, uax_1 (discussed subsequently) without any special designation (a feature of R).
- A uniform grid in x of 51 points is defined with the seq utility for the interval $x_l = 0 \leq x \leq x_u = 1$. Therefore, the vector x has the values $x = 0, 0.02, \ldots, 1$.

```
#
#       Spatial grid
        n=51;xl=0;xu=1;
        x=seq(from=xl,to=xu,by=(xu-xl)/(n-1));
```

- A uniform grid in t of 6 output points is defined with the seq utility for the interval $t_0 = 0 \leq t \leq t_f = 0.15$. Therefore, the vector tout has the values $t = 0, 0.03, \ldots, 0.15$.

```
#
#       Independent variable for ODE integration
        t0=0;tf=0.15;nout=6;
        tout=seq(from=t0,to=tf,by=(tf-t0)/(nout-1));
        ncall=0;
      }
```

At this point, the intervals of x and t in Eqs. (19.1)–(19.4) are defined. Also, the counter for the calls to the MOL/ODE routine pde_1a is initialized (and passed to pde_1a without a special designation).

- The programming is similar for ncase=2 (Eqs. (19.5) are programmed). A uniform grid in x of 101 points is defined with the seq utility for the interval $x_l = -10 \le x \le x_u = 20$ so that the vector x has the values $x = -10, -9.70, \dots, 20$.
- A uniform grid in t of 6 output points is defined with the seq utility for the interval $t_0 = 0 \le t \le t_f = 10$ so the vector tout has the values $t = 0, 2, \dots, 10$.

```
#
   if(ncase==2){
     alpha=1;beta=0;gamma=1;c0=0;
     k=(11/19)^0.5;
#
#    Spatial grid
     n=101;xl=-10;xu=20;
     x=seq(from=xl,to=xu,by=(xu-xl)/(n-1));
#
#    Independent variable for ODE integration
     t0=0;tf=10;nout=6;
     tout=seq(from=t0,to=tf,by=(tf-t0)/(nout-1));
     ncall=0;
   }
```

The number of MOL/ODEs is increased from 51 (ncase=1) to 101 (ncase=2) to improve the spatial resolution in x (this will be clear when the graphical output in Figures 19.1a and 19.2a is discussed). Also, the time intervals are different for ncase=1,2 because of the difference in the solutions for ncase=1,2. At this point, the intervals of x and t in Eqs. (19.1)–(19.3) and (19.5) are defined as $-10 \le x \le 20, 0 \le t \le 10$.

- The IC, Eq. (19.2), is taken as the analytical solution of Eqs. (19.4) or (19.5) (depending on ncase) with $t = 0$.

```
#
# Initial condition (IC)
  u0=rep(0,n);
  for(i in 1:n){
    u0[i]=ua_1(x[i],0);
  }
```

- The system of MOL/ODEs is integrated by the library integrator lsodes (available in deSolve) with the sparse matrix option specified. As expected, the inputs to lsodes are the ODE function, pde_1a, the IC vector u0, and the vector of output values of t, tout. The length of u0 (51 for ncase=1, 101 for ncase=2) informs lsodes how many ODEs are to be integrated. func,y,times are reserved names.

```
#
# ODE integration
  out=lsodes(y=u0,times=tout,func=pde_1a,
    sparsetype="sparseint",rtol=1e-6,atol=1e-6,
    maxord=5);
  nrow(out)
  ncol(out)
```

The numerical solution to the ODEs is returned in matrix out. In this case, out has the dimensions $nout \times (n+1) = 6 \times 52$ for ncase=1 or $nout \times (n+1) = 6 \times 102$ for ncase=2. The offset $n+1$ is required since the first element of each column has the output t (also in tout), and the $2, \ldots, n+1 = 2, \ldots, 52$ or 102 column elements have the ODE solutions. This indexing of out is used next.

- The numerical $u(x,t)$ is taken from the solution matrix out returned by lsodes. The intervals in t and x are covered by two nested fors (with indices it and i for t and x, respectively). The offset i+1 accounts for t placed in the first column position of out (t[it]=out[it,1]).

```
#
# Store numerical, analytical solutions for plotting
  u=matrix(0,nrow=n,ncol=nout);
  ua=matrix(0,nrow=n,ncol=nout);
  t=rep(0,nout);
  for(it in 1:nout){
    t[it]=out[it,1];
  for(i in 1:n){
    u[i,it]=out[it,i+1];
    ua[i,it]=ua_1(x[i],t[it]);
  }
  u[1,it]=ua_1(x[1],t[it]);
  u[n,it]=ua_1(x[n],t[it]);
  }
```

The analytical solution from ua_1 is placed in array ua for comparison with the numerical solution in u. The numerical solution at $x = x_l, x = x_u$ is placed in u with ua_1 since these boundary values are not returned from lsodes in out (only dependent variables from the integration of ODEs in pde_1a are returned, but the boundary values are set with ua_1 in pde_1a).

- The numerical and analytical solutions, and their difference, are displayed.

```
#
# Numerical solution
  for(it in 1:nout){
```

```
cat (sprintf ("\n        t        x    u (x, t)
                ua (x, t)      diff")) ;
if (ncase==1) {iv=seq (from=1, to=n, by=5) ; }
if (ncase==2) {iv=seq (from=1, to=n, by=10) ; }
for (i in iv) {
   diff=u [i, it] -ua [i, it] ;
   cat (sprintf ("\n %6.3f%8.3f%9.4f%9.4f%9.4f",
                t [it] , x [i] , u [i, it] , ua [i, it] , diff)) ;
}
   cat (sprintf ("\n")) ;
}
cat (sprintf ("\n ncall = %4d\n", ncall)) ;
```

Every fifth or tenth value is displayed for ncase=1, 2, respectively. Finally, the number of calls to the ODE/MOL routine pde_1a (considered next) is displayed as a measure of the computational effort required for computing the numerical solution.

- The numerical and analytical solutions are plotted in 2D with matplot, matpoints, and in 3D with persp.

```
#
# Plot numerical solution
  matplot (x, u, type="l", lwd=2, col="black",
           lty=1, xlab="x", ylab="u (x, t) ",
           main="Kuramoto-Sivashinsky") ;
    matpoints (x, ua, pch="o", col="black") ;
#
# Plot 3D numerical solution
    persp (x, t, u, theta=55, phi=45, xlim=c (xl, xu) ,
           ylim=c (t0, t [nout] ) , xlab="x", ylab="t",
           zlab="u (x, t) ") ;
```

This concludes the discussion of the main program in Listing 19.1. The ODE/MOL routine pde_1a called by lsodes in the main program of Listing 19.1 is considered next.

19.3 ODE Routine

pde_1a for Eq. (19.1) follows.

```
  pde_1a=function (t, u, parms) {
#
# Function pde_1a computes the t derivative vector
```

```
# for u(x,t)
#
# Dirichlet BCs at x = xl,xu
  u[1]=ua_1(x[1],t);
  u[n]=ua_1(x[n],t);
#
# ux
  tablex=splinefun(x,u);
  ux=tablex(x,deriv=1);
#
# Neumann BCs at x = xl,xu
  ux[1]=uax_1(x[1],t);
  ux[n]=uax_1(x[n],t);
#
# uxx
  tablexx=splinefun(x,ux);
  uxx=tablexx(x,deriv=1);
#
# uxxx
  tablexxx=splinefun(x,uxx);
  uxxx=tablexxx(x,deriv=1);
#
# uxxxx
  tablexxxx=splinefun(x,uxxx);
  uxxxx=tablexxxx(x,deriv=1);
#
# PDE
  ut=rep(0,n);
  for(i in 2:(n-1)){
    ut[i]=-u[i]*ux[i]-alpha*uxx[i]-
          beta*uxxx[i]-gamma*uxxxx[i];
  }
  ut[1]=0;ut[n]=0;
#
# Increment calls to pde_1a
  ncall <<- ncall+1;
#
# Return derivative vector
  return(list(c(ut)))
}
```

Listing 19.2 ODE/MOL routine pde_1a for Eq. (19.1).

We can note the following details about pde_1a:

- The function is defined.

```
pde_1a=function(t,u,parms){
#
# Function pde_1a computes the t derivative vector
# for u(x,t)
```

t is the current value of t in Eq. (19.1). u is the 51 (ncase=1) or 101 (ncase=2) vector of ODE dependent variables. parm is an argument to pass parameters to pde_1a (unused, but required in the argument list). The arguments must be listed in the order stated to properly interface with lsodes called in the main program.

- BCs (19.3a,b) are implemented with ua_1 (discussed subsequently).

```
#
# Dirichlet BCs at x = xl,xu
  u[1]=ua_1(x[1],t);
  u[n]=ua_1(x[n],t);
```

The subscripts 1, n correspond to $x = x_l, x_u$ at the boundaries.

- $\dfrac{\partial u}{\partial x}$ in Eq. (19.1) is computed with splinefun.

```
#
# ux
  tablex=splinefun(x,u);
  ux=tablex(x,deriv=1);
```

- BCs (19.3c,d) are implemented with uax_1 (discussed subsequently).

```
#
# Neumann BCs at x = xl,xu
  ux[1]=uax_1(x[1],t);
  ux[n]=uax_1(x[n],t);
```

The subscripts 1, n correspond to $x = x_l, x_u$ at the boundaries.

- $\dfrac{\partial^2 u}{\partial x^2}, \dfrac{\partial^3 u}{\partial x^3}, \dfrac{\partial^4 u}{\partial x^4}$ in Eq. (19.1) are computed with splinefun by successive (stagewise) differentiation.

```
#
# uxx
  tablexx=splinefun(x,ux);
  uxx=tablexx(x,deriv=1);
#
# uxxx
```

```
    tablexxx=splinefun(x,uxx);
    uxxx=tablexxx(x,deriv=1);
#
# uxxxx
    tablexxxx=splinefun(x,uxxx);
    uxxxx=tablexxxx(x,deriv=1);
```

- Equation (19.1) is programmed at the interior points (using $2 : (n-1)$).

```
#
# PDE
    ut=rep(0,n);
    for(i in 2:(n-1)){
      ut[i]=-u[i]*ux[i]-alpha*uxx[i]-
             beta*uxxx[i]-gamma*uxxxx[i];
    }
    ut[1]=0;ut[n]=0;
```

The derivatives in t are set to zero at the boundaries to avoid having lsodes move them away from the values prescribed by BCs (19.3a,b).

- The counter for the calls to pde_1a is incremented and its value is returned to the calling program (of Listing 19.1) with the <<- operator.

```
#
# Increment calls to pde_1a
    ncall <<- ncall+1;
```

- The derivative vector ut is returned to lsodes, which requires a list. c is the R vector utility. The combination of return, list, c gives lsodes the required derivative vector for the next step along the solution.

```
#
# Return derivative vector
    return(list(c(ut)))
}
```

The final } concludes pde_1a.

This completes the programming of Eqs. (19.1)–(19.3). The subordinate routines for Eqs. (19.4) and (19.5) follow.

19.4 Subordinate Routines

The following routines are straightforward implementations of the corresponding equations noted in the titles (captions) of the listings (with switching based on ncase). The various parameters are defined in the main program of Listing 19.1. The arguments (x, t) are the independent variables (x, t) in Eq. (19.1). The analytical solutions are returned numerically to the calling routines with return, c (a list is not required as in Listing 19.2).

```
ua_1=function(x,t){
#
# Function ua_1 computes two analytical solutions
# of the Kuramoto-Sivashinsky equation
#
# Analytical solution
  if(ncase==1){
    F=k/(1+exp(-k*x-lambda*t));
    ua=c0+c1*F+c2*F^2+c3*F^3;
  }
  if(ncase==2){
    H=tanh(0.5*k*x-(15/19)*k^2*t);
    ua=(15/19)*k*(11*H^3-9*H+2);
  }
#
# Return analytical solution
  return(c(ua))
}
```

Listing 19.3a ua_1 for Eqs. (19.4a,b) and (19.5a).

```
uax_1=function(x,t){
#
# Function uax_1 computes the derivative of two
# analytical solutions of the Kuramoto-Sivashinsky
# equation
#
# Derivative of analytical solution
  if(ncase==1){
    F=k/(1+exp(-k*x-lambda*t));
    Fx=(k^2)*(1+exp(-k*x-lambda*t))^(-2)*
        exp(-k*x-lambda*t);
    uax=c1*Fx+2*c2*F*Fx+3*c3*F^2*Fx;
  }
  if(ncase==2){
    arg=0.5*k*x-(15/19)*k^2*t;
    H=tanh(arg);
```

```
    Hx=(1-tanh(arg)^2)*0.5*k;
    uax=(15/19)*k*(33*H^2*Hx-9*Hx);
  }
#
# Return analytical solution derivative
  return(c(uax))
}
```

Listing 19.3b uax_1 for Eqs. (19.4c,d) and (19.5b,c).

This completes the programming of Eqs. (19.1)–(19.5). The numerical and graphical output from the R routines is considered next.

19.5 Model Output

Abbreviated numerical output for ncase=1 is in Table 19.1.

Table 19.1 Abbreviated numerical output for ncase=1.

```
[1]  6

[1]  52

      t        x     u(x,t)    ua(x,t)      diff
   0.000    0.000    8.5000     8.5000    0.0000
   0.000    0.100    6.5339     6.5339    0.0000
   0.000    0.200    4.3974     4.3974    0.0000
   0.000    0.300    2.1103     2.1103    0.0000
   0.000    0.400   -0.3048    -0.3048    0.0000
   0.000    0.500   -2.8237    -2.8237    0.0000
   0.000    0.600   -5.4212    -5.4212    0.0000
   0.000    0.700   -8.0717    -8.0717    0.0000
   0.000    0.800  -10.7503   -10.7503    0.0000
   0.000    0.900  -13.4329   -13.4329    0.0000
   0.000    1.000  -16.0975   -16.0975    0.0000
             .                    .
             .                    .
             .                    .
```

Table 19.1 (Continued)

```
      Output for t = 0.03 to 0.120 removed

                   .                  .
                   .                  .
                   .                  .
      t        x      u(x,t)    ua(x,t)       diff
   0.150    0.000    17.0247    17.0247     0.0000
   0.150    0.100    17.1082    17.1012     0.0070
   0.150    0.200    17.0943    17.0244     0.0699
   0.150    0.300    16.8925    16.7775     0.1150
   0.150    0.400    16.4935    16.3457     0.1478
   0.150    0.500    15.8574    15.7165     0.1409
   0.150    0.600    15.0006    14.8805     0.1201
   0.150    0.700    13.9043    13.8319     0.0723
   0.150    0.800    12.6011    12.5689     0.0322
   0.150    0.900    11.0847    11.0939    -0.0093
   0.150    1.000     9.4138     9.4138     0.0000

   ncall =   462
```

We can note the following details about this output:

- The solution array out has the dimensions 6 x 52 for 6 output points and 51 ODEs, with t included as an additional element, that is, $51 + 1 = 52$.
- The interval in x is $((1 - (-0))/(51 - 1) = 0.02$ (with every fifth value in Table 19.1).
- The output interval in t is 0.03 as specified in the main program of Listing 19.1.
- The ICs for the numerical and analytical solutions are the same since ua_1 was used in Listing 19.1 to define the numerical IC (with ncase = 1).
- The numerical solution is in acceptable agreement with the analytical solution (from the values of diff at $t = 0.15$). Thus, the succession of four first-order differentiations in pde_1 of Listing 19.2 is valid.
- The computational effort is modest, ncall = 462.

The agreement between the numerical and analytical solutions is clear in Figure 19.1a (with numerical as solid lines and analytical as points).

The 3D graphical output from persp is in Figure 19.1b.

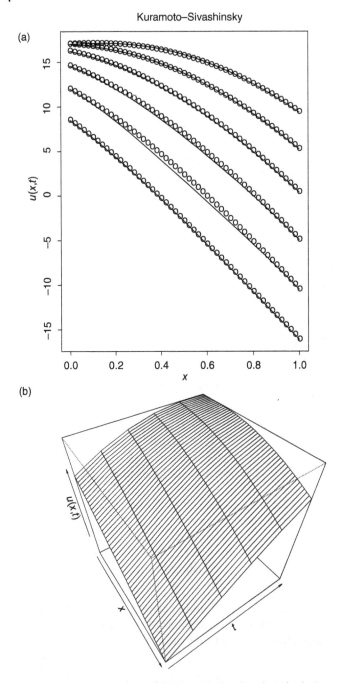

Figure 19.1 (a) Comparison of the numerical and analytical solutions of Eq. (19.1) ncase = 1, solid – num, points – anal. (b) Numerical solution to Eq. (19.1), ncase=1.

Abbreviated numerical output for ncase=2 is in Table 19.2.

Table 19.2 Abbreviated numerical output for ncase=2.

```
[1] 6

[1] 102

      t         x     u(x,t)   ua(x,t)      diff
  0.000  -10.000   0.0143    0.0143    0.0000
  0.000   -7.000   0.1377    0.1377    0.0000
  0.000   -4.000   1.1527    1.1527    0.0000
  0.000   -1.000   2.8481    2.8481    0.0000
  0.000    2.000  -0.5221   -0.5221    0.0000
  0.000    5.000   1.8117    1.8117    0.0000
  0.000    8.000   2.3378    2.3378    0.0000
  0.000   11.000   2.3961    2.3961    0.0000
  0.000   14.000   2.4021    2.4021    0.0000
  0.000   17.000   2.4027    2.4027    0.0000
  0.000   20.000   2.4028    2.4028    0.0000
                      .                     .
                      .                     .
                      .                     .
       Output for t = 2 to t = 8 removed
                      .                     .
                      .                     .
                      .                     .
      t         x     u(x,t)   ua(x,t)      diff
 10.000  -10.000   0.0000    0.0000    0.0000
 10.000   -7.000  -0.0002    0.0000   -0.0002
 10.000   -4.000   0.0002    0.0001    0.0000
 10.000   -1.000   0.0018    0.0014    0.0003
 10.000    2.000   0.0144    0.0141    0.0002
 10.000    5.000   0.1365    0.1362    0.0003
 10.000    8.000   1.1424    1.1426   -0.0002
 10.000   11.000   2.8616    2.8608    0.0008
 10.000   14.000  -0.5302   -0.5306    0.0004
 10.000   17.000   1.8054    1.8059   -0.0005
 10.000   20.000   2.3372    2.3372    0.0000

 ncall =   652
```

We can note the following details about this output:

- The solution array out has the dimensions 6 x 102 for 6 output points and 101 ODEs, with t included as an additional element, that is, 101 + 1 = 102.
- The interval in x is $((20 - (-10))/(101 - 1) = 0.3$ (with every tenth value in Table 19.1).
- The output interval in t is 2 as specified in the main program of Listing 19.1.
- The ICs for the numerical and analytical solutions are the same since ua_1 was used in Listing 19.1 to define the numerical IC (with ncase = 2).
- The numerical solution is in good agreement with the analytical solution (from the values of diff at $t = 10$). Thus, the succession of four first-order differentiations in pde_1 of Listing 19.2 is valid.
- The computational effort is modest, ncall = 652.

The agreement between the numerical and analytical solutions is clear in Figure 19.2a (with numerical as solid lines and analytical as points).

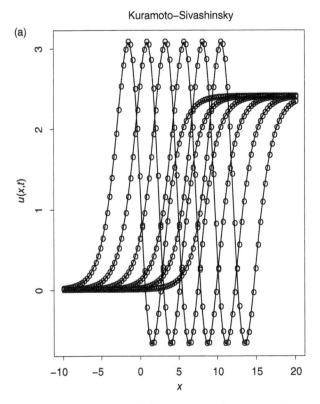

(a)

Kuramoto–Sivashinsky

Figure 19.2 (a) Comparison of the numerical and analytical solutions of Eq. (19.1) ncase = 2, solid – num, points – anal. (b) Numerical solution to Eq. (19.1), ncase=2.

(b)

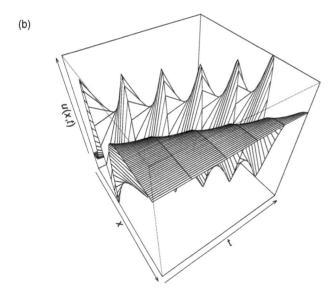

Figure 19.2 *(Continued)*

In particular, the maximum and minimum (peak) values of $u(x, t)$ are essentially constant with increasing t. This is a stringent test of the use of spline-fun in pde_1a. These peak values could be studied in detail by considering the numerical output.

The 3D graphical output from persp is in Figure 19.2b.

19.6 Summary and Conclusions

The SCMOL can accommodate the higher order derivatives in x in the Kuramoto–Sivashinsky equation (19.1). For ncase=2, the traveling wave solution of Eqs. (19.5) is clear in Figure 19.2a (the solution for ncase=1 is also a traveling wave, but this is not as apparent in Figure 19.1a). In both cases, the solutions are a function of a Lagrangian variable (a linear combination of x and t), as defined in Eqs. (19.4) and (19.5).

The choice of $n = 51,101$ for ncase=1, 2 gives acceptable spatial resolution in x, but these grid specifications could be changed to observe the effect on the accuracy of the numerical solution (the difference between the numerical and analytical solutions in the tabular numerical output). This is an example of h refinement to study the errors in numerical solutions.

However, the analytical solution is not actually required, but rather, the effect of changes in the number of spatial grid points on the numerical solution can be observed (by comparing numerical solutions for differing numbers of grid points). This is an important method for assessing the accuracy of a numerical solution when an analytical solution is not available (the usual case in most applications).

References

1 Kuramoto, Y. (1978), Diffusion-induced chaos in reaction systems, *Supplement of the Progress of Theoretical Physics*, **64**, 346–367.

2 Polyanin, A.D., and V.F. Zaitsev (2004), *Handbook of Nonlinear Partial Differential Equations*, Chapman & Hall/CRC, Boca Raton, FL.

3 Sivashinsky, G.I. (1977), Nonlinear analysis of hydrodynamic instability in laminar flames - I. Derivation of basic equations, *Acta Astronautica*, **4**, 1177–1206.

4 Sivashinsky, G.I. (1983), Instabilities, pattern formation, and turbulence in flames, *Annual Reviews of Fluid Mechanics*, **15**, 179–99.

20

Einstein–Maxwell Equations

The Einstein–Maxwell (EM) equations are a 2×2 (two equations in two unknowns) system of hyperbolic PDEs in cylindrical coordinates with angular and axial symmetry [1]. They have the form of a damped wave equation, so we consider an extension of the analysis in Chapter 6.

20.1 PDE Model

The EM equations are ([1], eqs. (22.61))

$$\frac{\partial^2 u}{\partial t^2} = \frac{\partial^2 u}{\partial r^2} + \frac{1}{r}\frac{\partial u}{\partial r} - e^{2u}\left[\left(\frac{\partial v}{\partial t}\right)^2 - \left(\frac{\partial v}{\partial r}\right)^2\right] \qquad (20.1a)$$

$$\frac{\partial^2 v}{\partial t^2} = \frac{\partial^2 v}{\partial r^2} + \frac{1}{r}\frac{\partial v}{\partial r} - 2\left[\frac{\partial u}{\partial r}\frac{\partial v}{\partial r} - \frac{\partial u}{\partial t}\frac{\partial v}{\partial t}\right] \qquad (20.1b)$$

where

u, v	field variables
r	radial coordinate
t	time.

Equations (20.1) are second order in r and t, and, therefore, each requires two initial conditions (ICs) and two boundary conditions (BCs).

$$u(r, t = 0) = f_1(r); \quad \frac{\partial u(r, t = 0)}{\partial t} = f_2(r) \qquad (20.2a,b)$$

$$v(r, t = 0) = g_1(r); \quad \frac{\partial v(r, t = 0)}{\partial t} = g_2(r) \qquad (20.2c,d)$$

$$\frac{\partial u(r = 0, t)}{\partial r} = 0; \quad \frac{\partial u(r = r_0, t)}{\partial r} = 0 \qquad (20.3a,b)$$

$$\frac{\partial v(r = 0, t)}{\partial r} = 0; \quad \frac{\partial v(r = r_0, t)}{\partial r} = 0 \qquad (20.3c,d)$$

Equations (20.1)–(20.3) are the 2×2 PDE system for $u(r, t), v(r, t)$ computed by the following routines.

Spline Collocation Methods for Partial Differential Equations: With Applications in R, First Edition.
William E. Schiesser.
© 2017 John Wiley & Sons, Inc. Published 2017 by John Wiley & Sons, Inc.
Companion website: www.wiley.com/go/Spline_Collocation

20.2   Main Program

A main program for Eqs. (20.1)–(20.3) follows.

```
#
# Einstein-Maxwell
#
# Delete previous workspaces
  rm(list=ls(all=TRUE))
#
# Access ODE integrator
  library("deSolve");
#
# Access functions for numerical solution
  setwd("f:/collocation/chap20");
  source("pde_1.R");
#
# Grid in r
  n=51;rl=0;ru=1;
  r=seq(from=rl,to=ru,by=(ru-rl)/(n-1));
#
# Parameters
  ncase=1;
  if(ncase==1){c=10;c1=0;c2=0;}
  if(ncase==2){c=10;c1=1;c2=1;}
#
# Independent variable for ODE integration
  nout=6;t0=0;tf=1;
  tout=seq(from=t0,to=tf,by=(tf-t0)/(nout-1));
#
# ICs
  u0=rep(0,4*n);
  for(i in 1:n){
    u0[i]     =exp(-c*r[i]^2);
    u0[i+n]   =0;
    u0[i+2*n] =exp(-c*r[i]^2);
    u0[i+3*n] =0;
  }
  ncall=0;
#
# ODE integration
  out=lsodes(y=u0,times=tout,func=pde_1,
      sparsetype="sparseint",rtol=1e-6,atol=1e-6,
      maxord=5);
```

```
  nrow(out)
  ncol(out)
#
# Arrays for numerical solution
  u1=matrix(0,nrow=n,ncol=nout);
  u2=matrix(0,nrow=n,ncol=nout);
  v1=matrix(0,nrow=n,ncol=nout);
  v2=matrix(0,nrow=n,ncol=nout);
  t=rep(0,nout);
  for(it in 1:nout){
  for(i  in 1:n){
    u1[i,it]=out[it,i+1];
    u2[i,it]=out[it,i+1+n];
    v1[i,it]=out[it,i+1+2*n];
    v2[i,it]=out[it,i+1+3*n];
       t[it]=out[it,1];
  }
  }
#
# Display selected output
  cat(sprintf("\n    c = %4.2f,   c1 = %4.2f,
              c2 = %4.2f\n",c,c1,c2));
  for(it in 1:nout){
    iv=seq(from=1,to=n,by=10);
    for(i in iv){
    cat(sprintf("\n        t         r      u1(r,t)
              u2(r,t)"));
    cat(sprintf("\n        t         r      v1(r,t)
              v2(r,t)\n"));
      cat(sprintf("%6.2f%8.3f%12.6f%12.6f\n",
            t[it],r[i],u1[i,it],u2[i,it]));
      cat(sprintf("%6.2f%8.3f%12.6f%12.6f\n",
            t[it],r[i],v1[i,it],v2[i,it]));
    }
   }
  cat(sprintf(" ncall = %4d\n",ncall));
#
# Plot 2D numerical solution
  matplot(r,u1,type="l",lwd=2,col="black",
         lty=1,xlab="r",ylab="u1(r,t)",
         main="Einstein-Maxwell");
  matplot(r,u2,type="l",lwd=2,col="black",
```

```
            lty=1,xlab="r",ylab="u2(r,t)",
            main="Einstein-Maxwell");
    matplot(r,v1,type="l",lwd=2,col="black",
            lty=1,xlab="r",ylab="v1(r,t)",
            main="Einstein-Maxwell");
    matplot(r,v2,type="l",lwd=2,col="black",
            lty=1,xlab="r",ylab="v2(r,t)",
            main="Einstein-Maxwell");
```

Listing 20.1 Main program for Eqs. (20.1)–(20.3).

We can note the following details about Listing 20.1:

- Previous workspaces are removed. Then the ODE integrator library deS-olve is accessed. Note that the setwd (set working directory) uses / rather than the usual \. Also, setwd requires editing for the local computer (to specify the directory for the R routines).

```
#
# Einstein-Maxwell
#
# Delete previous workspaces
  rm(list=ls(all=TRUE))
#
# Access ODE integrator
  library("deSolve");
#
# Access functions for numerical solution
  setwd("f:/collocation/chap20");
  source("pde_1.R");
```

pde_1 is the routine for the MOL/ODEs (discussed subsequently).
- A uniform grid in r with 51 points is defined with the seq utility for the interval $r_l = 0 \le r \le r_u = 1$. Therefore, the vector r has the values $r = 0, 0.02, \ldots, 1$.

```
#
# Grid in r
  n=51;rl=0;ru=1;
  r=seq(from=rl,to=ru,by=(ru-rl)/(n-1));
```

- The model parameters (constants) are defined numerically.

```
#
# Parameters
  ncase=1;
```

```
if(ncase==1){c=10;c1=0;c2=0;}
if(ncase==2){c=10;c1=1;c2=1;}
```

The use of c, c1, c2 is discussed subsequently.

- A uniform grid in t of 6 output points is defined with the seq utility for the interval $0 \le t \le t_f = 1$. Therefore, the vector tout has the values $t = 0, 0.2, \dots, 1$.

```
#
# Independent variable for ODE integration
  nout=6;t0=0;tf=1;
  tout=seq(from=t0,to=tf,by=(tf-t0)/(nout-1));
```

At this point, the intervals of r and t in Eqs. (20.1) are defined.

- The IC for Eqs. (20.1), Eqs. (20.2) is placed in a vector u0 of length $4(51) = 204$.[1] Specifically, from ICs (20.2a–d)

$$u(r, t = 0) = f_1(r) = e^{-cr^2}; \quad \frac{\partial u(r, t = 0)}{\partial t} = f_2(r) = 0$$

$$v(r, t = 0) = g_1(r) = e^{-cr^2}; \quad \frac{\partial v(r, t = 0)}{\partial t} = g_2(r) = 0$$

```
#
# ICs
  u0=rep(0,4*n);
  for(i in 1:n){
    u0[i]     =exp(-c*r[i]^2);
    u0[i+n]   =0;
    u0[i+2*n] =exp(-c*r[i]^2);
    u0[i+3*n] =0;
  }
  ncall=0;
```

The counter for the number of calls to pde_1 is initialized.

- The system of 204 ODEs is integrated by the library integrator lsodes (available in deSolve) with the sparse matrix option specified. As expected, the inputs to lsodes are the ODE function, pde_1, the IC vector u0, and the vector of output values of t, tout. The length of u0 (204) informs lsodes how many ODEs are to be integrated. func, y, times are reserved names.

---

1 This IC is based on the use of four variables, u_1, u_2, v_1, v_2, defined as

$$u_1(r, t) = u(r, t); \quad v_1(r, t) = v(r, t)$$

$$u_2(r, t) = \frac{\partial u(r, t)}{\partial t}; \quad v_2(r, t) = \frac{\partial v(r, t)}{\partial t}$$

```
#
# ODE integration
  out=lsodes(y=u0,times=tout,func=pde_1,
      sparsetype="sparseint",rtol=1e-6,atol=1e-6,
      maxord=5);
  nrow(out)
  ncol(out)
```

The numerical solution to the ODEs is returned in matrix out. In this case, out has the dimensions 6×205 (displayed in the numerical output). The offset $204 + 1 = 205$ is required since the first element of each column has the output t (also in tout), and the $2, \ldots, 205$ column elements have the 204 ODE solutions. This indexing of out is used next.

- The numerical solution for $u(r, t)$, $v(r, t)$, consisting of u_1, u_2, v_1, v_2, is taken from the solution matrix out returned by lsodes. The intervals in t and r are covered by two nested fors (with indices it and i for t and r, respectively). The offset i+1 accounts for t returned in the first column position of out and placed in the vector t[it].

```
#
# Arrays for numerical solution
  u1=matrix(0,nrow=n,ncol=nout);
  u2=matrix(0,nrow=n,ncol=nout);
  v1=matrix(0,nrow=n,ncol=nout);
  v2=matrix(0,nrow=n,ncol=nout);
  t=rep(0,nout);
  for(it in 1:nout){
  for(i  in 1:n){
    u1[i,it]=out[it,i+1];
    u2[i,it]=out[it,i+1+n];
    v1[i,it]=out[it,i+1+2*n];
    v2[i,it]=out[it,i+1+3*n];
        t[it]=out[it,1];
  }
  }
```

- The parameters of the numerical solution are displayed, followed by every tenth value of the solution.

```
#
# Display selected output
  cat(sprintf("\n    c = %4.2f,    c1 = %4.2f,
                c2 = %4.2f\n",c,c1,c2));
  for(it in 1:nout){
    iv=seq(from=1,to=n,by=10);
```

```
    for (i in iv) {
  cat (sprintf ("\n        t           r        u1 (r,t)
              u2 (r,t) ")) ;
  cat (sprintf ("\n        t           r        v1 (r,t)
              v2 (r,t) \n")) ;
      cat (sprintf ("%6.2f%8.3f%12.6f%12.6f\n",
                t [it] ,r [i] ,u1 [i,it] ,u2 [i,it])) ;
      cat (sprintf ("%6.2f%8.3f%12.6f%12.6f\n",
                t [it] ,r [i] ,v1 [i,it] ,v2 [i,it])) ;
  }
  }
  cat (sprintf (" ncall = %4d\n",ncall)) ;
```

The number of calls to pde_1 is displayed at the end of the solution.
- The numerical solutions are plotted with matplot.

```
#
# Plot 2D numerical solution
  matplot (r,u1,type="l",lwd=2,col="black",
          lty=1,xlab="r",ylab="u1 (r,t) ",
          main="Einstein-Maxwell") ;
  matplot (r,u2,type="l",lwd=2,col="black",
          lty=1,xlab="r",ylab="u2 (r,t) ",
          main="Einstein-Maxwell") ;
  matplot (r,v1,type="l",lwd=2,col="black",
          lty=1,xlab="r",ylab="v1 (r,t) ",
          main="Einstein-Maxwell") ;
  matplot (r,v2,type="l",lwd=2,col="black",
          lty=1,xlab="r",ylab="v2 (r,t) ",
          main="Einstein-Maxwell") ;
```

The ODE routine pde_1 called by lsodes in the main program of Listing 20.1 is considered next.

20.3 ODE Routine

pde_1 for Eqs. (20.1) follows.

```
  pde_1=function (t,u,parm) {
#
# Function pde_1 computes the t derivative vectors
# for u1 (r,t), u2 (r,t), v1 (r,t), v2 (r,t)
#
# One vector to four vectors
```

```
  u1=rep(0,n);u2=rep(0,n);
  v1=rep(0,n);v2=rep(0,n);
  for(i in 1:n){
    u1[i]=u[i];
    u2[i]=u[i+n];
    v1[i]=u[i+2*n];
    v2[i]=u[i+3*n];
  }
#
# u1r
  tabler=splinefun(r,u1);
  u1r=tabler(r,deriv=1);
#
# v1r
  tabler=splinefun(r,v1);
  v1r=tabler(r,deriv=1);
#
# BCs
  u1r[1]=0;u1r[n]=0;
  v1r[1]=0;v1r[n]=0;
#
# u1rr
  tablerr=splinefun(r,u1r);
  u1rr=tablerr(r,deriv=1);
#
# v1rr
  tablerr=splinefun(r,v1r);
  v1rr=tablerr(r,deriv=1);
#
# PDEs
  u1t=rep(0,n);u2t=rep(0,n);
  v1t=rep(0,n);v2t=rep(0,n);
  for(i in 1:n){
    u1t[i]=u2[i];
    v1t[i]=v2[i];
    term1=-exp(-2*u1[i])*(v2[i]^2-v1r[i]^2);
    term2=-2*(u1r[i]*v1r[i]-u2[i]*v2[i]);
    if(i==1){
       u2t[i]=2*u1rr[i]+c1*term1;
       v2t[i]=2*v1rr[i]+c2*term2;}
    if(i> 1){
       u2t[i]=u1rr[i]+(1/r[i])*u1r[i]+c1*term1;
       v2t[i]=v1rr[i]+(1/r[i])*v1r[i]+c2*term2;}
```

```
  }
#
# Four vectors to one vector
  ut=rep(0,4*n);
  for(i in 1:n){
    ut[i]    =u1t[i];
    ut[i+n]  =u2t[i];
    ut[i+2*n]=v1t[i];
    ut[i+3*n]=v2t[i];
  }
#
# Increment calls to pde_1
  ncall <<- ncall+1;
#
# Return derivative vector
  return(list(c(ut)));
  }
```

Listing 20.2 ODE/MOL routine pde_1 for Eqs. (20.1).

We can note the following details about pde_1:

- The function is defined.

```
  pde_1=function(t,u,parm) {
#
# Function pde_1 computes the t derivative vectors
# for u1(r,t),  u2(r,t),  v1(r,t),  v2(r,t)
```

t is the current value of t in Eqs. (20.1). u is the dependent variable 204-vector. parm is an argument to pass parameters to pde_1 (unused but required in the argument list). The arguments must be listed in the order stated to properly interface with lsodes called in the main program. The derivative vector of the LHS of Eqs. (20.1) is calculated next and returned to lsodes.
- u is placed in four 51-vectors, u1, u2, v1, v2, to facilitate the programming of Eqs. (20.1).

```
#
# One vector to four vectors
  u1=rep(0,n);u2=rep(0,n);
  v1=rep(0,n);v2=rep(0,n);
  for(i in 1:n){
    u1[i]=u[i];
```

```
    u2[i]=u[i+n];
    v1[i]=u[i+2*n];
    v2[i]=u[i+3*n];
}
```

- The first derivatives in r, $\dfrac{\partial u}{\partial r} = \dfrac{\partial u_1}{\partial r}, \dfrac{\partial v}{\partial r} = \dfrac{\partial v_1}{\partial r}$, are computed with `spline-fun`.

```
#
# u1r
    tabler=splinefun(r,u1);
    u1r=tabler(r,deriv=1);
#
# v1r
    tabler=splinefun(r,v1);
    v1r=tabler(r,deriv=1);
```

- Homogeneous Neumann BCs (20.3) are set. The subscripts 1, n correspond to $r = 0, 1$.

```
#
# BCs
    u1r[1]=0;u1r[n]=0;
    v1r[1]=0;v1r[n]=0;
```

- The second derivatives $\dfrac{\partial^2 u}{\partial r^2} = \dfrac{\partial^2 u_1}{\partial r^2}, \dfrac{\partial^2 v}{\partial r^2} = \dfrac{\partial^2 v_1}{\partial r^2}$ in Eqs. (20.1) are computed with `splinefun` by differentiating the first derivative (i.e., stagewise differentiation).

```
#
# u1rr
    tablerr=splinefun(r,u1r);
    u1rr=tablerr(r,deriv=1);
#
# v1rr
    tablerr=splinefun(r,v1r);
    v1rr=tablerr(r,deriv=1);
```

- The LHS derivatives of Eqs. (20.1) $\dfrac{\partial^2 u(r,t)}{\partial t^2} = \dfrac{\partial u_2(r,t)}{\partial t}, \dfrac{\partial^2 v(r,t)}{\partial t^2} = \dfrac{\partial v_2(r,t)}{\partial t}$ are computed using u_1, u_2, v_1, v_2.[2] The four 51-derivative vectors are placed in `u1t, u2t, v1t, v2t`.

---

2 For the singularity at $r = 0$ in Eqs. (20.1), by l'Hospital's rule and BCs (20.3a,c),

$$\frac{1}{r}\frac{\partial u}{\partial r} = \frac{\partial^2 u}{\partial r^2}; \quad \frac{1}{r}\frac{\partial v}{\partial r} = \frac{\partial^2 v}{\partial r^2}$$

```
#
# PDEs
  u1t=rep(0,n);u2t=rep(0,n);
  v1t=rep(0,n);v2t=rep(0,n);
  for(i in 1:n){
    u1t[i]=u2[i];
    v1t[i]=v2[i];
    term1=-exp(-2*u1[i])*(v2[i]^2-v1r[i]^2);
    term2=-2*(u1r[i]*v1r[i]-u2[i]*v2[i]);
    if(i==1){
        u2t[i]=2*u1rr[i]+c1*term1;
        v2t[i]=2*v1rr[i]+c2*term2;}
    if(i> 1){
        u2t[i]=u1rr[i]+(1/r[i])*u1r[i]+c1*term1;
        v2t[i]=v1rr[i]+(1/r[i])*v1r[i]+c2*term2;}
  }
```

- `u1t,u2t,v1t,v2t` are placed in a 204-vector `ut`.

```
#
# Four vectors to one vector
  ut=rep(0,4*n);
  for(i in 1:n){
    ut[i]    =u1t[i];
    ut[i+n]  =u2t[i];
    ut[i+2*n]=v1t[i];
    ut[i+3*n]=v2t[i];
  }
```

- The counter for the calls to `pde_1` is returned to the main program of Listing 20.1 by the `<<-` operator.

```
#
# Increment calls to pde_1
  ncall <<- ncall+1;
```

- The derivative in `ut` is returned to `lsodes` (the ODE integrator called in the main program of Listing 20.1).

```
#
# Return derivative vector
```

so that

$$\frac{\partial^2 u}{\partial r^2} + \frac{1}{r}\frac{\partial u}{\partial r} = 2\frac{\partial^2 u}{\partial r^2}; \quad \frac{\partial^2 v}{\partial r^2} + \frac{1}{r}\frac{\partial v}{\partial r} = 2\frac{\partial^2 v}{\partial r^2}$$

These forms for $r = 0$ appear in `2*u1rr[i]`, `2*v1rr[i]`.

```
return(list(c(ut)));
}
```

Note that a `list` is used as required by `lsodes`. The final `}` concludes `pde_1`.

20.4   Model Output

Execution of the preceding main program and ODE routine `pde_1` for `ncase=1` produces the following abbreviated numerical output in Table 20.1. We can note the following details about this output:

- As expected, the solution array `out` has the dimensions 6×205 for 6 output points and 204 ODEs, with t included as an additional element, that is, $204 + 1 = 205$.
- The interval in r is $((1-0)/(51-1) = 0.02$ (with every tenth value in Table 20.1).
- For `ncase=1`, $u(r, t)$ and $v(r, t)$ are the same for the same ICs.[3]
- The computational effort is modest, `ncall = 650`.

The solutions are displayed graphically in Figure 20.1a–d.[4]

For `ncase=2` (`c1=c2=1`), abbreviated output is in Table 20.2. We can note the following details about this output:

---

3 For `ncase=1`, `c1=c2=0` so that the nonlinear coupling terms in Eqs. (20.1),

$$- e^{2u} \left[\left(\frac{\partial v}{\partial t} \right)^2 - \left(\frac{\partial v}{\partial r} \right)^2 \right]$$

$$- 2 \left[\frac{\partial u}{\partial r} \frac{\partial v}{\partial r} - \frac{\partial u}{\partial t} \frac{\partial v}{\partial t} \right]$$

are dropped. Then Eqs. (20.1) reduce to the same 1D diffusion equation in cylindrical coordinates for $u(r, t)$ and $v(r, t)$.

Also, for the ICs

$$u(r, t = 0) = f_1(r) = e^{-cr^2}; \quad \frac{\partial u(r, t = 0)}{\partial t} = f_2(r) = 0$$

$$v(r, t = 0) = g_1(r) = e^{-cr^2}; \quad \frac{\partial v(r, t = 0)}{\partial t} = g_2(r) = 0$$

and the problems in $u(r, t)$, $v(r, t)$ are the same.

This may seem like a trivial test case, but it is worth executing since if the solutions are not the same, a programming error is indicated. Also, for `ncase=2`, `c1=c2=1` and the nonlinear coupling terms are included. Therefore, any difference in $u(r, t)$, $v(r, t)$ results from the coupling terms. This use of `ncase=1, 2` demonstrates how the effects of particular RHS terms in simultaneous PDEs can be identified.

4 The solutions are rather detailed and complicated. Their evolution in t can be discerned to some extent by observing the departure from the ICs. The plots could also be extended with legends and identification of the individual curves (in t). This can be readily accomplished with graphical utilities (options) available with `matplot`.

Table 20.1 Abbreviated output for Eqs. (20.1)–(20.3), ncase=1.

```
[1]  6

[1]  205
```

c = 10.00,   c1 = 0.00,   c2 = 0.00

t	r	u1(r,t)	u2(r,t)
t	r	v1(r,t)	v2(r,t)
0.00	0.000	1.000000	0.000000
0.00	0.000	1.000000	0.000000

t	r	u1(r,t)	u2(r,t)
t	r	v1(r,t)	v2(r,t)
0.00	0.200	0.670320	0.000000
0.00	0.200	0.670320	0.000000

t	r	u1(r,t)	u2(r,t)
t	r	v1(r,t)	v2(r,t)
0.00	0.400	0.201897	0.000000
0.00	0.400	0.201897	0.000000

t	r	u1(r,t)	u2(r,t)
t	r	v1(r,t)	v2(r,t)
0.00	0.600	0.027324	0.000000
0.00	0.600	0.027324	0.000000

t	r	u1(r,t)	u2(r,t)
t	r	v1(r,t)	v2(r,t)
0.00	0.800	0.001662	0.000000
0.00	0.800	0.001662	0.000000

t	r	u1(r,t)	u2(r,t)
t	r	v1(r,t)	v2(r,t)
0.00	1.000	0.000045	0.000000
0.00	1.000	0.000045	0.000000

.
.
.

(Continued)

Table 20.1 (Continued)

```
     Output for t = 0.2 to 0.8 removed

                   .                    .
                   .                    .
                   .                    .
     t         r       u1(r,t)        u2(r,t)
     t         r       v1(r,t)        v2(r,t)
   1.00     0.000     -0.060626       0.158822
   1.00     0.000     -0.060626       0.158822

     t         r       u1(r,t)        u2(r,t)
     t         r       v1(r,t)        v2(r,t)
   1.00     0.200     -0.066068       0.222558
   1.00     0.200     -0.066068       0.222558

     t         r       u1(r,t)        u2(r,t)
     t         r       v1(r,t)        v2(r,t)
   1.00     0.400     -0.074901       0.496411
   1.00     0.400     -0.074901       0.496411

     t         r       u1(r,t)        u2(r,t)
     t         r       v1(r,t)        v2(r,t)
   1.00     0.600     -0.021122       0.731395
   1.00     0.600     -0.021122       0.731395

     t         r       u1(r,t)        u2(r,t)
     t         r       v1(r,t)        v2(r,t)
   1.00     0.800      0.183552      -0.167318
   1.00     0.800      0.183552      -0.167318

     t         r       u1(r,t)        u2(r,t)
     t         r       v1(r,t)        v2(r,t)
   1.00     1.000      0.331441      -1.140403
   1.00     1.000      0.331441      -1.140403

   ncall =   650
```

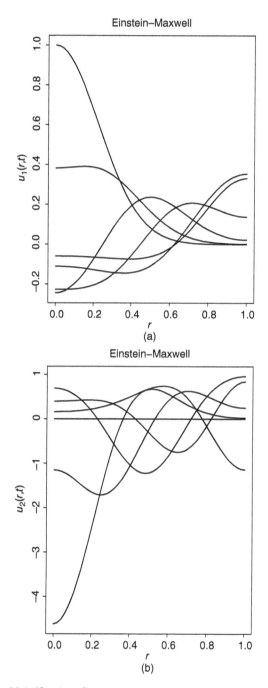

Figure 20.1 (Continued)

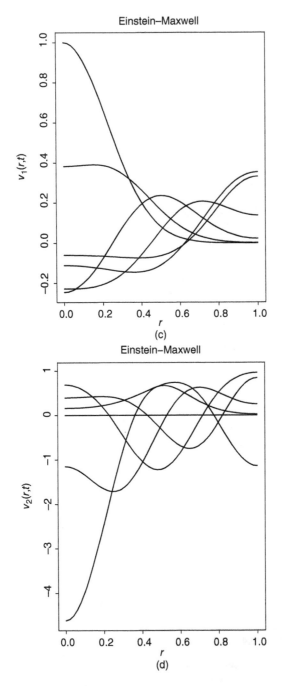

Figure 20.1 (a) $u_1(r, t) = u(r, t)$ from Eqs. (20.1)–(20.3), ncase=1. (b) $u_2(r, t) = \dfrac{\partial u(r, t)}{\partial t}$ from Eqs. (20.1)–(20.3), ncase=1. (c) $v_1(r, t) = v(r, t)$ from Eqs. (20.1)–(20.3), ncase=1. (d) $v_2(r, t) = \dfrac{\partial v(r, t)}{\partial t}$ from Eqs. (20.1)–(20.3), ncase=1.

Table 20.2 Abbreviated output for Eqs. (20.1)–(20.3), ncase=2.

```
[1]  6

[1]  205

   c = 10.00,   c1 = 1.00,   c2 = 1.00

      t          r        u1(r,t)        u2(r,t)
      t          r        v1(r,t)        v2(r,t)
   0.00       0.000      1.000000       0.000000
   0.00       0.000      1.000000       0.000000

      t          r        u1(r,t)        u2(r,t)
      t          r        v1(r,t)        v2(r,t)
   0.00       0.200      0.670320       0.000000
   0.00       0.200      0.670320       0.000000

      t          r        u1(r,t)        u2(r,t)
      t          r        v1(r,t)        v2(r,t)
   0.00       0.400      0.201897       0.000000
   0.00       0.400      0.201897       0.000000

      t          r        u1(r,t)        u2(r,t)
      t          r        v1(r,t)        v2(r,t)
   0.00       0.600      0.027324       0.000000
   0.00       0.600      0.027324       0.000000

      t          r        u1(r,t)        u2(r,t)
      t          r        v1(r,t)        v2(r,t)
   0.00       0.800      0.001662       0.000000
   0.00       0.800      0.001662       0.000000

      t          r        u1(r,t)        u2(r,t)
      t          r        v1(r,t)        v2(r,t)
   0.00       1.000      0.000045       0.000000
   0.00       1.000      0.000045       0.000000
                  .                         .
                  .                         .
                  .                         .
```

(Continued)

Table 20.2 (Continued)

```
      Output for t = 0.2 to 0.8 removed
                  .                      .
                  .                      .
                  .                      .
       t          r         u1(r,t)        u2(r,t)
       t          r         v1(r,t)        v2(r,t)
      1.00      0.000      -0.067592       0.179669
      1.00      0.000      -0.034443       0.094420

       t          r         u1(r,t)        u2(r,t)
       t          r         v1(r,t)        v2(r,t)
      1.00      0.200      -0.073954       0.247583
      1.00      0.200      -0.036981       0.137539

       t          r         u1(r,t)        u2(r,t)
       t          r         v1(r,t)        v2(r,t)
      1.00      0.400      -0.085504       0.554943
      1.00      0.400      -0.036644       0.310384

       t          r         u1(r,t)        u2(r,t)
       t          r         v1(r,t)        v2(r,t)
      1.00      0.600      -0.031970       0.851850
      1.00      0.600       0.025539       0.248361

       t          r         u1(r,t)        u2(r,t)
       t          r         v1(r,t)        v2(r,t)
      1.00      0.800       0.195956      -0.177160
      1.00      0.800       0.153584      -0.410411

       t          r         u1(r,t)        u2(r,t)
       t          r         v1(r,t)        v2(r,t)
      1.00      1.000       0.358601      -1.214713
      1.00      1.000       0.186315      -0.493650

      ncall =   703
```

- As expected, $u(r, t)$ and $v(r, t)$ are different (from the effect of the nonlinear coupling terms).
- The computational effort is modest, ncall = 703 (the inclusion of the nonlinear coupling terms did not add appreciably to ncall).

The general features of the solutions for $u(r, t), v(r, t)$ can be observed in Figures 20.2a–d.

20.5 Summary and Conclusions

The system of Eqs. (20.1)–(20.3) indicates that the SCMOL can accommodate simultaneous, nonlinear, variable coefficient, second order in space and time PDEs.

The choice of $n = 51$ appears to give acceptable spatial resolution in r, but n can be easily changed to observe the effect on the numerical solution of the discretization in r (h refinement). The error tolerances for lsodes can also be changed to observe the effect on the accuracy of the t integration. This type of experimentation (numerical error analysis) should be part of any new PDE study (omitted here to conserve space).

An important feature of splines is that a spatial interval does not have to be defined on a grid with uniform spacing.[5] The grid can be variable (nonuniform) to place grid points where the solution changes rapidly in space to improve the spatial resolution.

The overall intention of all of the preceding PDE examples is to demonstrate the general applicability and flexibility of SCMOL analysis. However, a successful result for a new PDE application cannot be guaranteed in advance, that is, SCMOL is not a mechanical procedure from beginning to end, but rather, may require the application of the analyst's experience and insights to arrive at a solution of acceptable accuracy with a reasonable computational effort. We hope the routines and detailed discussions can facilitate a successful outcome.

---

5 In most of the preceding examples, the spatial intervals were defined numerically with a uniform (equally spaced) grid through the use of the seq utility. But this procedure could be replaced with a definition of a variable grid. In principle, the only limitation is that the grid spacing should vary smoothly, for example, discontinuities in the grid spacing are not permitted. But otherwise, the grid points can be concentrated where they are most needed, that is, where the solution spatial variation is greatest.

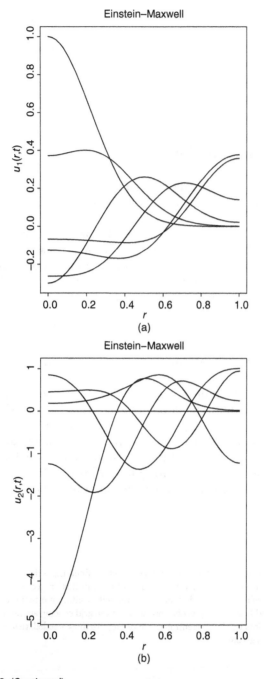

Figure 20.2 (Continued)

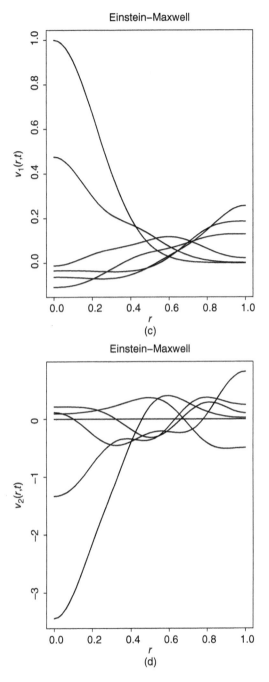

Figure 20.2 (a) $u_1(r, t) = u(r, t)$ from Eqs. (20.1)–(20.3), ncase=2. (b) $u_2(r, t) = \dfrac{\partial u(r, t)}{\partial t}$ from Eqs. (20.1)–(20.3), ncase=2. (c) $v_1(r, t) = v(r, t)$ from Eqs. (20.1)–(20.3), ncase=2. (d) $v_2(r, t) = \dfrac{\partial v(r, t)}{\partial t}$ from Eqs. (20.1)–(20.3), ncase=2.

Reference

1 Stephani, H., et al. (2000), *Exact Solutions of Einstein's Field Equations*, Cambridge University Press, Cambridge, UK.

A

Differential Operators in Three Orthogonal Coordinate Systems

Differential operators that can be used in the formulation of PDEs in Cartesian, cylindrical, and spherical coordinates [1, 2] are summarized in this appendix.

∇ is a vector differential operator that has three components in the three coordinate systems. For the divergence of a vector,

Table A1.1 Divergence of a vector, $\nabla \cdot$ [1, 2].

Coordinate system	Components
Cartesian	$\begin{bmatrix} [\nabla]_x = \dfrac{\partial}{\partial x} \\[2mm] [\nabla]_y = \dfrac{\partial}{\partial y} \\[2mm] [\nabla]_z = \dfrac{\partial}{\partial z} \end{bmatrix}$

Coordinate system	Components
cylindrical	$\begin{bmatrix} [\nabla]_r = \dfrac{1}{r}\dfrac{\partial}{\partial r}(r) \\[2mm] [\nabla]_\theta = \dfrac{1}{r}\dfrac{\partial}{\partial \theta} \\[2mm] [\nabla]_z = \dfrac{\partial}{\partial z} \end{bmatrix}$

Coordinate system	Components
spherical	$\begin{bmatrix} [\nabla]_r = \dfrac{1}{r^2}\dfrac{\partial}{\partial r}(r^2) \\[2mm] [\nabla]_\theta = \dfrac{1}{r\sin\theta}\dfrac{\partial}{\partial \theta}(\sin\theta) \\[2mm] [\nabla]_\phi = \dfrac{1}{r\sin\theta}\dfrac{\partial}{\partial \phi} \end{bmatrix}$

Spline Collocation Methods for Partial Differential Equations: With Applications in R, First Edition.
William E. Schiesser.
© 2017 John Wiley & Sons, Inc. Published 2017 by John Wiley & Sons, Inc.
Companion website: www.wiley.com/go/Spline_Collocation

For the gradient of a scalar,

Table A1.2 Gradient of a scalar, ∇ [1, 2].

Coordinate system	Components
Cartesian	$\begin{bmatrix} [\nabla]_x = \dfrac{\partial}{\partial x} \\[2mm] [\nabla]_y = \dfrac{\partial}{\partial y} \\[2mm] [\nabla]_z = \dfrac{\partial}{\partial z} \end{bmatrix}$

Coordinate system	Components
cylindrical	$\begin{bmatrix} [\nabla]_r = \dfrac{\partial}{\partial r} \\[2mm] [\nabla]_\theta = \dfrac{1}{r}\dfrac{\partial}{\partial \theta} \\[2mm] [\nabla]_z = \dfrac{\partial}{\partial z} \end{bmatrix}$

Coordinate system	Components
spherical	$\begin{bmatrix} [\nabla]_r = \dfrac{\partial}{\partial r} \\[2mm] [\nabla]_\theta = \dfrac{1}{r}\dfrac{\partial}{\partial \theta} \\[2mm] [\nabla]_\phi = \dfrac{1}{r\sin\theta}\dfrac{\partial}{\partial \phi} \end{bmatrix}$

$\Delta = \nabla \cdot \nabla = \nabla^2$ (the *Laplacian*) follows directly from the preceding components of $\nabla \cdot$ (divergence of a vector in Table A1.1) and ∇ (gradient of a scalar in Table A1.2).

Cartesian coordinates (with unit vector $(\mathbf{i}, \mathbf{j}, \mathbf{k})$):

$$\nabla \cdot \nabla = \left(\mathbf{i}\frac{\partial}{\partial x} + \mathbf{j}\frac{\partial}{\partial y} + \mathbf{k}\frac{\partial}{\partial z}\right) \cdot \left(\mathbf{i}\frac{\partial}{\partial x} + \mathbf{j}\frac{\partial}{\partial y} + \mathbf{k}\frac{\partial}{\partial z}\right) = \frac{\partial^2}{\partial x^2} + \frac{\partial^2}{\partial y^2} + \frac{\partial^2}{\partial z^2}$$

Cylindrical coordinates (with unit vector $(\mathbf{i}_r, \mathbf{j}_\theta, \mathbf{k}_z)$):

$$\begin{aligned}
\nabla \cdot \nabla &= \left(\mathbf{i}_r\frac{1}{r}\frac{\partial}{\partial r}(r) + \mathbf{j}_\theta\frac{1}{r}\frac{\partial}{\partial \theta} + \mathbf{k}_z\frac{\partial}{\partial z}\right) \cdot \left(\mathbf{i}_r\frac{\partial}{\partial r} + \mathbf{j}_\theta\frac{1}{r}\frac{\partial}{\partial \theta} + \mathbf{k}_z\frac{\partial}{\partial z}\right) \\
&= \frac{1}{r}\frac{\partial}{\partial r}\left(r\frac{\partial}{\partial r}\right) + \frac{1}{r}\frac{\partial}{\partial \theta}\left(\frac{1}{r}\frac{\partial}{\partial \theta}\right) + \frac{\partial}{\partial z}\left(\frac{\partial}{\partial z}\right) \\
&= \frac{1}{r}\left(\frac{\partial}{\partial r} + r\frac{\partial^2}{\partial r^2}\right) + \frac{1}{r^2}\frac{\partial}{\partial \theta}\frac{\partial}{\partial \theta} + \frac{\partial}{\partial z}\frac{\partial}{\partial z} \\
&= \left(\frac{\partial^2}{\partial r^2} + \frac{1}{r}\frac{\partial}{\partial r}\right) + \frac{1}{r^2}\frac{\partial^2}{\partial \theta^2} + \frac{\partial^2}{\partial z^2}
\end{aligned}$$

Spherical coordinates (with unit vector $(\mathbf{i}_r, \mathbf{j}_\theta, \mathbf{k}_\phi)$):

$$\nabla \cdot \nabla = \left(\mathbf{i}_r \frac{1}{r^2} \frac{\partial}{\partial r}(r^2) + \mathbf{j}_\theta \frac{1}{r \sin \theta} \frac{\partial}{\partial \theta}(\sin \theta) + \mathbf{k}_\phi \frac{1}{r \sin \theta} \frac{\partial}{\partial \phi} \right)$$
$$\cdot \left(\mathbf{i}_r \frac{\partial}{\partial r} + \mathbf{j}_\theta \frac{1}{r} \frac{\partial}{\partial \theta} + \mathbf{k}_\phi \frac{1}{r \sin \theta} \frac{\partial}{\partial \phi} \right)$$
$$= \frac{1}{r^2} \frac{\partial}{\partial r} \left(r^2 \frac{\partial}{\partial r} \right) + \frac{1}{r \sin \theta} \frac{\partial}{\partial \theta} \left(\sin \theta \frac{1}{r} \frac{\partial}{\partial \theta} \right) + \frac{1}{r \sin \theta} \frac{\partial}{\partial \phi} \left(\frac{1}{r \sin \theta} \frac{\partial}{\partial \phi} \right)$$
$$= \frac{1}{r^2} \frac{\partial}{\partial r} \left(r^2 \frac{\partial}{\partial r} \right) + \frac{1}{r^2 \sin \theta} \frac{\partial}{\partial \theta} \left(\sin \theta \frac{\partial}{\partial \theta} \right) + \frac{1}{r^2 \sin^2 \theta} \frac{\partial^2}{\partial \phi^2}$$

The preceding information may be useful for some of the PDE application systems stated in coordinate-free format, then restated for a particular coordinate system.

References

1 Bird, R.B., W.E. Stewart, and E.N. Lightfoot (2002), *Transport Phenomena*, Second Edition, John Wiley & Sons, Inc., Hoboken, NJ, p. 836.
2 Soetaert, K., J. Cash, and F. Mazzia (2012), *Solving Differential Equations in R*, Springer-Verlag, Heidelberg, pp. 140–141.

Index

Spline Collocation Methods for Partial Differential Equations: With Applications in R, First Edition.
William E. Schiesser.
© 2017 John Wiley & Sons, Inc. Published 2017 by John Wiley & Sons, Inc.
Companion website: www.wiley.com/go/Spline_Collocation